Modern Methods in Partial Differential Equations

Martin Schechter
University of California, Irvine

Dover Publications, Inc.
Mineola, New York

Bibliographical Note

This Dover edition, first published in 2014, is an unabridged republication of the work originally published in 1977 by the McGraw-Hill Book Company, New York.

Library of Congress Cataloging-in-Publication Data

Schechter, Martin.
Modern methods in partial differential equations / Martin Schechter.—Dover edition.
pages cm.
Summary: "When first published in 1977, this volume made recent accomplishments in its field available to advanced undergraduates and beginning graduate students of mathematics. Requiring only some familiarity with advanced calculus and rudimentary complex function theory, it covered discoveries of the previous three decades, a particularly fruitful era. Now it remains a permanent, much-cited contribution to the ever-expanding literature on partial differential equations"—Provided by publisher.
Includes bibliographical references and index.
ISBN-13: 978-0-486-49296-4 (pbk.)
ISBN-10: 0-486-49296-6 (pbk.)
1. Differential equations, Partial. I. Title.

QA374.S35 2014
515'.353—dc23

2013026076

Manufactured in the United States by Courier Corporation
49296601 2014
www.doverpublications.com

BS"D
To Deborah, with love

ACKNOWLEDGMENTS

I want to thank Magdalene (Dilly) McNamara, Marie McDivite and Harriet Nachmann for transforming my illegible scribble into a work of art. Bobbe Friedman deserves much credit for her invaluable help. Also, many thanks are due to Ronald Levin for useful suggestions and Alex Gelman for help in correcting the proofs. Special thanks are due to my special daughter, Sarah, for typing parts of the manuscript. Most of all I am indebted to my wife, Deborah, whose contribution to this effort is beyond measure.

Martin Schechter

CONTENTS

LIST OF SYMBOLS

The numbers following the symbols refer to the pages where they are introduced or explained

PREFACE

The subject of partial differential operators is very broad and encompasses many aspects of analysis. We can convince ourselves of this fact merely by noting that the study of analytic functions of a complex variable is, in reality, the study of solutions of a simple system of partial differential equations (the Cauchy–Riemann equations). Moreover, in our present state of knowledge, we have only scratched the surface. The more involved scientific theory becomes, the more and varied the partial differential equations that arise. Other branches of mathematics also contribute their share. The types of partial differential equations and systems that are possible can stagger the imagination.

There seems to be a dichotomy in the types of books published on the topic of partial differential equations. One kind of presentation studies the equations of classical physics (the wave, heat, and potential energy equations) and employs mostly separation of variables and Fourier series. This part of the theory has been thoroughly studied and is near completion. There are many excellent books written on this topic. It can be referred to as the classical theory of partial differential equations, because the major part of it was known more than fifty years ago.

In the past thirty years there has been a surge of activity in new directions. Higher order equations and systems have been studied, including those that do not fall into previously defined categories. Even the concept of a solution has been broadened, and operators more general than partial differential operators have become popular.

The purpose of this book is to introduce the student to the modern techniques and methods that have been used in the newer theory. I want the student to get a taste and feeling for the powerful tools that are used, but I do not want him to be engrossed in the minute details that can hide the basic ideas behind the methods. I try to take the middle of the road—to attack problems of greater generality than those considered in the classical theory, but not to require the most general and refined machinery available. With respect to the latter, I try to introduce new tools sparingly, and to use only what is needed to obtain a meaningful (albeit not necessarily the most powerful) result. In this way I hope the student will be able to distinguish the trees from the forest.

Another reason for my keeping the background material to a minimum is to allow students to learn the subject at an earlier stage. The material in this book has been taught to first year graduate students at Yeshiva University over a ten year period. Moreover, I have endeavored to present it in such a way as to make it accessible to undergraduates as well. The reader should have a basic knowledge of advanced calculus. Lebesgue integration theory is not actually used. It is only noted that L^2 is complete. Anyone who is willing to accept this fact needs no more background in integration theory. The theory of analytic functions of a complex variable is used only in a few places, and even then only in an elementary way. All background material in Hilbert space and linear algebra is given where necessary. It is actually surprising that one can accomplish so much with so little.

A book on partial differential equations can hope to cover only a small fraction of the basic material that is known. The author must select, but there are no logical grounds upon which to base such a selection. Many authors choose what they feel are the most important topics, but the subjectivity of their choices is fairly obvious.

The present book is no different in this respect. I have chosen subject matter that I feel will motivate the student and introduce him to techniques that have wide applicability to many other problems, in partial differential equations as well as other branches of analysis. Also, I want to give the reader a fair idea of what is to be expected in other situations, and what methods can be used. In addition, I want a uniform theme and outlook; throughout the book I consider a single linear partial differential operator of arbitrary order. In each problem I look for existence, uniqueness, estimates, and regularity of solutions. I try to pick analytical tools that can be used throughout the book, and not only in one or two isolated instances. Whenever possible I try to use the same spaces of functions throughout (basically the L^2 spaces).

I have tried to give proper credit to various research articles used in obtaining material, but I am sure that I have benefited directly or indirectly from countless others. In addition I have added, to the bibliography, several books of interest on the modern theory of partial differential equations.

June, 1976 Martin Schechter

CHAPTER

ONE

EXISTENCE OF SOLUTIONS

1-1 INTRODUCTION

A partial differential equation is, as the name implies, an equation containing a partial derivative. Of course, the derivative is to be taken of an unknown function of more than one variable (if the function were known, we could take the derivative and it would disappear; if it depended only on one variable, we would call the equation an ordinary differential equation). The simplest partial differential equation is

$$\frac{\partial u(x, y)}{\partial x} = 0 \tag{1-1}$$

where the unknown function u depends on two variables x, y. The solution of Eq. (1-1) is obviously

$$u(x, y) = g(y) \tag{1-2}$$

where $g(y)$ is any function of y alone. Although this example is fairly simple, we should examine it a bit more closely. First of all, what do we mean by a "solution" of Eq. (1-1)? You say, "That is obvious; we mean simply a function $u(x, y)$, which, when substituted into Eq. (1-1), makes the equation hold." However, a little reflection shows immediately that certain problems arise, albeit that for this particular equation they are easily solved.

We cannot just substitute any proposed solution into Eq. (1-1): we must start by differentiating it. Therefore, the first requirement we must impose on $u(x, y)$ in order that it be a solution of Eq. (1-1) is that it possess a derivative with respect to x. Second, for what values of x, y is Eq. (1-1) to hold? All real values, or just some? This certainly has to be specified. Next, let us examine our "solution," Eq. (1-2). What kind of function is $g(y)$? Must it possess a derivative with respect to y, or can it even be discontinuous? Or perhaps it need not be a function at all in the usual sense, but a so-called "distribution" (ignore this last statement if you have never heard the term).

Another observation is that no matter what kind of functions we admit, Eq. (1-1) will have many solutions. If a particular solution is desired, then we must prescribe additional restrictions, or "side conditions."

The upshot of all this is that with a partial differential equation, we must also be told where the equation applies and what kind of functions are acceptable as solutions. This information is usually supplied from the application where the equation originated. However, there are important cases when the "side conditions" are not clear from the application, and have to be determined by studying the equation. They are then used to determine "meaningful" situations in the application.

Needless to say, the number of partial differential equations (and systems of equations) that can be dreamt up is infinite. The number of equations arising in applications is not much smaller. To complicate matters, experience has shown us that a slight modification of an equation (such as the change in sign of a term) may cause solutions to be completely different in nature, with entirely different methods required for solving them. It should come as no surprise, therefore, that as yet we are nowhere near a systematic treatment of partial differential equations. At best, the present state of knowledge can be described as a conglomeration of particular methods (the word "tricks" may even be more appropriate) which work in special cases. Thus, any treatment of partial differential equations, no matter how extensive, must necessarily restrict itself to a relatively small area of the subject.

We have chosen to deal with *linear* partial differential equations primarily because they are the easiest to deal with. The most general linear partial differential equation involving one unknown function $u(x_1, \ldots, x_n)$ can be written in the form

$$\sum_{\mu_1 + \cdots + \mu_n \leq m} a_{\mu_1, \ldots, \mu_n}(x_1, \ldots, x_n) \frac{\partial^{\mu_1 + \cdots + \mu_n} u}{\partial x_1^{\mu_1} \cdots \partial x_n^{\mu_n}} = f(x_1, \ldots, x_n) \tag{1-3}$$

where summation is taken over all nonnegative integers $\mu_1, \ldots, \mu_n$, and the a's and f are given functions. (Since we have not as yet defined what we mean by a linear equation, you might as well take Eq. (1-3) to be the definition.)

One look at Eq. (1-3) should be sufficient to discourage anyone from studying partial differential equations. (If it does not accomplish this effect, I shall do better later on.) However, once we have survived the initial impact, we see that a bit of shorthand will do a lot of good. For instance, if we let μ stand for the *multi-index* $(\mu_1, \ldots, \mu_n)$ with *norm* $|\mu| = \mu_1 + \cdots + \mu_n$, and let x stand for the vector $(x_1, \ldots, x_n)$ and write

$$D^\mu = \frac{\partial^{|\mu|}}{\partial x_1^{\mu_1} \cdots \partial x_n^{\mu_n}}$$

then Eq. (1-3) becomes

$$\sum_{|\mu| \leq m} a_\mu(x) D^\mu u(x) = f(x) \tag{1-4}$$

which looks much better.

Let us examine Eq. (1-4) a little more closely. The left-hand side consists of a sum of terms, each of which is a product of a coefficient and a derivative of u. We may consider it as a *differential operator* A acting on u. We can then write Eq. (1-4) more simply as

$$Au = f \tag{1-5}$$

where

$$A = \sum_{|\mu| \leq m} a_\mu(x) D^\mu \tag{1-6}$$

The operator A is called linear because

$$A(\alpha_1 u_1 + \alpha_2 u_2) = \alpha_1 A u_1 + \alpha_2 A u_2 \tag{1-7}$$

holds for all functions u_1, u_2 and all numbers α_1, α_2. Equation (1-5) is called linear because the operator A is linear.

Now what do we mean by a solution of Eq. (1-4)? Since derivatives up to and including those of order m are involved, it seems quite natural to require that these derivatives exist and are continuous, and when they are substituted into Eq. (1-4) the equality holds. We take this as our present definition. Later on, we shall find it convenient, if not essential, to modify this definition quite drastically.

Where do we want Eq. (1-4) to hold? Obviously, it should hold in some subset Ω of $(x_1, \ldots, x_n)$ space. This subset has to be specified. Of course, $u(x)$ has to be defined in a neighborhood of each point of Ω in order that the appropriate derivatives be defined. Since we want our solutions to have continuous derivatives up to order m in Ω, we shall give this set of functions a name. We denote the n-dimensional coordinate space by E^n.

Definition 1-1 Let Ω be a set in E^n. We let $C^m(\Omega)$ denote the set of all functions defined in a neighborhood of each point of Ω, and having all derivatives of order $\leq m$ continuous in Ω. If a function u is in $C^m(\Omega)$ for each m, then it is called infinitely differentiable and said to be in $C^\infty(\Omega)$, i.e.,

$$C^\infty(\Omega) = \bigcap_{m=0}^{\infty} C^m(\Omega)$$

For $m = 0$, we write $C(\Omega) = C^0(\Omega)$. This is the set of functions continuous in Ω.

1-2 EQUATIONS WITHOUT SOLUTIONS

The first question that might be asked concerning Eq. (1-4) is whether or not it has a solution in a given set Ω. To make the environment as conducive as possible, let us be willing to take Ω as the sphere Σ_r consisting of those points $(x_1, \ldots, x_n)$ of E^n satisfying

$$x_1^2 + \cdots + x_n^2 < r^2$$

where r is some positive number. (The reason for calling Σ_r a sphere should be evident.) Let us even be willing to assume that the function f and the coefficients of A are infinitely differentiable in Σ_r (i.e., they are in $C^\infty(\Sigma_r)$). Under such circumstances one might reasonably expect a solution of Eq. (1-4) to be guaranteed. Unfortunately, this is decidedly not the case. A simple example was discovered by H. Lewy (1957) (pronounced Layvee), and a study of it is instructive.

The setting is three-dimensional space and we denote the coordinates by x, y, t (we save the letter z for another quantity). The equation is simple to write down:

$$u_x + iu_y + 2(ix - y)u_t = f(x, y, t) \tag{1-8}$$

A word of explanation is in order. The coefficients of this equation are complex-valued, while it was hitherto tacitly assumed that the functions and coefficients considered were real-valued. The following considerations will clarify the matter.

Suppose we allow f to be complex-valued in the sense that there are two bona fide, real-valued functions $f_1(x, y, t)$ and $f_2(x, y, t)$, such that $f = f_1 + if_2$. It is to be understood that there need not be any connection between the two functions f_1 and f_2. We assume the same for any solution u. Then Eq. (1-8) is equivalent to the system

$$u_{1x} - u_{2y} - 2xu_{2t} - 2yu_{1t} = f_1 \tag{1-9}$$

$$u_{2x} + u_{1y} + 2xu_{1t} - 2yu_{2t} = f_2 \tag{1-10}$$

which involves only real functions. It is a system of two equations in two unknowns. Thus, Eq. (1-8) is just a short way of writing Eqs. (1-9) and (1-10). The fact that Eq. (1-8) is a system, is not a factor in its lack of solutions. We shall also exhibit a single equation with real f and real coefficients which has no solution.

Now back to Eq. (1-8). To simplify it, we introduce the complex variable $z = x + iy$. Then $u(x, y, t)$ is a function of z and t. It is an analytic function of z only if it satisfies the Cauchy–Riemann equations

$$u_{1x} = u_{2y} \qquad u_{1y} = -u_{2x} \tag{1-11}$$

or their abbreviated form

$$u_x + iu_y = 0 \tag{1-12}$$

To abbreviate even further, set

$$2u_z = u_x - iu_y \qquad 2u_{\bar{z}} = u_x + iu_y \tag{1-13}$$

Then Eq. (1-12) becomes

$$u_{\bar{z}} = 0 \tag{1-14}$$

while Eq. (1-8) becomes

$$u_{\bar{z}} + izu_t = \tfrac{1}{2}f \tag{1-15}$$

Now let Ω be the set $x^2 + y^2 < a$, $|t| < b$, where a and b are any fixed positive numbers. We shall show that there is an $f \in C^\infty(\Omega)$ such that Eq. (1-8) has no solution in $C^1(\Omega)$. Since a and b are arbitrary, it will follow that Eq. (1-8) does not have a solution in Σ_r for any $r > 0$.

To carry out our proof, let $\psi(\sigma, \tau)$ be a continuously differentiable complex-valued function of two real variables σ, τ which vanishes outside the rectangle $0 < \sigma < a$, $|\tau| < b$. Set

$$\varphi(x, y, t) = \psi(\rho, t) \qquad \rho = x^2 + y^2$$

Note that φ has continuous derivatives in x, y, t space and vanishes outside Ω. By the chain rule, we have

$$\varphi_z(x, y, t) = \bar{z}\psi_\rho(\rho, t) \tag{1-16}$$

Now suppose there were a solution u of Eq. (1-8) in Ω. Then,

$$\iiint_\Omega (u_{\bar{z}} + izu_t)\bar{\varphi}\, dx\, dy\, dt = \frac{1}{2}\iiint_\Omega f\bar{\varphi}\, dx\, dy\, dt$$

where the bar denotes complex conjugation. Integrating the left-hand integral by parts (see Sec. 1–3), we have

$$-\iiint_\Omega u\overline{(\varphi_z - i\bar{z}\varphi_t)}\, dx\, dy\, dt = \frac{1}{2}\iiint_\Omega f\bar{\varphi}\, dx\, dy\, dt$$

(There are no boundary integrals because φ vanishes on the boundary of Ω). By Eq. (1-16) this becomes

$$-\iiint_\Omega zu\overline{(\psi_\rho - i\psi_t)}\, dx\, dy\, dt = \frac{1}{2}\iiint_\Omega f\bar{\varphi}\, dx\, dy\, dt \tag{1-17}$$

We now introduce coordinates ρ, θ in place of x and y, where

$$\tan\theta = \frac{y}{x}$$

Noting that $2d\rho\, d\theta = dx\, dy$, we see that Eq. (1-17) becomes

$$-\int_{-b}^{b}\int_0^{2\pi}\int_0^a zu\overline{(\psi_\rho - i\psi_t)}\, d\rho\, d\theta\, dt = \frac{1}{2}\int_{-b}^{b}\int_0^{2\pi}\int_0^a f\bar{\psi}\, d\rho\, d\theta\, dt \tag{1-18}$$

We now set

$$U(\rho, t) = \int_0^{2\pi} zu\, d\theta \tag{1-19}$$

and assume that f does not depend on θ. Since ψ also does not depend on θ, we have

$$-\int_{-b}^{b}\int_0^a U\overline{(\psi_\rho - i\psi_t)}\, d\rho\, dt = \pi\int_{-b}^{b}\int_0^a f\bar{\psi}\, d\rho\, dt$$

We now integrate the left-hand side by parts, obtaining

$$\int_{-b}^{b}\int_{0}^{a}(U_\rho + iU_t - \pi f)\overline{\psi}\,d\rho\,dt = 0$$

The next step is to note that ψ was any continuously differentiable function which vanished outside of $0 < \rho < a$, $|t| < b$. It follows from well-known arguments (see Sec. 1–3) that

$$U_\rho + iU_t = \pi f \qquad 0 < \rho < a \qquad |t| < b \tag{1-20}$$

Next take $f = g'(t)$, where g is a smooth, real-valued function of t alone, and set

$$V(\rho, t) = U + \pi i g \tag{1-21}$$

Then
$$V_\rho + iV_t = 0 \qquad 0 < \rho < a \qquad |t| < b$$

and hence V is an analytic function of $\rho + it$ on this set. Since $u(x, y, t)$ is continuous on $0 \le \rho < a$, $|t| < b$, so is $U(\rho, t)$. Moreover, $U(0, t) = 0$ by Eq. (1-19). Thus

$$\text{Re } V(0, t) = 0 \qquad |t| < b \tag{1-22}$$

Since V is analytic in $0 < \rho < a$, $|t| < b$, and its real part vanishes for $\rho = 0$, we know that we can continue V analytically across the line $\rho = 0$ (see any good book on complex variables). In particular, $V(0, t)$ is an analytic function of t in $|t| < b$ (in the sense of power series). But $V(0, t) = \pi i g(t)$. Thus, we have shown that in order for Eq. (1-8) to have a solution when f depends on t alone, it is necessary that f be an analytic function of t. If we take, for example

$$g(t) = \begin{cases} e^{-1/t} & t > 0 \\ 0 & t \le 0 \end{cases} \tag{1-23}$$

then f has continuous derivatives of all orders, but is not analytic in any neighborhood of $t = 0$. Hence, Eq. (1-8) can have no solution for such an f.

Now we can give an example of a "real" equation without solutions. Let Au stand for the left-hand side of Eq. (1-15). Let $\bar{A}$ represent the operator obtained from A by taking the complex conjugate of all of the coefficients in A.

Then
$$A\bar{A}u = \tfrac{1}{4}(u_{xx} + u_{yy}) + (xu_{yt} - yu_{xt}) + (x^2 + y^2)u_{tt} - iu_t \tag{1-24}$$

This, unfortunately, does not quite make the grade because of the last term. But we do have

$$A\bar{A} = B - i\frac{\partial}{\partial t}$$

where B is a linear operator with real coefficients.

Thus
$$A\bar{A}(\overline{A\bar{A}}) = \left(B - i\frac{\partial}{\partial t}\right)\left(B + i\frac{\partial}{\partial t}\right) = B^2 + \frac{\partial^2}{\partial t^2}$$

Now, I claim that the equation

$$B^2u + u_{tt} = f \tag{1-25}$$

cannot have a solution when $f = g'$ and g is given by Eq. (1-23). For if u were a solution of Eq. (1-25), then $v = \frac{1}{2}\bar{A}(\overline{A\bar{A}})u$ would be a solution of Eq. (1-8), contradicting our previous result. This example was given by F. Treves (1962).

It might be noted that Eq. (1-25) would be much harder to deal with directly. The fact that we were allowed to use complex-valued functions brought about a great savings. This is true in many other situations in the study of partial differential equations.

1-3 INTEGRATION BY PARTS

In Sec. 1-2 we employed an elementary but very useful technique, which we will review here for the benefit of anyone who is a bit rusty. It is integration by parts. Let Ω be an open, connected set (domain) in E^n with a piecewise smooth boundary. This means that the boundary $\partial\Omega$ of Ω consists of a finite number of surfaces each of which can be expressed in the form

$$x_j = h(x_1, \ldots, x_{j-1}, x_{j+1}, \ldots, x_n)$$

for some j, with the function h having continuous first derivatives. The closure $\bar{\Omega}$ of Ω is the union of Ω and its boundary $\partial\Omega$. Assume that Ω is bounded, i.e., that it is contained in some Σ_R for R sufficiently large. If $f \in C^1(\bar{\Omega})$, then

$$\int_\Omega \frac{\partial f}{\partial x_k}\, dx = \int_{\partial\Omega} f\gamma_k\, d\sigma \qquad 1 \le k \le n \tag{1-26}$$

where $dx = dx_1 \cdots dx_n$, γ_k is the cosine of the angle between the x_k-axis and the outward normal to $\partial\Omega$, and $d\sigma$ is the surface element on $\partial\Omega$. (Note that we use only one integral sign for a volume integral; it would not be easy to write n of them.) Equation (1-26) has many names attached to it, including Gauss, Green, Stokes, divergence, etc. For a proof we can refer to any good book on advanced calculus, e.g., Spivak (1965).

Now suppose u and v have continuous derivatives in $\bar{\Omega}$ and their product vanishes on $\partial\Omega$. Then, by Eq. (1-26), we have

$$\int_\Omega u\, \frac{\partial v}{\partial x_k}\, dx = -\int_\Omega \frac{\partial u}{\partial x_k}\, v\, dx \qquad 1 \le k \le n \tag{1-27}$$

This is the formula employed in Sec. 1-2. It is a very convenient one, since it allows us to "throw" derivatives from one function to another. It is so convenient that the first general rule for all people studying partial differential equations is: when you do not know what to do next, integrate by parts.

There is one feature of Eq. (1-27) which appears harmless, but which has done more to fill mental institutions with partial differential equations people than any other single factor, namely, the minus sign. However, there is a way of avoiding it. The method is as follows. As agreed before, we can allow complex-valued functions provided we understand that there need not be any connection between their real and imaginary parts. Moreover, it is easy to check that Eqs. (1-26) and (1-27) hold for such functions. Thus, if we take $v = \bar{w}$ in Eq. (1-27) and set

$$D_k = -i \frac{\partial}{\partial x_k} \tag{1-28}$$

then we have

$$\int_\Omega u \, \overline{D_k w} \, dx = \int_\Omega D_k u \bar{w} \, dx \tag{1-29}$$

Presto, the minus sign has disappeared.

This calls for a slight change in the notation used in Sec. 1-1. We now write:

$$D^\mu = D_1^{\mu_1} \cdots D_n^{\mu_n} = \frac{(-i)^{|\mu|} \partial^{|\mu|}}{\partial x_1^{\mu_1} \cdots \partial x_n^{\mu_n}} \tag{1-30}$$

As before, every linear operator can be written in the form of Eq. (1-4), but in converting from Eq. (1-3) to Eq. (1-4) we must now multiply the coefficients by powers of i.

In Sec. 1-2, we considered an expression of the form

$$\int_\Omega Au\bar{\varphi} \, dx$$

and integrated by parts. Now suppose A is given by Eq. (1-4) with each coefficient $a_\mu(x)$ in $C^m(\bar{\Omega})$. Assume also that the function φ is in $C^m(\bar{\Omega})$ as well, and vanishes on and near the boundary $\partial\Omega$. Then we can, by repeated use of Eq. (1-29), throw all derivatives in A over on to φ. This gives

$$\int_\Omega Au\bar{\varphi} \, dx = \int_\Omega u\overline{A'\varphi} \, dx \tag{1-31}$$

where

$$A'\varphi = \sum_{|\mu| \le m} D^\mu(\bar{a}_\mu \varphi) \tag{1-32}$$

(See, no minus signs!) We are too lazy to carry out the differentiations, but we know that after they are all carried out we can write A' in the form

$$A' = \sum_{|\mu| \le m} b_\mu(x) D^\mu \tag{1-33}$$

Thus, A' is a linear partial differential operator just like A. Its coefficients, b_μ, depend only on the coefficients of A and their derivatives. We call A' the *formal adjoint* of A. As we saw in Sec. 1-2, the formal adjoint of $\partial/\partial\bar{z} + iz \, \partial/\partial t$ is $\partial/\partial z - i\bar{z} \, \partial/\partial t$.

Now let u be a function continuous in Ω and suppose

$$\int_\Omega u\bar{\varphi} \, dx = 0 \tag{1-34}$$

for all functions $\varphi \in C^\infty(\Omega)$ which vanish near $\partial\Omega$. Then I claim that u vanishes identically in Ω. For suppose there were a point $x_0 \in \Omega$ such that $\operatorname{Re} u(x_0) > 0$. Since

u is continuous, Re $u(x) > 0$ in some neighborhood of x_0, say for $|x - x_0| < r$,

where
$$|x|^2 = x_1^2 + \cdots + x_n^2$$

We claim that we can find a function $\varphi \in C^\infty(\Omega)$, such that

$$\varphi(x) > 0 \qquad \text{for} \qquad |x - x_0| < r$$

$$\varphi(x) = 0 \qquad \text{for} \qquad |x - x_0| \geq r$$

Assuming this for the moment, we note that $u\varphi$ has the same properties.

Hence
$$\text{Re} \int_\Omega u\varphi \, dx > 0$$

But this contradicts Eq. (1-34). Similarly, we cannot have Re $u(x) < 0$ anywhere, and the same holds for Im u as well. Hence $u \equiv 0$.

To construct our function φ, we set

$$\begin{aligned} j(x) &= a \exp\left[(|x|^2 - 1)^{-1}\right] \qquad &|x| < 1 \\ &= 0 &|x| \geq 1 \end{aligned} \tag{1-35}$$

where a is a constant $\neq 0$. It is left as an exercise to verify that $j(x) \in C^\infty(E^n)$. We now merely take $\varphi(x) = j[(x - x_0)/r]$.

In later chapters it will be useful to know the following fact. Let Ω be a bounded domain, and let K be any bounded closed subset of Ω. Then there is a $\psi \in C^\infty(\Omega)$ which vanishes near $\partial\Omega$, and such that

$$0 \leq \psi(x) \leq 1 \qquad x \in \Omega \tag{1-36}$$

$$\psi(x) = 1 \qquad x \in K \tag{1-37}$$

To construct ψ, note that there is an $\varepsilon > 0$ such that the distance from any point in K to any point in $\partial\Omega$ is always $> 3\varepsilon$ (the proof is left as an exercise). Let K_ε be the set of all $x \in \Omega$ such that there is a $y \in K$ satisfying $|x - y| < \varepsilon$. Choose a in Eq. (1-35) so that

$$\int_{|x|<1} j(x) \, dx = 1 \tag{1-38}$$

Then define ψ to be

$$\psi(x) = \varepsilon^{-n} \int_{K_\varepsilon} j\left(\frac{x - y}{\varepsilon}\right) dy \tag{1-39}$$

By differentiating under the integral sign, one verifies easily that $\psi \in C^\infty(E^n)$ (this is also left as an exercise). Now, if x is within a distance of ε of $\partial\Omega$, then it is a distance $\geq 2\varepsilon$ from K and, hence, a distance $\geq \varepsilon$ from K_ε. In this case, the integral vanishes identically in Eq. (1-39) showing that $\psi(x) = 0$. Thus, ψ vanishes near $\partial\Omega$. If $x \in K$, then

$$\psi(x) = \varepsilon^{-n} \int_{|x-y|<\varepsilon} j\left(\frac{x-y}{\varepsilon}\right) dy$$

$$= \int_{|z|<1} j(z)\, dz = 1$$

Since $j(x) \geq 0$, in general we have $\psi(x) \geq 0$ and

$$\psi(x) \leq \varepsilon^{-n} \int_{|x-y|<\varepsilon} j\left(\frac{x-y}{\varepsilon}\right) dy = 1$$

This proves the desired properties.

Recall that we have assumed throughout that Ω has a piecewise smooth boundary. We shall continue to do so in the future unless otherwise stated.

1-4 A NECESSARY CONDITION

Now that we have seen that a linear partial differential equation need not have a solution, it is natural to ask which equations have solutions.

To tackle this problem, let Ω be a domain in E^n and let A be a linear operator of the form of Eq. (1-6). Suppose we are given a function f defined on Ω. We want to know whether or not the equation

$$Au = f \tag{1-40}$$

has a solution in Ω. Since we are novices, we are willing to assume as much smoothness of $\partial\Omega$, f, and the coefficients of A, as necessary to obtain a solution. This is not enough, however, as the example in Sec. 1-2 shows.

Since we do not know how to begin, let us start from the other end. Let us assume that Eq. (1-40) has a solution, and see if this implies anything concerning A, f, or Ω. To apply the first general principle of partial differential equations (see Sec. 1-3), we let φ be any function in $C^\infty(\Omega)$ which vanishes near $\partial\Omega$ and for $|x|$ large (in case Ω is not bounded). The reason for these restrictions on φ is that we want to apply Eq. (1-31). Note that Eq. (1-31) is known to hold for bounded domains. However, if Ω is unbounded, we can still use it if we make φ vanish outside some sphere Σ_R for R sufficiently large. For then we are, effectively, integrating over a bounded domain. Note that R can vary depending on the function φ, as long as it is finite for each φ.

The set of functions in $C^\infty(\Omega)$ which vanish near $\partial\Omega$ and outside some Σ_R have a special name because of their usefulness. They are called *test* functions. In general, the *support* of a function is the closure of the set of points where it does not vanish. Thus, a test function in Ω is a function in $C^\infty(\Omega)$ having bounded support in Ω. Since closed bounded sets in E^n are compact, the term *compact support* is sometimes used.

Let us introduce the notation

$$(u, v) = \int_\Omega u(x)\overline{v(x)}\,dx \tag{1-41}$$

Thus, Eq. (1-29) becomes

$$(u, D_k v) = (D_k u, v) \tag{1-42}$$

Now returning to Eq. (1-41), we have by Eq. (1-31)

$$(f, \varphi) = (Au, \varphi) = (u, A'\varphi) \tag{1-43}$$

(note that we are assuming that the coefficients of A are in $C^m(\overline{\Omega})$). Thus, by the Schwarz inequality (see Sec. 1-5)

$$|(f, \varphi)| = |(u, A'\varphi)| \le \|u\| \; \|A'\varphi\|$$

where
$$\|u\|^2 = (u, u) = \int_\Omega |u|^2\,dx$$

From this we see immediately that a necessary condition for Eq. (1-40) to have a solution is that

$$|(f, \varphi)| \le C\,\|A'\varphi\| \tag{1-44}$$

hold for all test functions φ in Ω and some constant C. Is it sufficient? In order to find out, let us try to work backwards from inequality (1-44). In doing so we shall need some facts about Hilbert space: these are reviewed in Sec. 1-5 (those familiar with the facts can leave out that section, skipping straight to Sec. 1-6).

1-5 SOME NOTIONS FROM HILBERT SPACE

We recall the definition of a complex Hilbert space H. It consists of elements $u, v, w, \ldots$, for which are defined the operations of addition and multiplication by complex numbers (scalars) $\alpha, \beta, \gamma, \ldots$, and such that

$$u + v = v + u \tag{1-45}$$

$$(u + v) + w = u + (v + w) \tag{1-46}$$

$$(\alpha + \beta)u = \alpha u + \beta u \tag{1-47}$$

$$\alpha(u + v) = \alpha u + \alpha v \tag{1-48}$$

$$\alpha(\beta u) = (\alpha\beta)u \tag{1-49}$$

There is an element 0 of H, such that

$$u + 0 = u \tag{1-50}$$

For each pair u, v of elements of H there is defined a complex number called their scalar product and denoted by (u, v), such that

$$(\alpha u + \beta v, w) = \alpha(u, w) + \beta(v, w) \tag{1-51}$$

$$(u, v) = \overline{(v, u)} \tag{1-52}$$

$$(u, u) > 0 \quad \text{when} \quad u \neq 0 \tag{1-53}$$

This list is not complete, but let us stop for a moment and examine (1-45) to (1-53). First, we note that

$$0u = 0 \tag{1-54}$$

For we have, by Eqs. (1-47) and (1-51)

$$(0u, 0u) = (1u + (-1)u, 0u) = (u, 0u) - (u, 0u) = 0$$

showing that $0u = 0$ by (1-53). From this we see that

$$(0, u) = 0 \tag{1-55}$$

For by Eqs. (1-54) and (1-51)

$$(0, u) = (0u, u) = 0(u, u) = 0$$

Similarly, we have

$$u + (-1)u = 0 \tag{1-56}$$

For if u, v are any elements of H, we have

$$\begin{aligned}(u + (-1)u, v) &= 1(u + (-1)u, v) \\ &= (1u + (-1)u, v) \\ &= (0, v) = 0\end{aligned}$$

by Eqs. (1-51), (1-48), (1-49), (1-47), (1-54), and (1-55). Taking $v = u + (-1)u$, we see that $u + (-1)u = 0$, by (1-53). Adding $1u$ to both sides of Eq. (1-56) and applying Eqs. (1-46), (1-47), (1-54), and (1-50), we get

$$u = 1u \tag{1-56a}$$

From Eq. (1-56) we see that we can define subtraction. For if $u + v = w$, then $u + v + (-1)v = w + (-1)v$ and, thus, $u = w + (-1)v$ by Eqs. (1-46), (1-56), and (1-50). We write $-v$ for $(-1)v$ and $w - v$ for $w + (-1)v$. From Eqs. (1-49) and (1-54), it follows that

$$\alpha 0 = 0 \tag{1-57}$$

If we set

$$\| u \| = (u, u)^{1/2}$$

we have

$$\| \alpha u \| = |\alpha| \; \| u \| \tag{1-58}$$

$$\| u \| > 0 \quad \text{if} \quad u \neq 0 \tag{1-59}$$

$$\| u + v \| \leq \| u \| + \| v \| \tag{1-60}$$

$$\| u + v \|^2 + \| u - v \|^2 = 2 \| u \|^2 + 2 \| v \|^2 \tag{1-61}$$

$$|(u, v)| \leq \| u \| \; \| v \| \tag{1-62}$$

We call $\| u \|$ the *norm* of u. Statements (1-60), (1-61), and (1-62) are called the *triangle inequality*, the *parallelogram law*, and the *Schwarz inequality*, respectively. The parallelogram law is the most easily proved of the three. It follows from Eqs. (1-51) and (1-52) since

$$\| u + v \|^2 = (u + v, u + v) = \| u \|^2 + (u, v) + (v, u) + \| v \|^2$$

$$\| u - v \|^2 = (u - v, u - v) = \| u \|^2 - (u, v) - (v, u) + \| v \|^2$$

To prove statement (1-62), consider the expression

$$\| \alpha u + v \|^2 = (\alpha u + v, \alpha u + v) = |\alpha|^2 \;\| u \|^2 + 2 \operatorname{Re} \alpha(u, v) + \| v \|^2$$

$$= \left| \alpha \| u \| + \frac{(v, u)}{\| u \|} \right|^2 + \| v \|^2 - \frac{|(u, v)|^2}{\| u \|^2}$$

where we have assumed that $u \neq 0$. [If $u = 0$, then statement (1-62) follows from Eq. (1-55).] Take α to be such that the first term on the right vanishes, i.e., $\alpha = -(v, u)/\| u \|^2$.

Hence
$$\| v \|^2 - \frac{|(u, v)|^2}{\| u \|^2} \geq 0$$

which is precisely (1-62). Once we have the Schwarz inequality, (1-60) follows easily from

$$\begin{aligned} \| u + v \|^2 &= \| u \|^2 + 2 \operatorname{Re} (u, v) + \| v \|^2 \\ &\leq \| u \|^2 + 2 \| u \| \; \| v \| + \| v \|^2 \\ &= (\| u \| + \| v \|)^2 \end{aligned}$$

There is, finally, one more axiom which must be added to the list (1-45) to (1-53), namely

$$H \text{ is complete} \tag{1-63}$$

By this we mean that for any sequence $\{u_k\}$ of elements of H which satisfy

$$\| u_j - u_k \| \to 0 \quad \text{as} \quad j, k \to \infty \tag{1-64}$$

there is an element $u \in H$, such that

$$\| u - u_k \| \to 0 \quad \text{as} \quad k \to \infty \tag{1-65}$$

A sequence satisfying statement (1-64) is called a *Cauchy sequence*. Note that (1-65) implies (1-64) by (1-60). Completeness is very important in applications. We shall sometimes abbreviate (1-65) to

$$u_k \to u \text{ in } H \quad \text{as} \quad k \to \infty \tag{1-66}$$

We now turn to some important properties which all Hilbert spaces have. A subset S of H is called a *subspace* if $\alpha u + \beta v$ is in S for all scalars α, β whenever u and v are in S. It is called *closed* if every Cauchy sequence of elements in S converges to an element of S. Clearly, a closed subspace of a Hilbert space is itself a Hilbert space.

Lemma 1-2 Let M be a closed subspace of H. Then for every $u \in H$ not in M there is a $v \in M$, such that

$$\| u - v \| = \operatorname*{glb}_{w \in M} \| u - w \| \tag{1-67}$$

PROOF Set $d = \text{glb} \| u - w \|$, $w \in M$. Then there is a minimizing sequence $\{w_k\} \subseteq M$, such that $\| u - w_k \| \to d$, as $k \to \infty$. From Eq. (1-61), we see that

$$4 \| u - \tfrac{1}{2}(w_k + w_j) \|^2 + \| w_k - w_j \|^2 = 2(\| u - w_k \|^2 + \| u - w_j \|^2) \to 4d^2$$

$$\text{as} \quad j, k \to \infty$$

Since $\frac{1}{2}(w_k + w_j) \in M$ (it was for this reason that the hypotheses were made on M; it shows where they can be weakened):

$$4 \| u - \tfrac{1}{2}(w_k + w_j) \|^2 \geq 4d^2$$

and hence $\| w_k - w_j \| \to 0$ as $j, k \to \infty$. By the completeness of H, there is a $v \in M$, such that $\| w_k - v \| \to 0$. This means that

$$\| u - v \| = \lim \| u - w_k \| = d$$

and the proof is complete.

Theorem 1-3: Projection theorem Let M be a closed subspace of H. Then for every $u \in H$ there is a $v \in M$, such that $(u - v, M) = 0$ (i.e., such that $(u - v, w) = 0$ for all $w \in M$).

PROOF If $u \in M$, set $v = u$. Otherwise, we know by Lemma 1-2 that there is a $v \in M$ such that $\| u - v \| = d$, the "distance" from u to M. Now, if $w \neq 0$ is any element of M

$$\begin{aligned} \| u - v \|^2 &\leq \| u - v - \alpha w \|^2 \\ &= \| u - v \|^2 - 2 \operatorname{Re} \bar{\alpha}(u - v, w) + |\alpha|^2 \| w \|^2 \end{aligned}$$

for all complex α. In particular, this holds if we take

$$\alpha = \frac{(u - v, w)}{\| w \|^2}$$

Thus $$\| u - v \|^2 \leq \| u - v \|^2 - 2 \frac{|(u - v, w)|^2}{\| w \|^2} + \frac{|(u - v, w)|^2}{\| w \|^2}$$

and hence $$|(u - v, w)|^2 \leq 0$$

This, of course, can be true only if

$$(u - v, w) = 0$$

Hence $(u - v, M) = 0$, and the proof is complete.

Corollary 1-4 If $M \neq H$, there is an element $w \neq 0$ of H, such that $(w, M) = 0$.

PROOF By assumption, there is a $u \in H$ which is not in M. By Theorem 1-3, there is a $v \in M$, such that $(u - v, M) = 0$. Set $w = u - v$. Clearly $w \neq 0$.

A bounded linear functional F on H is a complex-valued function on H, which satisfies the following conditions:

1. $$F(\alpha u + \beta v) = \alpha F u + \beta F v$$
   for all complex α, β and $u, v \in H$.
2. There is a constant K, such that
   $$|Fu| \leq K \| u \| \qquad u \in H \tag{1-68}$$
   We define
   $$\| F \| = \underset{u \in H}{\text{lub}} \frac{|Fu|}{\| u \|} \tag{1-69}$$

Note that a bounded linear functional is continuous in the sense that $u_k \to u$ in H implies $Fu_k \to Fu$. For by (1-68)

$$|F(u_k - u)| \leq K \| u_k - u \| \to 0 \quad \text{as} \quad k \to \infty$$

Theorem 1-5: Fréchet–Riesz For every bounded linear functional F on H there is an element $f \in H$, such that

$$Fu = (u, f) \qquad u \in H \tag{1-70}$$

and

$$\| F \| = \| f \| \tag{1-71}$$

PROOF Let N be the set of all $v \in H$, such that $Fv = 0$. N is a subspace of H. For if u and v are in N, $F(\alpha u + \beta v) = \alpha F u + \beta F v = 0$ for all scalars α, β and hence $\alpha u + \beta v \in N$. Moreover, N is a closed subspace of H. For if $v_k \in N$ and $v_k \to v$, then

$$|Fv| = |F(v - v_k)| \leq K \| v - v_k \| \to 0 \quad \text{as} \quad k \to \infty$$

Hence $v \in N$.

Now if $N = H$, the theorem is easily proved by setting $f = 0$. Otherwise there is a $w \neq 0$ in H, such that $(w, N) = 0$ (Corollary 1-4). Therefore, $Fw \neq 0$ and

$$F\left(u - \frac{Fu}{Fw} w\right) = Fu - \frac{Fu}{Fw} Fw = 0$$

Thus, $u - (Fu/Fw)w$ is in N for all $u \in H$. In particular, this means that

$$\left(u - \frac{Fu}{Fw} w, w\right) = 0$$

i.e., that

$$(u, w) = \frac{Fu}{Fw} \| w \|^2$$

Thus
$$Fu = \left(u, \frac{w}{\|w\|^2} \overline{Fw}\right)$$

This proves Eq. (1-70) if we take $f = w\overline{Fw}/\|w\|^2$. To obtain Eq. (1-71), we note that, by Eq. (1-70) and (1-62)

$$|Fu| \leq \|f\| \; \|u\|$$

This shows that $\|F\| \leq \|f\|$. On the other hand, if we take $u = f$ in Eq. (1-70), we have

$$Ff = \|f\|^2$$

or
$$\|f\| = \frac{|Ff|}{\|f\|} \leq \|F\|$$

Thus, the theorem is proved.

Let S be a subspace of H. We define S (the *closure* of S) to be the set of all $f \in H$ which are the limits of elements of S. Thus, $f \in \overline{S}$ if there is a sequence $\{v_k\} \leq S$, such that

$$\|v_k - f\| \to 0$$

Clearly $S \subset \overline{S}$ and $S = \overline{S}$, if S is closed. It is easily checked that $\overline{S}$ is a closed subspace of H and is the "smallest" one containing S.

Theorem 1-6: Hahn–Banach Let S be a subspace of H, and let F be a bounded linear functional on S, satisfying

$$|Fu| \leq K_0 \|u\| \qquad u \in S \tag{1-72}$$

Then there is a bounded linear functional G on the whole of H, such that

$$Gu = Fu \qquad u \in S \tag{1-73}$$

and
$$|Gv| \leq K_0 \|v\| \qquad v \in H \tag{1-74}$$

PROOF We first extend F to $\overline{S}$. This is done as follows. If $\{v_k\} \subset S$ and $\|v_k - f\| \to 0$, set

$$Ff = \lim Fv_k$$

This limit exists since

$$|F(v_k - v_j)| \leq K_0 \|v_k - v_j\| \to 0$$

It is independent of the particular sequence $\{v_k\}$ for if $\{v_k'\}$ is another, then

$$|Fv_k - Fv_k'| = |F(v_k - v_k')| \leq K_0 \|v_k - v_k'\|$$
$$\leq K_0(\|v_k - f\| + \|f - v_k'\|) \to 0$$

Moreover, $\quad |Ff| = \lim |Fv_k| \leq \lim K_0 \|v_k\| = K_0 \|f\|$

Hence, F can be extended to be a bounded linear functional on $\bar{S}$. Now, for every $w \in H$ there is a $w_1 \in \bar{S}$, such that $(w - w_1, \bar{S}) = 0$ (Theorem 1-3). We set

$$Gw = Fw_1$$

Clearly G satisfies condition 1 of the definition of a bounded linear functional (see Corollary 1-4). That it satisfies condition 2 follows from the fact that

$$\| w \|^2 = \| w_1 \|^2 + \| w - w_1 \|^2$$

Thus
$$\| w_1 \| \le \| w \|$$

and
$$|Gw| = |Fw_1| \le K_0 \| w_1 \| \le K_0 \| w \|$$

and the proof is complete.

This is all we need for the present chapter. Now we shall give some further results which we want to use later on.

Lemma 1-7 Let S be a subspace of H. Denote the set of all $u \in H$ satisfying $(u, S) = 0$ by $S^\perp$. Clearly $S^\perp$ is a closed subspace of H. Moreover, we have

$$(S^\perp)^\perp = \bar{S}$$

PROOF Suppose $u \in \bar{S}$. Then there is a sequence $\{u_k\}$ of elements of S, such that $u_k \to u$ in H. Thus, if $w \in S^\perp$

$$(u, w) = \lim\,(u_k, w) = 0$$

Hence $u \in (S^\perp)^\perp$. Conversely, suppose $u \in (S^\perp)^\perp$. If $u \notin \bar{S}$, then $u = v + w$, where $v \in \bar{S}$ and $w \in \bar{S}^\perp$ (Theorem 1-3). Since both u and v are in $(S^\perp)^\perp$, so is w. Thus, $(w, S^\perp) = 0$. But $w \in S^\perp$ and, hence, $w = 0$. Thus, $u = v \in \bar{S}$.

We shall call a sequence $\{v_k\}$ of H *weakly convergent* to an element $v \in H$, if

$$(v_k - v, w) \to 0$$

for each $w \in H$.

Theorem 1-8 If $\{v_k\}$ is a sequence, such that

$$\| v_k \| \le C \tag{1-75}$$

then it has a subsequence which converges weakly.

PROOF Now for fixed j the complex numbers (v_k, v_j) are bounded. Hence, by the usual diagonal procedure we can find a subsequence $\{\hat{v}_k\}$ of $\{v_k\}$, such that $(\hat{v}_k, v_j)$ converges for each fixed j. Thus, $(\hat{v}_k, v_j)$ converges if f is in the set S of all elements of H which can be written in the form

$$f = \sum_{i=1}^{N} \alpha_i v_i$$

for any integer N. It also converges if f is the limit of a sequence $\{f_n\}$ of elements of S. For

$$|(\hat{v}_k - \hat{v}_l, f)| \leq |(\hat{v}_k - \hat{v}_l, f - f_n)| + |(\hat{v}_k - \hat{v}_l, f_n)|$$
$$\leq 2C \|f - f_n\| + |(\hat{v}_k - \hat{v}_l, f_n)|$$

For any $\varepsilon > 0$ we can take n so large that $2C\|f - f_n\| < \varepsilon/2$ and, once n is chosen, we take k and l so large that $|(\hat{v}_k - \hat{v}_l, f_n)| < \varepsilon/2$. This shows that $(\hat{v}_k, f)$ converges for each $f \in \overline{S}$. By Theorem 1-3 for every $w \in H$ there is a $w_1 \in \overline{S}$, such that

$$(w - w_1, \overline{S}) = 0$$

This means that

$$(\hat{v}_k, w) = (\hat{v}_k, w - w_1) + (\hat{v}_k, w_1) = (\hat{v}_k, w_1)$$

which shows that $(\hat{v}_k, w)$ converges for each $w \in H$.

If this is the case, we can set $Fw = \lim (w, \hat{v}_k)$. Then, F is a bounded linear functional on H. Applying Theorem 1-5, we see that there is a $v \in H$, such that

$$Fw = (w, v)$$

i.e., such that

$$(\hat{v}_k - v, w) \to 0 \tag{1-76}$$

and the proof is complete.

Theorem 1-9: Banach–Saks Under the hypotheses of Theorem 1-8, there is a subsequence $\{\tilde{v}_k\}$ of $\{v_k\}$ and a $v \in H$, such that

$$\left\| \frac{\tilde{v}_1 + \cdots + \tilde{v}_k}{k} - v \right\| \to 0 \quad \text{as} \quad k \to \infty$$

Proof By Theorem 1-8 there is a subsequence $\{\hat{v}_k\}$ of $\{v_k\}$ and a $v \in H$, such that (1-76) holds. Set $u_k = \hat{v}_k - v$. Then $\|u_k\| \leq C_1$ for some constant C_1 and $(u_k, w) \to 0$ for each $w \in H$. We now pick a subsequence $\{\tilde{u}_k\}$ of $\{u_k\}$, such that

$$|(\tilde{u}_1, \tilde{u}_{k+1})| \leq \frac{1}{k}, \ldots, |(\tilde{u}_k, \tilde{u}_{k+1})| \leq \frac{1}{k}$$

This is easily done by induction.

Thus
$$\sum_{i=1}^{k} |(\tilde{u}_i, \tilde{u}_{k+1})| \leq 1 \qquad k = 1, 2, \ldots,$$

and, consequently

$$\sum_{j=2}^{k} \sum_{i=1}^{j-1} |(\tilde{u}_i, \tilde{u}_j)| < k$$

Thus
$$\left\| \frac{(\tilde{u}_1 + \cdots + \tilde{u}_k)}{k} \right\|^2 = k^{-2} \sum_{j=1}^{k} \sum_{i=1}^{k} (\tilde{u}_i, \tilde{u}_j)$$
$$\leq k^{-2} \left\{ kC_1^2 + 2 \sum_{j=2}^{k} \sum_{i=1}^{j-1} |(\tilde{u}_i, \tilde{u}_j)| \right\}$$
$$< k^{-1}(C_1^2 + 2) \to 0 \quad \text{as} \quad k \to \infty$$

we now set $\tilde{v}_k = \tilde{u}_k + v$ to complete the proof.

Theorem 1-10: Banach–Steinhaus Let $\{F_n\}$ be a sequence of bounded linear functionals on H, such that, for each $u \in H$

$$\operatorname*{lub}_n |F_n(u)| < \infty \tag{1-77}$$

Then there is a constant C such that

$$|F_n(u)| \leq C \|u\| \qquad u \in H \tag{1-78}$$

PROOF We first note that if there is a $v_0 \in H$ and constants $\delta > 0$, K, such that

$$|F_n(v)| \leq K \quad \text{for} \quad \|v - v_0\| < \delta \tag{1-79}$$

then statement (1-78) holds. For let u be any element of H, and set $v = v_0 + \delta u/2\|u\|$. Then $\|v - v_0\| < \delta$. Therefore, by (1-79)

$$|F_n(v)| \leq K$$

But
$$F_n(v) = F_n(v_0) + \frac{\delta F_n(u)}{2\|u\|}$$

Hence
$$|F_n(u)| \leq \frac{2\|u\|(|F_n(v_0)| + |F_n(v)|)}{\delta}$$
$$\leq \frac{4K\|u\|}{\delta}$$

showing that (1-78) holds with $C = 4K/\delta$.

If statement (1-79) did not hold, there would be a $u_1 \in H$ and an integer n_1, such that

$$\|u_1\| = 1 \qquad |F_{n_1}(u_1)| > 1$$

Since F_{n_1} is continuous, there is a $\delta_1 > 0$, such that

$$|F_{n_1}(u)| > 1 \quad \text{for} \quad \|u - u_1\| < \delta_1$$

We take $\delta_1 < 1$. We now claim that there must also be a $u_2 \in H$ and an integer $n_2 > n_1$, such that

$$\|u_2 - u_1\| < \delta_1 \qquad |F_{n_2}(u_2)| > 2$$

for otherwise (1-79) will hold with $K = 2$, $v_0 = v_1$, and $\delta = \delta_1$. By continuity, there is a $\delta_2 > 0$, such that

$$|F_{n_2}(u)| > 2 \quad \text{for} \quad \|u - u_2\| < \delta_2$$

Take $\delta_2 < \min[\delta_1 - \|u_1 - u_2\|, \frac{1}{2}]$. This is to guarantee that $\|u - u_2\| < \delta_2$ implies $\|u - u_1\| < \delta_1$. Continuing, we see that there are a $u_3 \in H$ and an integer $n_3 > n_2$, such that

$$\|u_3 - u_2\| < \delta_2 \qquad |F_{n_3}(u_3)| > 3$$

Thus, there is a $\delta_3 > 0$, such that

$$|F_{n_3}(u)| > 3 \quad \text{for} \quad \|u - u_3\| < \delta_3$$

We take $\delta_3 < \min[\delta_2 - \|u_3 - u_2\|, \frac{1}{3}]$. Continuing inductively, we find that there are a $u_k \in H$, an $n_k > n_{k-1}$, and a $\delta_k > 0$, satisfying

$$\delta_k < \min\left[\delta_{k-1} - \|u_k - u_{k-1}\|, \frac{1}{k}\right]$$

such that
$$|F_{n_k}(u)| > k \quad \text{for} \quad \|u - u_k\| < \delta_k \tag{1-80}$$

Now, if $j > k$, we have

$$\begin{aligned}\|u_j - u_k\| &\le \sum_{i=k}^{j-1} \|u_{i+1} - u_i\| \\ &\le \sum_{i=k}^{j-1} \delta_i - \delta_{i+1} \\ &= \delta_k - \delta_j < \frac{1}{k}\end{aligned} \tag{1-81}$$

Hence $\{u_k\}$ is a Cauchy sequence in H. By the completeness of H, there is a $u_0 \in H$, such that $u_k \to u_0$ in H. Holding k fixed and letting $j \to \infty$ in Eq. (1-81), we have

$$\|u_0 - u_k\| \le \delta_k$$

This shows, by (1-80), that

$$|F_{n_k}(u_0)| \ge k$$

Hence
$$\operatorname*{lub}_n |F_n(u_0)| = \infty$$

contrary to assumption. This proves the theorem.

Corollary 1-11 If $\{v_k\}$ is a weakly convergent sequence in H, then there is a constant C, such that (1-75) holds.

PROOF Set $F_k u = (u, v_k)$. Then $\{F_k\}$ is a sequence of bounded linear functionals on H. Moreover, for each u, statement (1-77) holds since the complex numbers (u, v_n) are convergent and, hence, bounded. Thus, there is a constant C, such that

$$|(u, v_n)| \leq C \| u \| \qquad u \in H \tag{1-82}$$

Taking $u = v_n$, we have

$$\| v_n \|^2 \leq C \| v_n \|$$

or

$$\| v_n \| \leq C$$

This completes the proof.

It sometimes arises, in practice, that a space H satisfies statements (1-45) to (1-53) but is not complete. If the property of completeness is desired (and it usually is), the following "trick" may be used. Let $\{u_k\}$ be any Cauchy sequence in H. If there is a $u \in H$ satisfying (1-65), all well and good. If not, we have located a "hole" in H. We fill this hole by inventing a new element u. We define u to be the limit of $\{u_k\}$ in H. We then let $\bar{H}$ be the set of all limits of Cauchy sequences in H. $\bar{H}$ is called the *completion* of H. It is not immediately clear that $\bar{H}$ is a Hilbert space, but a careful analysis shows that it satisfies all of the axioms including (1-63). Even so, it is not quite clear what is the nature of the new elements. In some cases we do not care; we only want to apply theorems from Hilbert space theory. In others we are able to describe the "ideal" elements quite accurately.

We now give another method of constructing $\bar{H}$. Let H be a space satisfying (1-45) to (1-53) but not (1-63). Let W be the collection of bounded linear functionals F which are of the form (1-70). Note that there is only one $f \in H$ that will work. For if $Fu = (u, f_1)$ as well, then $(u, f - f_1) = 0$ for all $u \in H$ and, consequently, $f_1 = f$. Note also that Eq. (1-71) holds, and note the reasoning at the end of the proof of Theorem 1-5.

If

$$Fu = (u, f) \qquad Gu = (u, g) \qquad u \in H$$

and we define

$$(F, G) = (g, f) \tag{1-83}$$

then it is easily verified that W satisfies (1-45) to (1-53). In fact, we can identify W with H through the correspondence $F \leftrightarrow f$.

Next, let Z be the set of all bounded linear functionals on H which are the limits in norm of functionals in W. By this we mean that $F \in Z$, if there is a sequence $\{F_n\}$ of functionals in W, such that

$$\| F_n - F \| \to 0 \quad \text{as} \quad n \to \infty \tag{1-84}$$

If (1-84) holds, and

$$F_n u = (u, f_n) \qquad u \in H$$

then $\{f_n\}$ is a Cauchy sequence in H; this follows from Eq. (1-71). If, in addition, $G_n u = (u, g_n)$ and $\| G_n - G \| \to 0$, we set

$$(F, G) = \lim (g_n, f_n)$$

From the fact that $\bar{W}$ satisfies (1-45) to (1-53) it follows that the same is true of Z. Moreover, Z has the additional property of being complete. To see this, let $\{F_n\}$ be a Cauchy sequence in Z. We know that for each n there is a $G_n \in W$, such that

$$\| F_n - G_n \| < \frac{1}{n} \tag{1-85}$$

It follows, therefore, that $\{G_n\}$ is a Cauchy sequence in W. If $G_n u = (u, g_n)$, then $\{g_n\}$ is a Cauchy sequence in H. In particular,

$$Gu = \lim (u, g_n) \qquad u \in H \tag{1-86}$$

exists, and is a bounded linear functional on H. Furthermore

$$\| G_n - G \| \to 0 \quad \text{as} \quad n \to \infty \tag{1-87}$$

To see this, note that for each $\varepsilon > 0$ there is an N so large, that

$$\frac{|(u, g_m - g_n)|}{\| u \|} < \varepsilon \qquad u \in H$$

whenever $m, n > N$. Letting $m \to \infty$, we have by Eq. (1-86)

$$\frac{|Gu - (u, g_n)|}{\| u \|} < \varepsilon, \qquad u \in H$$

for $n > N$. This gives (1-87). If we now combine (1-85) and (1-87), we obtain

$$\| F_n - G \| \to 0 \quad \text{as} \quad n \to \infty \tag{1-88}$$

Now (1-87) shows that $G \in Z$, while (1-88) shows that it is the limit in norm of the F_n. Thus, Z is complete. Clearly, W is dense in Z. Since W can be identified with H, our assertion is proved.

1-6 WEAK SOLUTIONS

We now return to Eq. (1-40). We have shown in Sec. 1-4 that a necessary condition for (1-40) to have a solution is that (1-44) hold for all test functions in Ω. Now, let us work back assuming that (1-44) holds. Does it follow that Eq. (1-40) has a solution?

Let us denote the set of test functions on Ω by $C_0^\infty(\Omega)$. Another important set of functions is the set of square integrable functions on Ω. These are the (complex-valued) measurable functions u defined on Ω, such that

$$\int_\Omega |u|^2 \, dx < \infty$$

Denote this set by $L^2(\Omega)$. It is a Hilbert space with scalar product given by (,). (For this fact we refer to any good book on real variables.) Clearly, $C_0^\infty(\Omega) \subset L^2(\Omega)$.

Now let W be the set of those $w \in L^2(\Omega)$ for which there is a $\varphi \in C_0^\infty$ satisfying

$$A'\varphi = w \tag{1-89}$$

Clearly, W is a subspace of $L^2(\Omega)$. For $w \in W$, we set

$$Fw = (\varphi, f) \tag{1-90}$$

where f is the given function in Eq. (1-40), and φ is any function in $C_0^\infty(\Omega)$ satisfying Eq. (1-89). For this definition to make sense, we must show that F depends only on w, and not on the particular φ chosen. So suppose φ_1 is another test function satisfying $A'\varphi_1 = w$. Then, by (1-44)

$$|(f, \varphi - \varphi_1)| \leq C \| A'(\varphi - \varphi_1) \| = C \| w - w \| = 0$$

showing that $(f, \varphi) = (f, \varphi_1)$. Thus, F depends only on w and not on φ. Now, F assigns to each $w \in W$ a complex number given by Eq. (1-90). Thus, it is a functional on W. It is clearly linear. Moreover, by (1-44)

$$|Fw| = |(\varphi, f)| \leq C \| A'\varphi \| = C \| w \| \tag{1-91}$$

which shows that F is bounded. By the Hahn–Banach theorem (Theorem 1-6), F can be extended to be a bounded linear functional on the whole of $L^2(\Omega)$. Hence, by the Fréchet–Riesz theorem (Theorem 1-5) there is a $u \in L^2(\Omega)$, such that $\| u \| = \| F \| \leq C$, and

$$Fw = (w, u) \tag{1-92}$$

for all $w \in L^2(\Omega)$. In particular, if $\varphi \in C_0^\infty(\Omega)$, then $A'\varphi \in W$ and, hence, by Eqs. (1-90) and (1-92)

$$(u, A'\varphi) = (f, \varphi) \tag{1-93}$$

This holds for all $\varphi \in C_0^\infty(\Omega)$. Note the similarity between (1-93) and (1-43). We would now like to integrate the left-hand side of Eq. (1-93) by parts, to obtain

$$(Au - f, \varphi) = 0 \tag{1-94}$$

Since this would be true for all $\varphi \in C_0^\infty(\Omega)$, it would follow that $Au = f$ (see the end of Sec. 1-3). This would give us what we want.

If we knew that $u \in C^m(\Omega)$, then we could indeed integrate Eq. (1-93) by parts, and then conclude that u is a solution of Eq. (1-40). But what do we know about u? Merely that it is in $L^2(\Omega)$ and satisfies Eq. (1-93). Does this imply that $u \in C^m(\Omega)$? Unfortunately not, as the example of Sec. 1-1 shows. For purpose Ω is a two-dimensional domain contained in the strip $|y| \leq R$, and let $g(y)$ be a continuous function of y in $|y| \leq R' > R$, which does not have a continuous derivative. As we know, one can find a sequence $\{g_n(y)\}$ of continuously differentiable functions which converge uniformly to $g(y)$ in $|y| \leq R$ (actually, we shall prove this later in Sec. 2-2). If φ is any test function in Ω, then

$$(g_n, \varphi_x) = -\left(\frac{\partial g_n}{\partial x}, \varphi\right) = 0$$

by integration by parts. Taking the limit as $n \to \infty$, we get

$$(g, \varphi_x) = 0 \qquad \varphi \in C_0^\infty(\Omega)$$

Clearly, this does not imply that $g \in C^1(\Omega)$.

Before we sit down to cry, let us try to salvage something from our work. We have shown that if u is a solution of $Au = f$, then (1-44) holds. However, if we examine our proof carefully, we will notice that we really did not use the fact that u is a solution, but only that it satisfies Eq. (1-93). A function $u \in L^2(\Omega)$ which satisfies (1-93) for all $\varphi \in C_0^\infty(\Omega)$ is called a *weak solution* of Eq. (1-40). Hence, we have proved:

Theorem 1-12 A necessary and sufficient condition that Eq. (1-40) have a weak solution is that inequality (1-44) hold for all $\varphi \in C_0^\infty(\Omega)$. If a weak solution is in $C^m(\Omega)$, then it is a solution. If (1-44) holds, there is a weak solution satisfying $\| u \| \le C$.

From all this we see that when we are looking for solutions of Eq. (1-40), there are two distinct questions we can ask:

1. When does Eq. (1-40) have a weak solution?
2. When is a weak solution sufficiently differentiable to be a solution?

Theorem 1-12 gives an answer to the first question. To make it useful, we must translate it into conditions involving A, f, and Ω. In Sec. 1-7 we discuss a large class of operators to which it can be applied.

1-7 OPERATORS WITH CONSTANT COEFFICIENTS

Let A be an operator with constant coefficients of the form

$$A = \sum_{|\mu| \le m} a_\mu D^\mu \tag{1-95}$$

We shall show that in any bounded domain Ω the equation $Au = f$ has a weak solution for any f. In fact, we shall prove

Theorem 1-13 If A has constant coefficients, then for each bounded domain Ω there is a constant C, such that

$$\| \varphi \| \le C \| A\varphi \| \qquad \varphi \in C_0^\infty(\Omega) \tag{1-96}$$

As a consequence of Theorem 1-13, we have

Corollary 1-14 If A has constant coefficients, Ω is any bounded domain, and $f \in L^2(\Omega)$, then $Au = f$ has a weak solution.

PROOF of Corollary 1-14 assuming Theorem 1-13 If A has constant coefficients, so does A'. In fact

$$A' = \sum_{|\mu| \le m} \bar{a}_\mu D^\mu \tag{1-97}$$

[see Eq. (1-32)]. Hence, Theorem 1-13 applies to A', and there is a constant C, such that

$$\| \varphi \| \leq C \| A'\varphi \| \qquad \varphi \in C_0^\infty(\Omega) \tag{1-98}$$

Since $f \in L^2(\Omega)$, we have by Schwarz's inequality [see statement (1-62)]

$$|(f, \varphi)| \leq \| f \| \ \| \varphi \| \leq C' \| A'\varphi \| \qquad \varphi \in C_0^\infty(\Omega)$$

Thus, (1-44) holds, and $Au = f$ has a weak solution, by Theorem 1-12.

The proof of Theorem 1-13 is not difficult. To carry it out, we introduce some convenient notation. Let $\xi_1, \ldots, \xi_n$ be n variables, and set

$$\xi^\mu = \xi_1^{\mu_1} \cdots \xi_n^{\mu_n}$$

for $\mu = (\mu_1, \ldots, \mu_n)$. Consider the polynomial

$$P(\xi) = \sum_{|\mu| \leq m} a_\mu \xi^\mu \tag{1-99}$$

in $\xi_1, \ldots, \xi_n$. If we replace $\xi_1, \ldots, \xi_n$ by $D_1, \ldots, D_n$, then we get

$$P(D) = \sum_{|\mu| \leq m} a_\mu D^\mu \tag{1-100}$$

showing that $P(D)$ is just the operator A. Now what good does this do us? Well, suppose we want to compute $A(uv)$. Since

$$D_k(uv) = u\, D_k v + v\, D_k u$$

we can write this in the form

$$D_k(uv) = (\overset{u}{D}_k + \overset{v}{D}_k) uv$$

if we agree that $\overset{u}{D}_k$ is to operate only on u and $\overset{v}{D}_k$ is to operate only on v. Thus

$$P(D)uv = P(\overset{u}{D} + \overset{v}{D})uv$$

Now, by Taylor's formula

$$P(\eta + \xi) = \sum_{|\mu| \leq m} \frac{1}{\mu!} P^{(\mu)}(\eta) \xi^\mu \tag{1-101}$$

where $\mu! = \mu_1! \cdots \mu_n!$ and

$$P^{(\mu)}(\eta) = \frac{\partial^{|\mu|} P(\eta)}{\partial \eta_1^{\mu_1} \cdots \partial \eta_n^{\mu_n}} \qquad P^{(0)}(\eta) = P(\eta)$$

Hence

$$A(uv) = \sum_{|\mu| \leq m} \frac{1}{\mu!} P^{(\mu)}(\overset{u}{D}) \overset{v}{D}{}^\mu (uv)$$

$$= \sum_{|\mu| \leq m} \frac{1}{\mu!} P^{(\mu)}(D)u\, D^\mu v \tag{1-102}$$

This is a very convenient formula, as we shall see later on several occasions.

Let $\bar{P}(\xi)$ be the polynomial with coefficients which are the complex conjugates of those of $P(\xi)$, namely

$$\bar{P}(\xi) = \sum_{|\mu| \le m} \bar{a}_\mu \xi^\mu \tag{1-103}$$

Then, by Eq. (1-97) we have $A' = \bar{P}(D)$. Moreover

$$\begin{aligned} \| A'\varphi \|^2 &= \| \bar{P}(D)\varphi \|^2 = [\bar{P}(D)\varphi, \bar{P}(D)\varphi] \\ &= [\varphi, P(D)\bar{P}(D)\varphi] = [P(D)\varphi, P(D)\varphi] \\ &= \| P(D)\varphi \|^2 = \| A\varphi \|^2 \qquad \varphi \in C_0^\infty(\Omega) \end{aligned} \tag{1-104}$$

We shall prove Theorem 1-13 by means of

Lemma 1-15 Suppose Ω is contained in the strip $|x_k - a| \le M/2$, and let $P^{(k)}(\xi) = \partial P(\xi)/\partial \xi_k$. Then

$$\| P^{(k)}(D)\varphi \| \le mM \| P(D)\varphi \| \qquad \varphi \in C_0^\infty(\Omega) \tag{1-105}$$

An immediate consequence of this is

Corollary 1-16 If Ω is contained in the region $|x_k - a_k| \le M_k/2$, $1 \le k \le n$, then for any multi-index μ

$$\| P^{(\mu)}(D)\varphi \| \le \frac{m!}{(m - |\mu|!} M^\mu \| P(D)\varphi \| \qquad \varphi \in C_0^\infty(\Omega) \tag{1-106}$$

where $M = (M_1, \ldots, M_n)$.

Corollary 1-16 is obtained by repeated applications of Lemma 1-15. It implies Theorem 1-13 because there always is a multi-index μ, such that $P^{(\mu)}(\xi) = \text{constant} \neq 0$.

It thus remains to give the

PROOF of Lemma 1-15 We employ induction on m. Suppose that

$$\| Q^{(k)}(D)\varphi \| \le (m-1)M \| Q(D)\varphi \| \qquad \varphi \in C_0^\infty(\Omega) \tag{1-107}$$

for all polynomials $Q(\xi)$ of the form

$$Q(\xi) = \sum_{|\mu| < m} c_\mu \xi^\mu \tag{1-108}$$

By making a simple translation, we may assume $a = 0$. By Eq. (1-102), we have

$$P(D)(x_k\varphi) = x_k P(D)\varphi + P^{(k)}(D)\varphi$$

$$\begin{aligned}
\text{Hence } \| P^{(k)}(D)\varphi \|^2 &= (P(D)(x_k\varphi) - x_kP(D)\varphi, P^{(k)}(D)\varphi) \\
&= (P(D)(x_k\varphi), P^{(k)}(D)\varphi) - (x_kP(D)\varphi, P^{(k)}(D)\varphi) \\
&= (\bar{P}^{(k)}(D)(x_k\varphi), \bar{P}(D)\varphi) - (x_kP(D)\varphi, P^{(k)}(D)\varphi) \\
&= (x_k\bar{P}^{(k)}(D)\varphi + \bar{P}^{(kk)}(D)\varphi, \bar{P}(D)\varphi) - (P(D)\varphi, x_kP^{(k)}(D)\varphi) \\
&\le \| P(D)\varphi \| [M \| P^{(k)}(D)\varphi \| + \| P^{(kk)}(D)\varphi \|] \qquad (1\text{-}109)
\end{aligned}$$

where $P^{(kk)}(\xi) = \partial^2 P(\xi)/\partial\xi_k^2$, and we have used Eq. (1-104). Since $P^{(k)}(\xi)$ is of the form (1-108), we have, by the induction hypothesis, that

$$\| P^{(kk)}(D)\varphi \| \le (m-1)M \| P^{(k)}(D)\varphi \|$$

Hence
$$\| P^{(k)}(D)\varphi \|^2 \le mM \| P(D)\varphi \| \; \| P^{(k)}(D)\varphi \|$$

Dividing both sides by $\| P^{(k)}(D)\varphi \|$ we obtain (1-105). To complete the proof, we note that for $m = 1$, $P^{(kk)}(\xi) \equiv 0$. In this case, (1-109) implies (1-105) immediately.

Inequalities (1-96) and (1-106) are due to Hörmander (1955). Our proofs follow those of Malgrange (1961).

Remark. We didn't really have to assume that $\partial\Omega$ was piecewise smooth for any of the results of this section. Since we were dealing with functions which vanished near $\partial\Omega$, we were actually integrating only over the supports of these functions.

PROBLEMS

1-1 Show that the function $g(t)$ given by Eq. (1-23) has continuous derivatives of all orders in $(-\infty, \infty)$, but is not analytic in a neighborhood of $t = 0$.

1-2 Show that the function $j(x)$ given by Eq. (1-35) is in $C^\infty(E^n)$.

1-3 Let K_1 and K_2 be two closed subsets of E^n which do not intersect. If K_1 is bounded, show that the distance between K_1 and K_2 is positive. Use this to prove the statement following (1-37).

1-4 By differentiating under the integral sign, show that the function $\psi(x)$ given by (1-39) is in $C^\infty(E^n)$.

1-5 Let $P(\xi)$ be a polynomial. Show that there exists a multi-index μ, such that $P^{(\mu)}(\xi) = \text{constant} \neq 0$.

1-6 Give an example of a space H satisfying (1-45) to (1-53) with the property that there is a sequence $\{F_k\}$ of bounded linear functionals on H, such that the numbers $F_k(v)$ are bounded for each $v \in H$, while $\| F_k \| \to \infty$ as $k \to \infty$.

1-7 Prove that the space $\bar{H}$ described in Sec. 1-5 is a Hilbert space.

1-8 Show that a subspace S is closed if and only if $S = \bar{S}$.

1-9 Show that the closure of a subspace is the smallest closed subspace containing it.

1-10 Let W be any subset of a Hilbert space H. Show that $W^\perp$ is a closed subspace of H.

1-11 Prove the second sentence in the proof of Theorem 1-8.

1-12 Show how to pick the sequence $\{\tilde{u}_k\}$ in the proof of Theorem 1-9.

1-13 Show that the space W described at the end of Sec. 1-5 satisfies (1-45) to (1-54).

CHAPTER

TWO

REGULARITY (CONSTANT COEFFICIENTS)

2-1 A NECESSARY CONDITION

In the first chapter we obtained an inequality which was necessary and sufficient for $Au = f$ to have a weak solution in a domain Ω. We then went on to show that, for bounded Ω, every operator with constant coefficients satisfies this inequality for every $f \in L^2(\Omega)$, and hence has a weak solution. However, we are still looking for solutions in $C^m(\Omega)$.

As usual when we are in unfamiliar territory, let us take as simple a case as possible. Assume that A has constant coefficients, and let us examine the case $f = 0$ first. Here we can always find smooth solutions. For let $\zeta = (\zeta_1, \ldots, \zeta_n)$ be a vector consisting of complex components, and set

$$(\zeta, x) = \zeta_1 x_1 + \cdots + \zeta_n x_n \tag{2-1}$$

where $x \in E^n$. Then, by differentiation

$$D_k\, e^{i(\zeta,x)} = \zeta_k\, e^{i(\zeta,x)} \tag{2-2}$$

and, if $P(\zeta)$ is a polynomial in the ζ_k, we have

$$P(D)\, e^{i(\zeta,x)} = P(\zeta)\, e^{i(\zeta,x)} \tag{2-3}$$

Thus, the function $u = e^{i(\zeta,x)}$ is a solution of

$$P(D)u = 0 \tag{2-4}$$

if ζ is a root of

$$P(\zeta) = 0 \tag{2-5}$$

Since Eq. (2-5) always has complex solutions, we see that Eq. (2-4) always has smooth solutions.

However, this does not rule out the possibility that Eq. (2-4) may have weak solutions which are not in $C^m(\Omega)$. This leads us to ask: what conditions must $P(\zeta)$ satisfy, in order that every weak solution of Eq. (2-4) be in $C^m(\Omega)$?

To examine this question, let W be the set of all weak solutions of Eq. (2-4), i.e., the set of all $u \in L^2(\Omega)$, such that

$$(u, \bar{P}(D)\varphi) = 0 \qquad \varphi \in C_0^\infty(\Omega) \tag{2-6}$$

Let us assume that each $u \in W$ is at least in $C^1(\Omega)$. Let Ω_1 be a domain such that $\bar{\Omega}_1 \subset \Omega$, and let B be the operator D_k, $1 \le k \le n$, applied to functions in $C^1(\bar{\Omega}_1)$. Now W is a closed subspace of $L^2(\Omega)$. For it is clearly a subspace, and if $u_n \in W$, $u_n \to u$ in $L^2(\Omega)$, then $u \in W$ by Eq. (2-6). Thus, W is itself a Hilbert space. Next note that B maps W into $L^2(\Omega_1)$ (in fact, it maps it into $C(\bar{\Omega}_1)$). B is linear and, by hypothesis, is defined on the whole of W. Moreover, it is a *closed* operator. This means that if $u_n \to u$ in W, and $Bu_n \to v$ in $L^2(\Omega_1)$, then $Bu = v$. To prove this we note that, if $\varphi \in C_0^\infty(\Omega_1)$, then

$$(Bu_n, \varphi) = (u_n, B\varphi)$$

and hence

$$(v, \varphi) = (u, B\varphi) \tag{2-7}$$

On the other hand, since $u \in W$, it is in $C^1(\Omega)$, and hence

$$(Bu, \varphi) = (u, B\varphi)$$

for any $\varphi \in C_0^\infty(\Omega_1)$. Comparing this with Eq. (2-7) we see that $Bu = v$. We now appeal to the important closed graph theorem (see Sec. 2-9) which tells us that B is a *bounded* operator from W to $L^2(\Omega_1)$. Thus, there is a constant C, such that

$$\int_{\Omega_1} |Bu|^2\, dx \le C \int_\Omega |u|^2\, dx \qquad u \in W$$

Since $B = D_k$ and k is any integer from 1 to n, we have

$$\sum_1^n \int_{\Omega_1} |D_k u|^2\, dx \le C' \int_\Omega |u|^2\, dx \qquad u \in W \tag{2-8}$$

Now, let ζ be any complex root of Eq. (2-5). Then $e^{i(\zeta,x)}$ is a smooth solution of Eq. (2-4), and certainly a weak solution. Substituting into (2-8) we get, by (2-2)

$$|\zeta|^2 \int_{\Omega_1} e^{-2(\operatorname{Im}\zeta, x)}\, dx \le C' \int_\Omega e^{-2(\operatorname{Im}\zeta, x)}\, dx$$

where $\operatorname{Im}\zeta = (\operatorname{Im}\zeta_1, \ldots, \operatorname{Im}\zeta_n)$. If Ω is contained in the sphere Σ_R, then

$$e^{2|(\operatorname{Im}\zeta, x)|} \le e^{2|\operatorname{Im}\zeta|\,|x|} \le e^{2R|\operatorname{Im}\zeta|}$$

and hence

$$|\zeta|^2 e^{-2R|\operatorname{Im}\zeta|} \int_{\Omega_1} dx \le C' e^{2R|\operatorname{Im}\zeta|} \int_\Omega dx$$

Thus, we have

$$|\zeta| \leq C_0\, e^{4R|\operatorname{Im}\zeta|} \qquad P(\zeta) = 0 \tag{2-9}$$

Hence, we have proved

Lemma 2-1 A necessary condition that every weak solution of Eq. (2-4) be in $C^1(\Omega)$, is that (2-9) holds.

We are going to investigate what (2-9) implies concerning the polynomial $P(\xi)$. To feel our way, let us consider first the case when $P(D)$ is homogeneous, i.e., when

$$P(\xi) = \sum_{|\mu|=m} a_\mu \xi^\mu \tag{2-10}$$

Now, suppose ξ_0 is a real root of $P(\xi) = 0$. Then, for real t,

$$\begin{aligned} P(t\xi_0) &= t^m P(\xi_0) \\ &= 0 \end{aligned}$$

so that $t\xi_0$ is a real root as well. By (2-9)

$$|t\xi_0| \leq C_0$$

since $\operatorname{Im}(t\xi_0) = 0$. Letting $t \to \infty$, we see that we must have $\xi_0 = 0$. Thus, the only real solution of $P(\xi) = 0$ is $\xi = 0$. A homogeneous operator having this property is called *elliptic*. Thus, we have proved

Corollary 2-2 If $P(D)$ is homogeneous and (2-9) holds, then $P(D)$ is elliptic.

Thus, if $P(\xi)$ is given by Eq. (2-10), we know that it must be elliptic in order that weak solutions of Eq. (2-4) be in $C^m(\Omega)$. In Sec. 2-4 we shall prove the converse. We shall show that if $P(D)$ is homogeneous and elliptic, then every weak solution of Eq. (2-4) is indeed in $C^\infty(\Omega)$. As a by-product, it will follow that every homogeneous elliptic operator $P(D)$ satisfies (2-9).

In order to carry out our plan, we must develop a few tools. We do this in Secs. 2-2 and 2-3.

The derivation of inequality (2-9) is due to Hörmander, (1955).

Since we used the concept of an operator in proving Lemma 2-1, we should give a precise definition. An operator A from a Hilbert space H_1 to a Hilbert space H_2 is a mapping, which assigns to each element χ of a subset V of H_1 an element $A\chi$ belonging to H_2. The set V is called the *domain of definition* of A, and is sometimes denoted by $D(A)$. The operator A is called *linear* if $D(A)$ is a subspace of H_1 and Eq. (1-7) of Sec. 1-1 holds. We say that A is defined everywhere in H_1 if $D(A) = H_1$. An operator A is called *bounded* if there is a constant K, such that $\|A\chi\| \leq K\|\chi\|$ for all χ in $D(A)$.

2-2 THE FRIEDRICHS MOLLIFIER

Set

$$j(x) = a \exp[(|x|^2 - 1)^{-1}] \qquad |x| < 1$$
$$= 0 \qquad |x| \geq 1$$

where

$$a = \left\{ \int_{|x|<1} \exp[(|x|^2 - 1)^{-1}]\, dx \right\}^{-1}$$

We choose this value of a in order to have

$$\int j(x)\, dx = 1$$

In Sec. 1-3, we noted that $j(x) \in C_0^\infty(E^n)$. For $\varepsilon > 0$, set

$$j_\varepsilon(x) = \varepsilon^{-n} j\left(\frac{x}{\varepsilon}\right)$$

Then for each $\varepsilon > 0$, we have

$$j_\varepsilon(x) \geq 0 \quad \text{in} \quad E^n \tag{2-11}$$

$$j_\varepsilon(x) = 0 \quad \text{for} \quad |x| \geq \varepsilon \tag{2-12}$$

$$\int j_\varepsilon(x)\, dx = 1 \tag{2-13}$$

For $u \in L^2(E^n)$, we set

$$J_\varepsilon u(x) = \int u(y) j_\varepsilon(x - y)\, dy = \int u(x - z) j_\varepsilon(z)\, dz \tag{2-14}$$

Theorem 2-3 For $u \in L^2(E^n)$

$$\| J_\varepsilon u \| \leq \| u \| \tag{2-15}$$

$$\| J_\varepsilon u - u \| \to 0 \quad \text{as} \quad \varepsilon \to 0 \tag{2-16}$$

PROOF By Eq. (2-14) and Schwarz's inequality (1-62)

$$|J_\varepsilon u|^2 \leq \left(\int |u(x - z)|^2 j_\varepsilon(z)\, dz \right) \left(\int j_\varepsilon(x)\, dx \right)$$

and hence

$$\int |J_\varepsilon u|^2\, dx \leq \int j_\varepsilon(z) \int |u(x - z)|^2\, dx\, dz$$
$$\leq \int j_\varepsilon(z)\, dz \int |u(x)|^2\, dx$$
$$= \| u \|^2$$

To prove (2-16), first assume that u is continuous. By Eq. (2-13)

$$J_\varepsilon u - u = \int j_\varepsilon(z)[u(x-z) - u(x)]\, dz$$

and hence $$|J_\varepsilon u - u|^2 \le \int j_\varepsilon(z)\,|u(x-z) - u(x)|^2\, dz$$

and $$\int |J_\varepsilon u - u|^2\, dx \le \int j_\varepsilon(z) \int |u(x-z) - u(x)|^2\, dx\, dz$$

$$\le \operatorname*{lub}_{|z|<\varepsilon} \int |u(x-z) - u(x)|^2\, dx$$

Now let $\rho > 0$ be given, and take R so large that

$$\int_{|x|>R} |u(x)|^2\, dx < \frac{\rho}{8}$$

Then for $|z| < R$

$$\int_{|x|>2R} |u(x-z)|^2\, dx \le \int_{|x|>R} |u(x)|^2\, dx < \frac{\rho}{8}$$

Since u is continuous, we can make

$$\max_{|z|\le\varepsilon} \int_{|x|<2R} |u(x-z) - u(x)|^2\, dx < \frac{\rho}{2}$$

by taking ε sufficient small. Combining these inequalities, we get

$$\int |J_\varepsilon u - u|^2\, dx < \rho$$

for ε sufficiently small. This proves (2-16) for u continuous. If u is not continuous, we can find a continuous function w in $L^2(E^n)$, such that

$$\| u - w \| < \frac{\rho}{3}$$

Hence $$\| J_\varepsilon u - u \| \le \| J_\varepsilon(u - w) \| + \| J_\varepsilon w - w \| + \| w - u \|$$

$$\le 2\| u - w \| + \| J_\varepsilon w - w \|$$

By what was just proved, we can make

$$\| J_\varepsilon w - w \| < \frac{\rho}{3}$$

by taking ε sufficiently small. This completes the proof.

$J_\varepsilon u$ is called a *mollifier*. It was first employed by Friedrichs (1953). I leave it as an exercise to verify the following simple properties

$$J_\varepsilon u \in C^\infty(E^n) \quad \text{for} \quad u \in L^2(E^n) \tag{2-17}$$

$$(J_\varepsilon u, v) = (u, J_\varepsilon v) \qquad u, v \in L^2(E^n) \tag{2-18}$$

$$D_k J_\varepsilon v = J_\varepsilon D_k v \qquad 1 \le k \le n \qquad v \in C^\infty(E^n) \tag{2-19}$$

In Sec. 1-6 we made use of the following result.

Lemma 2-4 Let Ω_1 and Ω_2 be bounded domains in E^n, such that $\bar{\Omega}_1 \subset \Omega_2$. Suppose u is a continuous function in Ω_2. Then there is a sequence $\{v_k\}$ of functions in $C^\infty(\Omega_2)$ which converge uniformly to u in Ω_1.

PROOF By shrinking Ω_2 a bit, we may assume that u is continuous on $\bar{\Omega}_2$. Let $\hat{u}$ be the function equal to u in Ω_2 and equal to 0 outside. Now u is uniformly continuous in $\bar{\Omega}_2$. Hence, if $\eta > 0$ is given, there is a $\delta > 0$, such that

$$|u(x - z) - u(x)| < \eta \quad \text{for} \quad |z| < \delta$$

Let $d > 0$ be the distance from $\bar{\Omega}_1$ to the boundary of Ω_2. Then if $\varepsilon < \min(\delta, d)$, we have

$$\begin{aligned} |J_\varepsilon \hat{u} - u(x)| &\le \int j_\varepsilon(z)\, |u(x - z) - u(x)|\, dz \\ &< \eta \int j_\varepsilon(z)\, dz \\ &= \eta \qquad x \in \Omega_1 \end{aligned}$$

This shows that $\{J_\varepsilon \hat{u}\}$ converges uniformly to u in Ω_1. Since $J_\varepsilon \hat{u} \in C^\infty(E^n)$, the proof is complete.

2-3 A FAMILY OF NORMS

Let $S = S(E^n)$ denote the set of functions $v(x) \in C^\infty(E^n)$, such that

$$(1 + |x|)^k |D^\mu v(x)|$$

are bounded for all k and μ. The Fourier transform of v is defined by

$$Fv(\xi) = \int e^{-i(\xi, x)} v(x)\, dx \tag{2-20}$$

If $v \in S$, we can differentiate Eq. (2-20) under the integral sign, to obtain

$$D^\mu Fv(\xi) = (-1)^{|\mu|} F(x^\mu v) \tag{2-21}$$

Moreover, if we integrate

$$F(D^\mu v) = \int e^{-i(\xi,x)} D^\mu v(x)\, dx$$

by parts, we obtain

$$F(D^\mu v) = \xi^\mu Fv \tag{2-22}$$

The Fourier transform has several useful properties, two of which will interest us. The first is the inversion formula

$$v(x) = (2\pi)^{-n} \int e^{i(x,\xi)} Fv(\xi)\, d\xi \tag{2-23}$$

and the second is Parseval's identity

$$\int v(x)\overline{w(x)}\, dx = (2\pi)^{-n} \int Fv(\xi)\overline{Fw(\xi)}\, d\xi \tag{2-24}$$

both holding for functions in S. We shall prove these in Sec. 2-5. Let us assume them here, and use the Fourier transform to define a family of scalar products.

For any real number s, set

$$(v, w)_s = \int (1 + |\xi|)^{2s} Fv\overline{Fw}\, d\xi \tag{2-25}$$

$$|v|_s^2 = (v, v)_s \tag{2-26}$$

Note that

$$(v, w) = (2\pi)^{-n}(v, w)_0 \tag{2-27}$$

by Eq. (2-24). Moreover, we have

$$|(v, w)_0| \le |v|_s\, |w|_{-s} \tag{2-28}$$

This follows from

$$(v, w)_0 = \int (1 + |\xi|)^s Fv \cdot (1 + |\xi|)^{-s}\overline{Fw}\, d\xi$$

and Schwarz's inequality (1-62). We even have

$$|v|_s = \operatorname*{lub}_{w\in S} \frac{|(v, w)_0|}{|w|_{-s}} \qquad v \in S \qquad s \text{ real} \tag{2-29}$$

Clearly, the left-hand side is $\ge$ the right-hand side, by (2-28). To show that they are equal, we note that if $v \in S$, so is $(1 + |\xi|)^{2s} Fv$. Hence, by the inversion formula Eq. (2-23), there is a $v_0 \in S$, such that $Fv_0 = (1 + |\xi|)^{2s} Fv$. For the function v_0, we have

$$|v_0|_{-s}^2 = \int (1 + |\xi|)^{-2s}(1 + |\xi|)^{4s} |Fv|^2\, d\xi$$

$$= |v|_s^2$$

and
$$(v, v_0)_0 = \int (1 + |\xi|)^{2s} |Fv|^2 \, d\xi$$
$$= |v|_s^2$$

Consequently,

$$\frac{|(v, v_0)_0|}{|v_0|_{-s}} = \frac{|v|_s^2}{|v|_s}$$
$$= |v|_s$$

showing that the least upper bound in Eq. (2-29) is a maximum, and equals $|v|_s$.

Another important observation is that

$$|D^\mu v|_s \leq |v|_{s+|\mu|} \qquad v \in S \tag{2-30}$$

This follows from the fact that

$$|\xi^\mu| \leq |\xi|^{|\mu|} \tag{2-31}$$

We shall need a method of telling when a function is in $C^k(\Omega)$. For this purpose we shall use the important

Lemma 2-5 Let k be a nonnegative integer and let s be a real number, such that $s - k > n/2$. Suppose u is a function in $L^2(E^n)$ and there is a sequence $\{v_n\}$ of functions in S, such that

$$|v_n|_s \leq C_1 \tag{2-32}$$

and
$$\| v_n - u \| \to 0 \quad \text{as} \quad n \to \infty \tag{2-33}$$

Then u is almost everywhere equal to a function in $C^k(E^n)$.

PROOF We shall prove the inequality

$$\max_{E^n} \sum_{|\mu| \leq k} |D^\mu v(x)| \leq C_2 |v|_s \qquad v \in S \tag{2-34}$$

Assuming (2-34) for the moment, we let H^s denote the completion of S with respect to the norm $|\ |_s$, and can check easily that H^s is a Hilbert space. Thus, by the Banach–Saks Lemma (see Theorem 1-9), $\{v_n\}$ has a subsequence whose arithmetical means $\{w_n\}$ converge in H^s. For convenience of notation, let us assume that $\{v_n\}$ is already the subsequence.

Then
$$w_n = \frac{(v_1 + \ldots + v_n)}{n}$$

and
$$|w_n - w_m|_s \to 0 \quad \text{as} \quad n, m \to \infty \tag{2-35}$$

It can be checked easily that

$$\| w_n - u \| \to 0 \quad \text{as} \quad n \to \infty \tag{2-36}$$

By statement (2-34) and (2-35), we have

$$\max_{E^n} \sum_{|\mu| \le k} |D^\mu[w_n(x) - w_m(x)]| \to 0 \quad \text{as} \quad n, m \to \infty \tag{2-37}$$

This shows that $\{w_n(x)\}$ converges uniformly to a function $w \in C^k(E^n)$. Now I claim that $u = w$ a.e. For, we have

$$\int_{|x|<R} |u - w|^2 \, dx \le 2 \int_{|x|<R} |u - w_n|^2 \, dx + 2 \int_{|x|<R} |w_n - w|^2 \, dx$$

$$\le 2 \| u - w_n \|^2 + C_R \max_{|x|<R} |w_n(x) - w(x)|^2 \to 0 \text{ as } n \to \infty$$

Since the left-hand side of this inequality does not depend on n, we have

$$\int_{|x|<R} |u - w|^2 \, dx = 0$$

for each R. This shows that $u = w$ a.e.

It, remains, therefore, to prove (2-34). By the inversion formula (2-23), we have

$$D^\mu v(x) = (2\pi)^{-n} \int e^{i(x,\xi)} \xi^\mu Fv \, d\xi$$

$$= (2\pi)^{-n} \int e^{i(x,\xi)} \xi^\mu (1 + |\xi|)^{s-k} Fv \cdot (1 + |\xi|)^{k-s} \, d\xi$$

By Schwarz's inequality (1-62) and (2-31)

$$|D^\mu v(x)|^2 \le \int |\xi^\mu|^2 (1 + |\xi|)^{2(s-k)} |Fv|^2 \, d\xi \int (1 + |\xi|)^{2(k-s)} \, d\xi$$

$$\le |v|_s^2 \int (1 + |\xi|)^{2(k-s)} \, d\xi \tag{2-38}$$

since $|\mu| \le k$. To complete the proof, all we need to show is that the last integral is finite when $s - k > n/2$. This follows from the fact that, for $R > 1$

$$\int_{|\xi|<R} (1 + |\xi|)^{2(k-s)} \, d\xi \le \left[\int_{|\xi|<1} d\xi + \int_{1<|\xi|<R} |\xi|^{2(k-s)} \, d\xi \right]$$

$$\le \text{constant} \left(1 + \int_1^R r^{2(k-s)+n-1} \, dr \right)$$

This last integral remains bounded as $R \to \infty$ since $2(k - s) + n < 0$. The proof is complete.

If a function u is almost everywhere equal to a function in $C^k(\Omega)$, we shall take the expedient of saying that u is in $C^k(\Omega)$. This will be true if we "correct" u on a set of measure zero.

Another observation, that we shall need, is

Lemma 2-6 If $v \in S$, then for each real s there is a constant c_s, such that

$$|J_\varepsilon v|_s \leq c_s \tag{2-39}$$

(The constant c_s does not depend on ε.)

PROOF Let $k \geq 0$ be an integer $\geq s$. Then the inequality

$$(1 + |\xi|)^s \leq K \sum_{|\mu| \leq k} |\xi^\mu|$$

implies
$$|J_\varepsilon v|_s \leq K \sum_{|\mu| \leq k} |D^\mu J_\varepsilon v|_0 \leq K' \sum_{|\mu| \leq k} \| D^\mu v \|$$

by (2-15), (2-19), and (2-27). This proves (2-39).

Lemma 2-5 is referred to as *Sobolev's lemma*. We shall prove a sharper form of inequality (2-39) in Sec. 2-5.

In future work it will be convenient to introduce the set of functions $S(\Omega)$ for an arbitrary domain Ω. It is the set of functions $v \in C^\infty(\bar{\Omega})$, such that

$$(1 + |x|)^k |D^\mu v(x)|$$

is bounded in Ω for each k and μ. Obviously, when Ω is a bounded domain, then $S(\Omega) = C^\infty(\bar{\Omega})$.

2-4 ELLIPTIC OPERATORS

We are now ready for the proof promised in Sec. 2-1. In that section we defined ellipticity for a homogeneous operator. We generalize this definition to include constant-coefficient operators of the form

$$P(D) = \sum_{|\mu| \leq m} a_\mu D^\mu \tag{2-40}$$

We say that $P(D)$ is of *order* m if not all of the a_μ with $|\mu| = m$ vanish. If $P(D)$ is of order m, set

$$p(D) = \sum_{|\mu| = m} a_\mu D^\mu \tag{2-41}$$

$p(D)$ is called the *principal part* of $P(D)$. We shall call $P(D)$ elliptic if the only real solution of $p(\xi) = 0$ is $\xi = 0$. Note that this definition coincides with the one given in Sec. 2-1 when $P(D)$ is homogeneous, i.e., when $P(D) = p(D)$.

We shall prove

Theorem 2-7 Let $P(D)$ be an elliptic operator with constant coefficients, and let Ω be any domain in E^n. If $f \in C^\infty(\Omega)$, then every weak solution of $P(D)u = f$ is in $C^\infty(\Omega)$.

Before beginning the proof of Theorem 2-7, we want to make an observation about elliptic operators. From the definition, we know that $p(\xi) \neq 0$ on the set $|\xi| = 1$ in E^n. Thus, the function $1/p(\xi)$ is continuous on this closed, bounded set and, hence, it is bounded there. Hence

$$\frac{1}{|p(\xi)|} \leq M \qquad |\xi| = 1$$

This means that for any real ξ

$$\frac{1}{|p(\xi/|\xi|)|} \leq M$$

or

$$|\xi|^m \leq M|p(\xi)| \qquad \xi \text{ real} \tag{2-42}$$

Since $P(D) - p(D)$ is of order $<m$, we have by (2-31)

$$|P(\xi) - p(\xi)| \leq C(1 + |\xi|)^{m-1} \tag{2-43}$$

These two inequalities imply

Lemma 2-8 If $P(D)$ is an elliptic operator of order m, then there is a constant C, such that

$$|v|_s \leq C(|P(D)v|_{s-m} + |v|_{s-1}) \qquad v \in S \tag{2-44}$$

PROOF First note that there is a constant K, such that

$$(1 + |\xi|)^{2m} \leq K(1 + |\xi|^{2m}) \tag{2-45}$$

In fact, if $|\xi| \geq 1$, then $(1 + |\xi|)^{2m} \leq (2|\xi|)^{2m}$ and $1 + |\xi|^{2m} \geq |\xi|^{2m}$. This shows that

$$\frac{(1 + |\xi|)^{2m}}{1 + |\xi|^{2m}} \leq 2^{2m} \qquad |\xi| \geq 1$$

For $|\xi| \leq 1$ this ratio is clearly bounded. Now, by (2-42) and (2-43)

$$\begin{aligned}(1 + |\xi|)^{2m} &\leq K(1 + |\xi|^{2m}) \leq K(1 + M^2|p(\xi)|^2) \\ &\leq K(1 + 2M^2|P(\xi)|^2 + 2M^2C^2(1 + |\xi|)^{2m-2}) \\ &\leq K(2M^2|P(\xi)|^2 + (2M^2C^2 + 1)(1 + |\xi|)^{2m-2}) \\ &\leq K'(|P(\xi)|^2 + (1 + |\xi|)^{2m-2})\end{aligned}$$

We now multiply both sides by $(1+|\xi|)^{2(s-m)}|Fv|^2$, and integrate with respect to ξ. If we note that $P(\xi)Fv = F(P(D)v)$, by Eq. (2-22), we immediately obtain inequality (2-44).

Employing (2-44), we can give the

PROOF of Theorem 2-7. Let φ be a function in $C_0^\infty(\Omega)$, and define φu to be zero outside Ω. We shall see that, for each $s \geq 0$, there is a constant c_s, such that

$$|J_\varepsilon(\varphi u)|_s \leq c_s \tag{2-46}$$

Since

$$\|J_\varepsilon(\varphi u) - \varphi u\| \to 0 \quad \text{as} \quad \varepsilon \to 0 \tag{2-47}$$

it will follow by Lemma 2-5 that $\varphi u \in C^\infty(E^n)$. Now, for any point $y \in \Omega$, we can find a $\varphi \in C_0^\infty(\Omega)$ which does not vanish in a neighborhood of y (see, for example, (1-36) and (1-37), Sec. 1-3). Consequently, the function $1/\varphi$ is in C^∞, in this neighborhood. Since $u = (\varphi u)\cdot(1/\varphi)$, we see that u is in C^∞ in this neighborhood. Since y was any point of Ω, it will follow that $u \in C^\infty(\Omega)$.

To carry out the argument, we note first that for any $\varphi \in C_0^\infty(\Omega)$ and each $\varepsilon > 0$, the function $J_\varepsilon(\varphi u)$ is in S. Thus, by Lemma 2-8

$$|J_\varepsilon(\varphi u)|_s \leq C(|P(D)J_\varepsilon(\varphi u)|_{s-m} + |J_\varepsilon(\varphi u)|_{s-1}) \tag{2-48}$$

To compute the first term on the right-hand side we note that for $v \in S$, we have, by Eq. (1-102) of Sec. 1-7

$$\begin{aligned}(P(D)J_\varepsilon(\varphi u), v) &= (u, \varphi \bar{P}(D)J_\varepsilon v)\\ &= (u, \bar{P}(D)(\varphi J_\varepsilon v)) - \sum_{|\mu|>0} \frac{1}{\mu!}(u, \bar{P}^{(\mu)}(D)J_\varepsilon v D^\mu\varphi)\\ &= (f, \varphi J_\varepsilon v) - \sum_{|\mu|>0} \frac{(J_\varepsilon(uD^\mu\varphi), \bar{P}^{(\mu)}(D)v)}{\mu!}\end{aligned}$$

Hence

$$\begin{aligned}|(P(D)J_\varepsilon(\varphi u), v)| &\leq |J_\varepsilon(\varphi f)|_{s-m}|v|_{m-s} + \sum_{|\mu|>0} \frac{(|J_\varepsilon(uD^\mu\varphi)|_{s-1}|\bar{P}^{(\mu)}(D)v|_{1-s})}{\mu!}\\ &\leq K_1|v|_{m-s}\left(|J_\varepsilon(\varphi f)|_{s-m} + \sum_{0<|\mu|\leq m} |J_\varepsilon(uD^\mu\varphi)|_{s-1}\right)\end{aligned} \tag{2-49}$$

Here we have employed (2-30) and the fact that $P^{(\mu)}(D)$ is of order $<m$ for $|\mu| > 0$. Since (2-49) holds for all $v \in S$, we have, by Eqs. (2-27) and (2-29)

$$|P(D)J_\varepsilon(\varphi u)|_{s-m} \leq (2\pi)^n K_1\left(|J_\varepsilon(\varphi f)|_{s-m} + \sum_{0\neq|\mu|\leq m} |J_\varepsilon(uD^\mu\varphi)|_{s-1}\right)$$

Substituting into (2-48), we have

$$|J_\varepsilon(\varphi u)|_s \leq C'\left(|J_\varepsilon(\varphi f)|_{s-m} + \sum_{|\mu|\leq m} |J_\varepsilon(uD^\mu\varphi)|_{s-1}\right) \tag{2-50}$$

Now $\varphi f \in C_0^\infty(\Omega) \subset S$. Hence, by Lemma 2-6

$$|J_\varepsilon(\varphi f)|_{s-m} \leq c_s \tag{2-51}$$

for each s. Now, suppose that s is fixed, and suppose that for each $\psi \in C_0^\infty(\Omega)$ there is a constant C, such that

$$|J_\varepsilon(\psi u)|_{s-1} \leq C \tag{2-52}$$

Let $\varphi \in C_0^\infty(\Omega)$ be given. Then, by (2-52)

$$\sum_{|\mu| \leq m} |J_\varepsilon(uD^\mu \varphi)|_{s-1} \leq C' \tag{2-53}$$

Applying inequalities (2-51) and (2-53), to (2-50), we see that

$$|J_\varepsilon(\varphi u)|_s \leq C' \tag{2-54}$$

This is true for each $\varphi \in C_0^\infty(\Omega)$. Thus, (2-52) holding for each $\psi \in C_0^\infty(\Omega)$ implies (2-54) for each $\varphi \in C_0^\infty(\Omega)$. We now employ induction on s. By hypothesis, (2-52) holds for $s = 1$. We have just shown that if it holds for $(s - 1)$, it also holds for s. This shows that (2-46) holds for each s, and completes the proof of the theorem.

2-5 FOURIER TRANSFORMS

In this section we shall give proofs of Eqs. (2-23) and (2-24) which were used in the proof of Theorem 2-7.

We first note that, in each case, it suffices to prove the identity for $n = 1$. For if $n > 1$, set

$$F_k v = \int_{-\infty}^{\infty} e^{-i\xi_k x_k}\, v\, dx_k$$

$$G_k w = \frac{1}{2\pi}\int_{-\infty}^{\infty} e^{ix_k\xi_k}\, w\, d\xi_k$$

Then the F_k and G_k commute and map S into itself. Moreover, $F = F_1 \cdots F_n$, and if we set $G = G_1 \cdots G_n$, then Eq. (2-23) claims that

$$v = GFv \qquad v \in S \tag{2-55}$$

If we can show that

$$v = G_k F_k v \qquad 1 \leq k \leq n \qquad v \in S \tag{2-56}$$

this will imply Eq. (2-55). But Eq. (2-56) is just the one-dimensional analog of Eq. (2-55). Similarly, Eq. (2-24) follows from

$$\int_{-\infty}^{\infty} v\bar{w}\, dx_k = \frac{1}{2\pi}\int_{-\infty}^{\infty} F_k v \overline{F_k w}\, d\xi_k \qquad 1 \leq k \leq n \qquad v, w \in S$$

applied for each k. Thus, we may assume $n = 1$.

To prove Eq. (2-23), set

$$G_R(x) = \frac{1}{2\pi}\int_{-R}^{R} e^{ix\xi}\, Fv\, d\xi$$
$$= \frac{1}{2\pi}\int_{-R}^{R} e^{ix\xi}\left[\int_{-\infty}^{\infty} e^{-i\xi y}\, v(y)\, dy\right] d\xi$$
$$= \frac{1}{2\pi}\int_{-\infty}^{\infty} v(y)\left[\int_{-R}^{R} e^{i\xi(x-y)}\, d\xi\right] dy \tag{2-57}$$

We are able to interchange the order of integration because the iterated integrals are absolutely convergent (any good text on advanced calculus may be consulted). Setting $t = y - x$, we have

$$G_R(x) = \frac{1}{2\pi}\int_{-\infty}^{\infty} v(x+t)\left[\int_{-R}^{R} e^{-i\xi t}\, d\xi\right] dt$$
$$= \frac{1}{\pi}\int_{-\infty}^{\infty} \frac{\sin Rt}{t}\, v(x+t)\, dt \tag{2-58}$$

where we have used

$$\int_{-R}^{R} e^{-i\xi t}\, d\xi = \int_{-R}^{R} (\cos \xi t - i \sin \xi t)\, d\xi$$
$$= 2t^{-1} \sin Rt$$

Now, it is well known that

$$\int_{-\infty}^{\infty} \frac{\sin Rt\, dt}{t} = \pi$$

Hence
$$G_R(x) - v(x) = \frac{1}{\pi}\int_{-\infty}^{\infty}\left[\frac{v(x+t) - v(x)}{t}\right] \sin Rt\, dt$$

Integrating by parts, we get

$$G_R(x) - v(x) = \frac{1}{\pi}\int_{-\infty}^{\infty}\left[\frac{v(x+t) - v(x)}{t}\right] d\left(-\frac{\cos Rt}{R}\right)$$
$$= \frac{1}{\pi R}\int_{-\infty}^{\infty} \cos Rt\, \frac{d}{dt}\left[\frac{v(x+t) - v(x)}{t}\right] dt \tag{2-59}$$

Now set

$$H(x, t) = \frac{d}{dt}\left[\frac{v(x+t) - v(x)}{t}\right]$$
$$= \frac{tv'(x+t) - v(x+t) + v(x)}{t^2}$$

Expanding the function $tv'(x+t) - v(x+t)$ in a Taylor series with remainder, we have

$$tv'(x+t) - v(x+t) + v(x) = \tfrac{1}{2}t^2(v''(x+t_1) + t_1 v'''(x+t_1))$$

where t_1 is between 0 and t. Since $v \in S$, this shows that $H(x,t)$ is continuous in (x,t) and that there is a constant K_1, such that

$$|H(x,t)| \le K_1$$

for all x, t. In addition, since $v(x)$, $v(x+t)$, and $(x+t)v'(x+t)$ are all bounded for $v \in S$, we have

$$|tv'(x+t) - v(x+t) + v(x)| \le K_2(1+|x|)$$

for some constant K_2. Hence, by Eq. (2-59)

$$|G_R(x) - v(x)| \le \frac{1}{\pi R}\int_{-\infty}^{\infty} |H(x,t)|\, dt$$

$$\le \frac{1}{\pi R}\left[2K_1 + (1+|x|)K_2 \int_{|t|>1} |t|^{-2}\, dt\right] \tag{2-60}$$

This shows that $G_R(x) \to v(x)$ as $R \to \infty$ and proves Eq. (2-23). It also shows that if we restrict x to some interval $|x| \le M$, then the convergence is uniform.

To prove Eq. (2-24), we write

$$\int_{-M}^{M} \overline{w(x)}v(x)\, dx = \int_{-M}^{M} \overline{w(x)}\left[\int_{-\infty}^{\infty} e^{ix\xi}\, Fv\, d\xi\right] dx$$

$$= \int_{-\infty}^{\infty} Fv\left[\int_{-M}^{M} e^{ix\xi}\, \overline{w(x)}\, dx\right] d\xi$$

$$= \int_{-\infty}^{\infty} Fv\overline{Fw}\, d\xi - \int_{-\infty}^{\infty} Fv\left[\int_{|x|>M} e^{ix\xi}\, \overline{w(x)}\, dx\right] d\xi$$

But

$$\left|\int_{-\infty}^{\infty} Fv\left[\int_{|x|>M} e^{ix\xi}\, \overline{w(x)}\, dx\right] d\xi\right| \le \int_{-\infty}^{\infty} |Fv|\, d\xi \int_{|x|>M} |w(x)|\, dx \to 0 \text{ as } M \to \infty$$

Hence
$$\int_{-\infty}^{\infty} v(x)\overline{w(x)}\, dx = \int_{-\infty}^{\infty} Fv\overline{Fw}\, d\xi$$

and the proof is complete.

Another important property of Fourier transforms is given by

Theorem 2-9 If $v, w \in S$ and

$$h(x) = \int v(y)w(x-y)\, dy \tag{2-61}$$

then $h \in S$, and

$$Fh = Fv \cdot Fw$$

PROOF First we note that Fv and Fw are in S and, hence, so is $Fv \cdot Fw$. Consequently, so is

$$
\begin{aligned}
G[Fv \cdot Fw] &= (2\pi)^{-n} \int e^{i(x,\xi)} Fv(\xi) \cdot Fw(\xi)\, d\xi \\
&= (2\pi)^{-n} \int e^{i(x,\xi)} Fw(\xi) \left[\int e^{-i(\xi,y)}\, v(y)\, dy \right] d\xi \\
&= (2\pi)^{-n} \int v(y) \left[\int e^{i(\xi,x-y)}\, Fw(\xi)\, d\xi \right] dy \\
&= \int v(y) w(x-y)\, dy
\end{aligned}
$$

by Eq. (2-23). Note that we were able to interchange the order of integration because the iterated integrals are absolutely convergent. This completes the proof.

As an application of this theorem we note that for $v \in S$

$$F(J_\varepsilon v) = Fj_\varepsilon Fv$$

Moreover,

$$
\begin{aligned}
Fj_\varepsilon(\xi) &= \int e^{-i(\xi,x)}\, j_\varepsilon(x)\, dx = \varepsilon^{-n} \int e^{-i(\xi,x)}\, j\left(\frac{x}{\varepsilon}\right) dx \\
&= \int e^{-i(\varepsilon\xi,y)}\, j(y)\, dy = Fj(\varepsilon\xi)
\end{aligned}
$$

Hence

$$F(J_\varepsilon v) = Fj(\varepsilon\xi) \cdot Fv \tag{2-62}$$

Note that

$$|Fj_\varepsilon(\xi)| = |Fj(\varepsilon\xi)| \le \int j(y)\, dy = 1$$

Thus, if $v \in S$, we have

$$
\begin{aligned}
|J_\varepsilon v|_s^2 &= \int (1 + |\xi|)^{2s} |F(J_\varepsilon v)|^2\, d\xi \\
&= \int (1 + |\xi|)^{2s} |Fj_\varepsilon(\xi)|^2\, |Fv|^2\, d\xi \\
&\le \int (1 + |\xi|)^{2s} |Fv|^2\, d\xi \\
&= |v|_s^2
\end{aligned}
$$

Hence, we have

$$|J_\varepsilon v|_s \le |v|_s \qquad v \in S, s \text{ real} \tag{2-63}$$

By completion, this also holds for $v \in H^s$. We also have

$$|J_\varepsilon v - v|_s \to 0 \quad \text{as} \quad \varepsilon \to 0 \qquad v \in H^s \tag{2-64}$$

Since S is dense in H^s and (2-63) holds, it suffices to prove (2-64) for $v \in S$. In fact, if $u \in H^s$ and $\rho > 0$ are given, then there is a $v \in S$, such that

$$|v - u|_s < \frac{\rho}{3}$$

Hence
$$|J_\varepsilon u - u|_s \leq |J_\varepsilon(u - v)|_s + |J_\varepsilon v - v|_s + |v - u|_s$$
$$< \frac{2\rho}{3} + |J_\varepsilon v - v|_s$$

If (2-64) holds for v, then we can take ε so small that $|J_\varepsilon v - v|_s < \rho/3$.

Thus
$$|J_\varepsilon u - u|_s < \rho$$

showing that (2-64) holds for u.

To prove (2-64) for $v \in S$, note that, by Eq. (2-62)

$$|J_\varepsilon v - v|_s^2 = \int |Fj(\varepsilon\xi) - 1|^2 \, |Fv|^2 (1 + |\xi|)^{2s} \, d\xi$$

Moreover
$$|Fj(\varepsilon\xi) - 1| \leq \int |e^{-i\varepsilon(\xi,y)} - 1| j(y) \, dy$$
$$\leq \varepsilon |\xi| \int |y| j(y) \, dy \tag{2-65}$$

by the inequality

$$|e^{-i\theta} - 1| \leq |\theta| \tag{2-66}$$

which follows from the fact that

$$2i \sin\left(\frac{\theta}{2}\right) = e^{i\theta/2} (1 - e^{-i\theta})$$

Hence

$$|J_\varepsilon v - v|_s^2 \leq \varepsilon^2 \left(\int |y| j(y) \, dy \right)^2 \int |\xi|^2 (1 + |\xi|)^{2s} |Fv|^2 \, d\xi \to 0 \quad \text{as} \quad \varepsilon \to 0$$

As a dual of Theorem 2-9, we have

Theorem 2-10

$$F(vw) = (2\pi)^{-n} \int Fv(\eta) Fw(\xi - \eta) \, d\eta \qquad v, w \in S$$

PROOF By Eq. (2-55)

$$\begin{aligned} F(vw) &= F(wGFv) \\ &= (2\pi)^{-n} \int e^{-i(\xi,x)}\, w(x) \left[\int e^{i(x,\eta)}\, Fv(\eta)\, d\eta \right] dx \\ &= (2\pi)^{-n} \int Fv(\eta) \left[\int e^{-i(\xi-\eta,x)}\, w(x)\, dx \right] d\eta \\ &= (2\pi)^{-n} \int Fv(\eta) Fw(\xi - \eta)\, d\eta \end{aligned}$$

2-6 HYPOELLIPTIC OPERATORS

Now let us see what inequality (2-9) implies for operators which are not homogeneous. To get an idea, let θ_j be the vector in E^n with zeros everywhere except in the j-th position, which has one as its entry. Thus, $\theta_1 = (1, 0, \ldots, 0)$, $\theta_2 = (0, 1, 0, \ldots, 0)$, etc. Let $P(\xi)$ be a polynomial satisfying (2-9), and set

$$g_j(t) = P(\xi + t\theta_j) \qquad 1 \le j \le n \tag{2-67}$$

For ξ fixed, $g_j(t)$ is a polynomial of degree $m_j \le m$ in t and, hence, has m_j complex roots $t_1, \ldots, t_{m_j}$. Thus

$$\frac{g_j'(t)}{g_j(t)} = \sum_1^{m_j} \frac{1}{t - t_k} \tag{2-68}$$

Since $P(\xi + t_k\theta_j) = 0$, we have by (2-9)

$$|\xi + t_k\theta_j| \le C_0\, e^{2R|\operatorname{Im} t_k|} \tag{2-69}$$

or

$$|\xi| \le |t_k| + C_0\, e^{2R|\operatorname{Im} t_k|}$$

This shows that $|t_k| \to \infty$ as $|\xi| \to \infty$. But

$$\left| \frac{P^{(j)}(\xi)}{P(\xi)} \right| = \left| \frac{g_j'(0)}{g_j(0)} \right| \le \sum_1^{m_j} \frac{1}{|t_k|} \tag{2-70}$$

where, as usual, $P^{(j)}(\xi) = \partial P(\xi)/\partial \xi_j$. Hence, (2-9) implies

$$\left| \frac{P^{(j)}(\xi)}{P(\xi)} \right| \to 0 \quad \text{as} \quad |\xi| \to \infty \qquad 1 \le j \le n \tag{2-71}$$

Thus, a necessary condition that every weak solution of $P(D)u = 0$ be in $C^1(\Omega)$ is that statement (2-71) holds. Now, (2-71) implies that there are constants $a > 0$, C_1, and C_2, such that

$$\left| \frac{P^{(j)}(\xi)}{P(\xi)} \right| \le \frac{C_1}{|\xi|^a} \qquad |\xi| > C_2 \qquad 1 \le j \le n \tag{2-72}$$

This should not be surprising, since the ratio of two polynomials is a rational function. If this rational function approaches zero as $|\xi| \to \infty$, it should do so in an "algebraic" way, i.e., in the sense of (2-72). We shall not prove (2-72), but rather assume it, since its proof is of an algebraic nature and would take us too far afield. We shall call polynomials (and their corresponding operators) which satisfy (2-72) *hypoelliptic*. We shall prove

Theorem 2-11 Let $P(D)$ be a hypoelliptic operator and let Ω be any domain. If $f \in C^\infty(\Omega)$, then every weak solution of $P(D)u = f$ is in $C^\infty(\Omega)$ and hence is a solution.

We shall carry out the proof in Sec. 2-8 after we develop a few preliminaries in Sec. 2-7.

Inequality (2-72) and Theorem 2-11 are due to Hörmander (1955).

2-7 COMPARISON OF OPERATORS

A polynomial $Q(\xi)$ is said to be *weaker* than a polynomial $P(\xi)$ if there is a constant C, such that

$$|Q(\xi)| \leq C \sum_\mu |P^{(\mu)}(\xi)| \tag{2-73}$$

In this case we have, by Eqs. (2-27) and (2-22), for $\varphi \in S$

$$\begin{aligned} \| Q(D)\varphi \|^2 &= (2\pi)^{-n} \int |F[Q(D)\varphi]|^2 \, d\xi \\ &= (2\pi)^{-n} \int |Q(\xi)F\varphi|^2 \, d\xi \leq C_1 \sum \int |P^{(\mu)}(\xi)F\varphi|^2 \, d\xi \\ &= C_1 \sum \int |F[P^{(\mu)}(D)\varphi]|^2 \, d\xi = (2\pi)^n C_1 \sum \| P^{(\mu)}(D)\varphi \|^2 \end{aligned}$$

If we now apply Corollary 1-16, Sec. 1-7, we obtain, for each bounded domain Ω

$$\| Q(D)\varphi \| \leq C_2 \| P(D)\varphi \| \qquad \varphi \in C_0^\infty(\Omega) \tag{2-74}$$

We also have the following

Theorem 2-12 If $Q(\xi)$ is weaker than $P(\xi)$, then for each real s and each bounded domain Ω, there is a constant C, such that

$$|Q(D)\varphi|_s \leq C |P(D)\varphi|_s \qquad \varphi \in C_0^\infty(\Omega) \tag{2-75}$$

In proving Theorem 2-12 we shall use the following two lemmas.

Lemma 2-13 If $Q(\xi)$ is weaker than $P(\xi)$, and $R(\xi) = (1 + |\xi|^2)^k$ for some integer $k \geq 0$, then QR is weaker than PR.

Lemma 2-14 If $s < 0$, then for each $b > 0$ there is a constant K, such that

$$K^{-1}|v|_s^2 \le \int_0^b \| J_\varepsilon v \|^2 \varepsilon^{-2s-1}\, d\varepsilon \le K|v|_s^2 \qquad v \in S \tag{2-76}$$

PROOF of Theorem 2-12 Let us assume these lemmas for the moment and show how they imply Theorem 2-12. If $s < 0$, we have by inequalities (2-76) and (2-74)

$$|Q(D)\varphi|_s^2 \le K \int_0^b \| J_\varepsilon Q(D)\varphi \|^2 \varepsilon^{-2s-1}\, d\varepsilon$$

$$\le KC' \int_0^b \| P(D)J_\varepsilon \varphi \|^2 \varepsilon^{-2s-1}\, d\varepsilon \le K^2C'|P(D)\varphi|_s^2$$

Note that we have used the fact that J_ε commutes with differentiation, and that $J_\varepsilon\varphi$ is in $C_0^\infty(\Omega_1)$ for some larger domain Ω_1.

Now, if $s > 0$, let k be an integer $\ge s/2$, and set $R(\xi) = (1 + |\xi|^2)^k$. Note that $R(\xi)$ is a polynomial in ξ. Set $Q_1(\xi) = Q(\xi)R(\xi)$, $P_1(\xi) = P(\xi)R(\xi)$. Then Q_1 is weaker than P_1, by Lemma 2-13. Moreover

$$|Q(D)\varphi|_s^2 = \int |Q_1(\xi)F\varphi|^2 (1 + |\xi|)^{2s}(1 + |\xi|^2)^{-2k}\, d\xi$$

$$\le C_3 \int |Q_1(\xi)F\varphi|^2 (1 + |\xi|)^{2s-4k}\, d\xi$$

$$= C_3|Q_1(D)\varphi|_{s-2k}^2 \le C_4|P_1(D)\varphi|_{s-2k}^2$$

$$= C_4 \int |P(\xi)F\varphi|^2 (1 + |\xi|)^{2s-4k}(1 + |\xi|^2)^{2k}\, d\xi$$

$$\le C_5 \int |P(\xi)F\varphi|^2 (1 + |\xi|)^{2s}\, d\xi = C_5|P(D)\varphi|_s^2$$

by the part already proved. We have employed the trivial inequality

$$\frac{(1 + |\xi|)^{2k}}{(1 + |\xi|^2)^k} + \frac{(1 + |\xi|^2)^k}{(1 + |\xi|^{2k})} \le \text{constant} \tag{2-77}$$

This completes the proof.

PROOF of Lemma 2-13 employs a simple result on polynomials which will be given in the next chapter (Lemma 3-14, of Sec. 3-4). There it will be shown that there are real vectors $\theta_1, \ldots, \theta_r$ in E^n, such that

$$\sum |P^{(\mu)}(\xi)| \le C \sum |P(\xi + \theta_j)| \tag{2-78}$$

where the constant C does not depend on ξ or $P(\xi)$. Now, for each fixed $\theta \in E^n$ there is a constant C_θ depending only on $|\theta|$, such that

$$1 + |\xi + \theta|^2 \le C_\theta(1 + |\xi|^2) \tag{2-79}$$

Hence, there is a constant C depending only on $|\theta_j|$ and k, such that

$$|R(\xi)| \le C|R(\xi + \theta_j)| \qquad 1 \le j \le r \tag{2-80}$$

Combining (2-78) and (2-80), we obtain

$$\sum |P^{(\mu)}(\xi)R(\xi)| \le C \sum |P_1(\xi + \theta_j)|$$

$$\le C \sum \frac{|P_1^{(\mu)}(\xi)\theta_j^{\mu}|}{\mu!} \le C' \sum |P_1^{(\mu)}(\xi)|$$

where $P_1(\xi) = P(\xi)R(\xi)$. Since $Q(\xi)$ is weaker than $P(\xi)$, we have

$$|Q(\xi)R(\xi)| \le C \sum |P^{(\mu)}(\xi)R(\xi)|$$

Combining the last two inequalities, we obtain the desired result. A proof of (2-78) will be given at the end of Sec. 3-4.

PROOF of Lemma 2-14 By Eqs. (2-27) and (2-62)

$$(2\pi)^n \| J_\varepsilon v \|^2 = \int |F(J_\varepsilon v)|^2 \, d\xi = \int |Fj(\varepsilon\xi)|^2 |Fv|^2 \, d\xi$$

Hence
$$\int_0^b \| J_\varepsilon v \|^2 \varepsilon^{-2s-1} \, d\varepsilon = (2\pi)^{-n} \int h(\xi) |Fv|^2 \, d\xi$$

where
$$h(\xi) = \int_0^b |Fj(\varepsilon\xi)|^2 \varepsilon^{-2s-1} \, d\varepsilon$$

Thus
$$h(\xi) \le \frac{-1}{2sb^{2s}}$$

For $|\xi| > 1$ we can get a better estimate. Set $t = \varepsilon(1 + |\xi|)$ and $\eta = \xi/(1 + |\xi|)$. Then $|\eta| \ge \frac{1}{2}$, and

$$h(\xi) \le (1 + |\xi|)^{2s} \int_0^\infty |Fj(t\eta)|^2 t^{-2s-1} \, dt$$

Since $j(x) \in S$, we have

$$|Fj(\xi)| \le \frac{K}{(1 + |\xi|)^k}$$

for k any integer. In particular, this holds for $k > 1 - 2s$. Hence

$$\int_0^\infty |Fj(t\eta)| t^{-2s-1} \, dt \le K \int_0^\infty \frac{t^{-2s-1}}{(1 + t|\eta|)^k} \, dt$$

$$= K|\eta|^{2s} \int_0^\infty \frac{r^{-2s-1}}{(1 + r)^k} \, dr \le K'$$

Thus, there is a constant M, such that

$$h(\xi) \leq M(1 + |\xi|)^{2s} \tag{2-81}$$

This proves the second inequality in (2-76). To prove the first, note that

$$Fj(0) = \int j(x)\, dx = 1$$

Hence, there is a $\delta > 0$, such that

$$|Fj(\xi)|^2 > \tfrac{1}{2} \quad \text{for} \quad |\xi| < \delta$$

Now, for $b|\xi| \leq \delta$, we have

$$h(\xi) \geq \frac{1}{2}\int_0^b \varepsilon^{-2s-1}\, d\varepsilon = \frac{-b^{-2s}}{4s} \tag{2-82}$$

If $b|\xi| > \delta$, then

$$\begin{aligned} h(\xi) &\geq \int_0^{\delta/|\xi|} |Fj(\varepsilon\xi)|^2 \varepsilon^{-2s-1}\, d\varepsilon \\ &\geq \frac{1}{2}\int_0^{\delta/|\xi|} \varepsilon^{-2s-1}\, d\varepsilon = -\frac{\delta^{-2s}}{4s}|\xi|^{2s} \end{aligned} \tag{2-83}$$

Combining (2-82) and (2-83) we see that there is a constant $m > 0$, such that

$$h(\xi) \geq m(1 + |\xi|)^{2s}$$

This completes the proof. Lemma 2-14 is due to Hörmander (1958).

2-8 PROOF OF REGULARITY

We are now ready to give the proof of Theorem 2-11. The proof of Theorem 2-7 will be followed closely.

Consider the norm

$$|||v|||_s = \sum_{|\mu| \neq 0} |P^{(\mu)}(D)v|_s$$

In terms of this norm, we have

$$|||v|||_{s+a} \leq C(|P(D)v|_s + |v|_s) \qquad v \in C_0^\infty(\Omega) \tag{2-84}$$

where $a > 0$ is the constant appearing in inequality (2-72) and Ω is a bounded domain. To see this, note that for $|\mu| \neq 0$, $P^{(\mu)}(\xi)$ is weaker than $P^{(j)}(\xi)$ for some j, $1 \leq j \leq n$. Thus, by Theorem (2-12)

$$|||v|||_{s+a} \leq C_1 \sum_1^n |P^{(j)}(D)v|_{s+a} \qquad v \in C_0^\infty(\Omega)$$

Furthermore, by (2-72)

$$(1 + |\xi|)^{2a} \sum_{1}^{n} |P^{(j)}(\xi)|^2 \leq C_2(|P(\xi)|^2 + 1) \qquad \xi \in E^n$$

These inequalities imply (2-84). Now, suppose u is a weak solution of $P(D)u = f$ in some domain, Ω_0. Let Ω be any bounded subdomain of Ω_0.

Then $$(u, \bar{P}(D)w) = (f, w) \qquad w \in C_0^\infty(\Omega) \tag{2-85}$$

We are going to show that for each real $\varphi \in C_0^\infty(\Omega)$, and each real s, there is a constant K, such that

$$|||J_\varepsilon(\varphi u)|||_s \leq K \tag{2-86}$$

It will then follow from Lemma 2-5 that $\varphi u \in C^\infty(\Omega)$ for each such φ. It thus follows that $u \in C^\infty(\Omega)$. We shall prove (2-86) by showing:

1. that if it holds for any φ when $s = t$, then it holds for any φ when $s = t + a$, and
2. that (2-86) holds when $s = -m$.

Since $a > 0$, it will then follow by induction that (2-86) holds for each s and φ.

To carry out the argument, let φ be any real-valued function in $C_0^\infty(\Omega)$, and v any function in S. Then

$$\begin{aligned}(P(D)J_\varepsilon(\varphi u), v) &= (\varphi u, \bar{P}(D)J_\varepsilon v) \\ &= (u, \bar{P}(D)(\varphi J_\varepsilon v)) - \sum_{|\mu|>0} \frac{(u, \bar{P}^{(\mu)}(D)J_\varepsilon v D^\mu \varphi)}{\mu!} \\ &= (f, \varphi J_\varepsilon v) - \sum_{|\mu|>0} \frac{(P^{(\mu)}(D)J_\varepsilon(uD^\mu\varphi), v)}{\mu!}\end{aligned}$$

where we have used Eq. (1-102) of Sec. 1-7. Hence

$$\begin{aligned}|(P(D)J_\varepsilon(\varphi u), v)| &\leq |v|_{-s}\left(|J_\varepsilon(\varphi f)|_s + \sum_{|\mu|>0} \frac{|P^{(\mu)}(D)J_\varepsilon(uD^\mu\varphi)|_s}{\mu!}\right) \\ &\leq |v|_{-s}\left(|J_\varepsilon(\varphi f)|_s + C \sum_{|\mu|\leq m} |||J_\varepsilon(uD^\mu\varphi)|||_s\right)\end{aligned}$$

Thus, by Eq. (2-29) and inequality (2-84)

$$|||J_\varepsilon(\varphi u)|||_{s+a} \leq C\left(|J_\varepsilon(\varphi f)|_s + \sum_{|\mu|\leq m} |||J_\varepsilon(uD^\mu\varphi)|||_s\right) \tag{2-87}$$

Now, by Lemma 2-6

$$|J_\varepsilon(\varphi f)|_s \leq \text{constant}$$

for each s and φ. Now, suppose

$$|||J_\varepsilon(\psi u)|||_s \leq \text{constant} \tag{2-88}$$

for each $\psi \in C_0^\infty(\Omega)$. Then (2-87) shows that

$$|||J_\varepsilon(\varphi u)|||_{s+a} \leq \text{constant}$$

for each $\varphi \in C_0^\infty(\Omega)$.

Moreover, (2-88) holds when $s = -m$. In fact, we have by inequalities (2-30) and (2-15)

$$\begin{aligned} |P^{(\mu)}(D)J_\varepsilon(\psi u)|_{-m} &\leq C|J_\varepsilon(\psi u)|_0 \\ &\leq C' \|\psi u\| \end{aligned}$$

which is finite since $u \in L^2(\Omega)$. Hence, we can apply induction to conclude that (2-88) holds for each s and ψ. Since there is a $\mu \neq 0$, such that $P^{(\mu)}(\xi) = \text{constant} \neq 0$, we have

$$|J_\varepsilon(\psi u)|_s \leq \text{constant}$$

for all s and ψ. This shows that $u \in C^\infty(\Omega)$. Since Ω was an arbitrary subdomain of Ω, the theorem follows.

2-9 THE CLOSED GRAPH THEOREM

In Sec. 2-1 we made use of the following theorem. We present its proof in this section.

Theorem 2-15: Closed graph theorem Let A be a linear operator from a Hilbert space χ to a Hilbert space Y defined everywhere on χ, and such that

$$x_n \to x \quad \text{in} \quad \chi \qquad Ax_n \to y \quad \text{in} \quad Y \tag{2-89}$$

implies that $Ax = y$. Then there is a constant C, such that

$$\|Ax\| \leq C\|x\| \qquad x \in \chi \tag{2-90}$$

Proof Let D be the set of those $y \in Y$ for which there is a $y^* \in \chi$, satisfying

$$(y, Ax) = (y^*, x) \qquad x \in \chi \tag{2-91}$$

Clearly D is a subspace of Y. Note that y^* is unique. For, if there were another element $z \in \chi$, satisfying

$$(y, Ax) = (z, x) \qquad x \in \chi$$

then

$$((y^* - z), x) = 0 \qquad x \in \chi$$

Take $x = y^* - z$. Then $\|y^* - z\| = 0$, showing that $z = y^*$.

We shall prove

1. There is a constant C, such that

$$\|y^*\| \leq C\|y\| \qquad y \in D \tag{2-92}$$

2. $\overline{D} = Y$

Assume these statements for the moment. We show how they imply the theorem. If (2-90) were false, there would be a sequence $\{x_n\}$ of elements of χ, such that

$$\| x_n \| = 1 \qquad \| Ax_n \| \to \infty \quad \text{in} \quad Y \quad \text{as} \quad n \to \infty \tag{2-93}$$

Then, by (2-91), (2-92), and (2-93)

$$|(y, Ax_n)| = |(y^*, x_n)| \leq \| y^* \| \; \| x_n \| \leq C \| y \|$$

for all $y \in D$. If y is any element of Y, there is a sequence $\{y_k\}$ of elements of D which converge to y by 2. Since

$$|(y_k, Ax_n)| \leq C \| y_k \| \qquad k = 1, 2, \ldots$$

we have, in the limit

$$|(y, Ax_n)| \leq C \| y \| \qquad y \in Y \tag{2-94}$$

for each n. Setting $y = Ax_n$, we have

$$\| Ax_n \|^2 \leq C \| Ax_n \|$$

or

$$\| Ax_n \| \leq C$$

which contradicts Eq. (2-93). It thus remains to prove 1 and 2.

To prove 1, suppose there is a sequence $\{y_n\}$ of elements of D, such that

$$\| y_n \| = 1 \qquad \| y_n^* \| \to \infty \quad \text{as} \quad n \to \infty$$

Set

$$F_n(x) = (Ax, y_n) = (x, y_n^*)$$

Since

$$|F_n(x)| \leq \| y_n^* \| \; \| x \|$$

we see that for each n, $F_n(x)$ is a bounded linear functional on χ. On the other hand

$$|F_n(x)| \leq \| Ax \| \; \| y_n \| = \| Ax \|$$

showing that

$$\operatorname*{lub}_n |F_n(x)| \leq \| Ax \|$$

$$< \infty$$

for each x. We can now apply the Banach–Steinhaus theorem (Theorem 1-10, Sec. 1-5) to the sequence $\{F_n\}$. Thus, there is a constant C, such that

$$|F_n(x)| \leq C \| x \| \qquad x \in \chi$$

In other words

$$|(x, y_n^*)| \leq C \| x \| \qquad x \in \chi$$

Take $x = y_n^*$. Then

$$\| y_n^* \|^2 \leq C \| y_n^* \|$$

or

$$\| y_n^* \| \leq C$$

contradicting Eq. (2-95).

To prove 2, let H be the set of ordered pairs $\{x, y\}$ of elements, where $x \in \chi$ and $y \in Y$. If we write

$$\alpha\{x, y\} = \{\alpha x, \alpha y\}$$

$$\{x_1, y_1\} + \{x_2, y_2\} = \{x_1 + x_2, y_1 + y_2\}$$

$$(\{x_1, y_1\}, \{x_2, y_2\}) = (x_1, x_2) + (y_1, y_2)$$

then H becomes a Hilbert space. It is called the *cartesian product* of χ and Y, and is denoted by $\chi \times Y$.

Now let G_A be the set of those elements of H of the form $\{x, Ax\}$. G_A is called the *graph* of A (for reasons not too difficult to explain). I claim that it is a closed subspace of H. In fact, if $\{x_n, Ax_n\} \to \{x, y\}$ in H, then (2-89) holds, and $Ax = y$ by hypothesis. Thus, $\{x, y\} \in G_A$.

We also note that $G_A^\perp$ consists of those elements of H of the form $\{-y^*, y\}$, where $y \in D$. Such elements are clearly in $G_A^\perp$ by Eq. (2-91), and if $\{z, y\} \in G_A^\perp$, then

$$(z, x) + (y, Ax) = 0 \qquad x \in \chi$$

This shows that $y \in D$ and $z = -y^*$.

Now suppose $\bar{D} \neq Y$. Then, by Corollary 1-4 of Sec. 1-5, there is a $w \neq 0$ in Y, such that $(w, \bar{D}) = 0$. Therefore,

$$(\{0, w\}, \{-y^*, y\}) = -(0, y^*) + (w, y) = 0$$

for every $y \in D$. This shows that the element $\{0, w\}$ is in $(G_A^\perp)^\perp$. But this latter subspace is just G_A by Lemma 1-7 of Sec. 1-5. Hence, there is an $x \in \chi$ such that $\{x, Ax\} = \{0, w\}$. But if $x = 0$, then $Ax = 0$, showing that $w = 0$. This contradicts the assumption that $\bar{D} \neq Y$, and completes the proof.

PROBLEMS

2-1 Prove statement (2-17) and Eqs. (2-18), (2-19).

2-2 Prove Eqs. (2-21) and (2-22).

2-3 Show that H^s is a Hilbert space for each s.

2-4 Prove statement (2-36).

2-5 Prove the identities following Eq. (2-58).

2-6 Prove inequality (2-66).

2-7 Prove inequality (2-77).

2-8 Prove inequality (2-79).

2-9 If $P(\xi)$ satisfies inequality (2-72), show that

$$|P^{(j)}(D)v|_{s+a} \leq C(|P(D)v|_s + |v|_s) \qquad v \in S$$

for each real s.

2-10 Show that (2-72) implies $a \leq 1$.

2-11 Show that the Fourier transform maps S into itself.

2-12 Prove that Parseval's identity in n dimensions follows from repeated applications of the one-dimensional case.

2-13 Verify the Taylor expansion following Eq. (2-59).

CHAPTER

THREE

REGULARITY (VARIABLE COEFFICIENTS)

3-1 FORMALLY HYPOELLIPTIC OPERATORS

The next thing that comes to mind is whether or not theorems similar to those of the preceding chapters, can be proved for equations with variable coefficients. The purpose of the present chapter is to introduce a class of operators with variable coefficients, which have regularity properties similar to the hypoelliptic constant-coefficient operators. To this end we introduce some terminology.

We shall call the operator

$$P(x, D) = \sum_{|\mu| \leq m} a_\mu(x)\, D^\mu \tag{3-1}$$

formally hypoelliptic in Ω if it satisfies the following conditions:

1. The coefficients $a_\mu(x)$ are in $C^\infty(\Omega)$
2. For each two points y and z in Ω, the polynomial $P(y, \xi)$ is weaker than the polynomial $P(z, \xi)$.
3. For each point $y \in \Omega$ the constant-coefficient operator $P(y, D)$ is hypoelliptic.

In the case of 3, the operator $P(y, D)$ is the constant-coefficient operator whose coefficients are the function a_μ, evaluated at the point y. We shall prove

Theorem 3-1 Assume that $P(x, D)$ is formally hypoelliptic in Ω. If $f \in C^\infty(\Omega)$, then every weak solution of

$$P(x, D)u = f \tag{3-2}$$

is in $C^\infty(\Omega)$, and hence is a solution.

As before, the theorem means that

$$(u, P'(x, D)\varphi) = (f, \varphi) \qquad \varphi \in C_0^\infty(\Omega) \tag{3-3}$$

implies that $u \in C^\infty(\Omega)$ (here $P'(x, D)$ is the formal adjoint of $P(x, D)$; see Sec. 1-3).

We shall prove the theorem in Sec. 3-2. Here we want to give some examples.

The operator $P(x, D)$ is called *elliptic at the point* x if the constant-coefficient operator, whose coefficients are the a_μ evaluated at x, is elliptic. It is called *elliptic in a domain* Ω if it is of the same order m and elliptic at each point of Ω. This is equivalent to saying that

$$p(x, \xi) \neq 0 \qquad x \in \Omega \qquad \xi \text{ real} \tag{3-4}$$

where

$$p(x, D) = \sum_{|\mu| = m} a_\mu(x) D^\mu \tag{3-5}$$

we have

Lemma 3-2 If $P(x, D)$ is elliptic in Ω and its coefficients $a_\mu(x)$ are in $C^\infty(\Omega)$, then $P(x, D)$ is formally hypoelliptic in Ω.

Proof By Theorem 2-7 of Sec. 2-4, every constant-coefficient elliptic operator must be hypoelliptic (actually, we could have verified this algebraically). Thus, 1 and 3 are immediately verified. To check 2, we note that any operator of order $\leq m$ is weaker than an elliptic operator of order m. For we showed in the proof of Lemma 2-8, Sec. 2-4, that there is a constant K, such that

$$(1 + |\xi|)^{2m} \leq K^2(|P(\xi)|^2 + (1 + |\xi|)^{2m-2}) \tag{3-6}$$

If $1 + |\xi| > 2^{1/2}K$, we have

$$K(1 + |\xi|)^{m-1} < \frac{(1 + |\xi|)^m}{2}$$

and hence

$$(1 + |\xi|)^m \leq 2^{1/2}K|P(\xi)| \tag{3-7}$$

Moreover, there is a constant K', such that

$$(1 + |\xi|)^n \leq K' \sum |P^{(\mu)}(\xi)| \qquad 1 + |\xi| \leq 2^{1/2}K \tag{3-8}$$

since the right-hand side does not vanish for real ξ. Combining inequalities (3-7) and (3-8), we get

$$(1 + |\xi|)^m \leq K'' \sum |P^{(\mu)}(\xi)| \qquad \xi \text{ real} \tag{3-9}$$

If $Q(\xi)$ is any polynomial of degree $\leq m$, then there is a constant K''', such that

$$|Q(\xi)| \leq K'''(1 + |\xi|)^m \tag{3-10}$$

Inequalities (3-9) and (3-10) prove the assertion.

Theorem 3-1 is due to Hörmander (1958) and Malgrange (1957).

3-2 PROOF OF REGULARITY

Let $P(D)$ be a constant-coefficient operator and Ω a bounded domain. Consider the expression

$$|w|_s^P = |P(D)w|_s \qquad w \in C_0^\infty(\Omega) \tag{3-11}$$

We note that this is a norm on $C_0^\infty(\Omega)$, i.e., it satisfies (1-58) to (1-60) of Sec. 1-5. The only thing that is not immediately obvious, is that it satisfies (1-59). This follows from Theorem 2-12 of Sec. 2-7, and the fact that the polynomial $Q(\xi) \equiv 1$ is weaker than $P(\xi)$. Thus, if $|w|_s^P = 0$, it follows that $|w|_s = 0$ and, hence, $w = 0$. Let $H_P^s(\Omega)$ be the completion of $C_0^\infty(\Omega)$ with respect to the norm (3-11). Then $H_P^s(\Omega)$ is a Hilbert space (the scalar product should be fairly obvious).

Now, suppose $u \in H_P^s(\Omega)$, and let $Q(\xi)$ be any polynomial weaker than $P(\xi)$. Then we can define $Q(D)u$ as an element of H^s. For, by Theorem 2-12 of Sec. 2-7.

$$|Q(D)w|_s \le C|w|_s^P \qquad w \in C_0^\infty(\Omega) \tag{3-12}$$

But if $u \in H_P^s(\Omega)$, there is a sequence $\{w_k\}$ of functions in $C_0^\infty(\Omega)$ which converges to u in $H_P^s(\Omega)$. Thus, by inequality (3-12), $\{Q(D)w_k\}$ is a Cauchy sequence in H^s. Hence, there is an $h \in H^s$ such that $Q(D)w_k \to h$ in H^s. Note that h is independent of the particular sequence $\{w_k\}$ chosen, as long as $w_k \to u$ in $H_P^s(\Omega)$. We now *define* $Q(D)u$ to be h. Applying (3-12) to the w_k and taking the limit, we get

$$|Q(D)u|_s \le C|u|_s^P \qquad u \in H_P^s(\Omega) \tag{3-13}$$

Note that

$$(u, \bar{Q}(D)v) = (Q(D)u, v) \qquad u \in H_P^s(\Omega) \qquad v \in S \tag{3-14}$$

Now let us turn to the

PROOF of Theorem 3-1 Assume that $P(x, D)$ is a formally hypoelliptic operator in Ω. Let y be any fixed point of Ω, and set

$$P(D) = P(y, D) \tag{3-15}$$

Let $a > 0$ be a constant, for which $P(D)$ satisfies inequality (2-72) of Sec. 2-6. Set $b = a/m$, where m is the order of $P(D)$. We are going to prove

Theorem 3-3 For each real s there is a neighborhood N_y of y, such that

$$|J_\varepsilon(\varphi u)|_s^P \le C\left(|\varphi f|_s + \sum_{|\mu|\le m} |uD^\mu_{\cdot}\varphi|_{s-b}^P\right) \qquad \varphi \in C_0^\infty(N_y) \tag{3-16}$$

whenever u, f satisfy Eq. (3-3), $\varphi f \in H^s(\Omega)$ and $\varphi u \in H_P^{s-b}(\Omega)$ for all $\varphi \in C_0^\infty(N_y)$. The constant C does not depend on ε, φ, f, or u.

Before embarking on the proof of Theorem 3-3, let us show how it implies Theorem 3-1. Suppose $f \in C^\infty(\Omega)$, $u \in L^2(\Omega)$, and Eq. (3-3) holds. Then $\varphi f \in C_0^\infty(\Omega)$ for each $\varphi \in C_0^\infty(\Omega)$. Now let y be any point of Ω and assume that there is an s and a neighboring Ω_1 of y, such that $\varphi u \in H_P^{s-b}(\Omega)$ for each $\varphi \in C_0^\infty(\Omega_1)$. By (3-16) there is a neighborhood $N_y \subseteq \Omega_1$, such that

$$|J_\varepsilon(\varphi u)|_s^P \leq \text{constant} \tag{3-17}$$

for each $\varphi \in C_0^\infty(N_y)$. Now, by inequality (3-17) and the Banach–Saks theorem (Theorem 1-9 of Sec. 1-5) there is a sequence $\varepsilon_k \to 0$ such that

$$\frac{[J_{\varepsilon_1}(\varphi u) + \cdots + J_{\varepsilon_k}(\varphi u)]}{k}$$

converges in $H_P^s(\Omega)$. Since this sequence converges to φu in H^s (see (2-64), Sec. 2-5), it must do so in $H_P^s(\Omega)$ as well. This shows that $\varphi u \in H_P^s(\Omega)$. Thus, we have shown that if there is a neighborhood Ω_1 of y such that $\varphi u \in H_P^{s-b}(\Omega)$ for each $\varphi \in C_0^\infty(\Omega_1)$, then there is a neighborhood $N_y \subseteq \Omega_1$ such that $\varphi u \in H_P^s(\Omega)$ for each $\varphi \in C_0^\infty(N_y)$. If we can find a t such that $\varphi u \in H_P^t(\Omega)$ for all $\varphi \in C_0^\infty(\Omega)$, it will follow by an induction argument that for each s there is a neighborhood N_y, such that $\varphi u \in H_P^s(\Omega)$ for all $\varphi u \in H_P^s(\Omega)$ for all $\varphi \in C_0^\infty(N_y)$. An application of Lemma 2-5 of Sec. 2-3 then shows that for each k there is a neighborhood N_y of y, such that $u \in C^k(N_y)$. Since y was any point of Ω, this means that $u \in C^\infty(\Omega)$.

Thus, to complete the proof, it sufficies to find a t, such that $\varphi u \in H_P^t(\Omega)$. This is easily done. We may take $t = -m$. For, by inequality (2-30), there is a constant C, such that

$$|P(D)w|_{-m} \leq C|w|_0 \qquad w \in C_0^\infty(\Omega) \tag{3-18}$$

Now we know that there is a sequence $\{w_k\}$ of functions in $C^\infty(\Omega)$ converging to u in $L^2(\Omega)$ (Theorem 2-3 of Sec. 2-2). By (3-18), $\{\varphi w_k\}$ is a Cauchy sequence in $H_P^{-m}(\Omega)$ for each $\varphi \in C_0^\infty(\Omega)$. Thus, $\varphi u \in H_P^{-m}(\Omega)$ for each such φ. This completes the proof.

Let us now turn to the proof of Theorem 3-3. By hypothesis, $P(D)$ is hypoelliptic. Let V be the set of all polynomials which are weaker than $P(\xi)$. Clearly, V is a vector space (if this term is unfamiliar, see Sec. 3-3). For if $Q_1(\xi)$ and $Q_2(\xi)$ are in V, so is $\alpha_1 Q_1(\xi) + \alpha_2 Q_2(\xi)$ for any scalars α_1, α_2. Moreover, V has finite dimension. To see this, note that V is a subspace of the vector space of all polynomials of degree $\leq m$. This follows from the fact that a polynomial of degree $> m$ cannot be weaker than one of degree m. Moreover, the monomials ξ^μ with $|\mu| \leq m$ are linearly independent, and every polynomial of degree $\leq m$ is a linear combination of them. Thus, V is finite dimensional. (All of the statements of this paragraph will be proved in the next section.)

Let $P_1(\xi), \ldots, P_N(\xi)$ be a basis for V. By assumptions 2 and 3, Sec. 3-1, for each $x \in \Omega$ the polynomial $P(x, \xi)$ is in V. Thus, we can write

$$P(x, \xi) = \sum_{j=1}^{N} c_j(x) P_j(\xi) \qquad x \in \Omega \tag{3-19}$$

We shall see that the coefficients $c_j(x)$ are in $C^\infty(\Omega)$. Let us postpone the proof of this fact until the end of the section. Assume it for the moment.

Since

$$P(D) = P(y, D) = \sum_{j=1}^{N} c_j(y) P_j(D) \tag{3-20}$$

we have, Eqs. (3-19), (3-20)

$$P(x, D) - P(D) = \sum_{j=1}^{N} b_j(x) P_j(D) \tag{3-21}$$

where $b_j(x) = c_j(x) - c_j(y)$. Hence

$$b_j(y) = 0 \qquad 1 \le j \le N \tag{3-22}$$

By Eq. (3-21), we have

$$P'(x, D) - \bar{P}(D) = \sum_{j=1}^{N} \bar{P}_j(D) \bar{b}_j(x) \tag{3-23}$$

To carry out the proof of Theorem 3-3, we shall need the following:

Lemma 3-4 If $\psi \in C_0^\infty(E^n)$ and $v \in H^s$, then $\psi v \in H^s$, and there is a constant C depending only on ψ and s, such that

$$|\psi v|_s \le \max_{E^n} |\psi(x)| \, |v|_s + C|v|_{s-1} \qquad v \in H^s \tag{3-24}$$

$$|J_\varepsilon(\psi v) - \psi J_\varepsilon v|_s \le C|v|_{s-1} \qquad v \in H^{s-1} \tag{3-25}$$

Lemma 3-5 Let $P(D)$ be a hypoelliptic operator of order m satisfying inequality (2-72), Sec. 2-6, with constant $a > 0$. Set $b = a/m$ and let $Q(\xi)$ be a polynomial weaker than $P(\xi)$. Then, for each real s and each bounded domain Ω

$$\sum_{|\mu|>0} |Q^{(\mu)}(D)\varphi|_{s+b} \le C(|P(D)\varphi|_s + |\varphi|_s) \qquad \varphi \in C_0^\infty(\Omega) \tag{3-26}$$

We shall prove these lemmas in Sec. 3-4. Assuming them for the present, we give the

PROOF of Theorem 3-3 Since the $P_j(\xi)$ are weaker than $P(\xi)$, there is a constant C_0, such that

$$|P_j(D)w|_s \le C_0 |P(D)w|_s \qquad w \in H_P^s(\Omega) \tag{3-27}$$

By Eq. (3-22) there is a sphere Ω_1 with center y, such that

$$\sum_{j=1}^{N} |b_j(x)| \le \frac{1}{2C_0} \qquad x \in \Omega_1 \tag{3-28}$$

Let N_y be a sphere with center y and radius less than that of Ω_1. Then there is a ψ in $C_0^\infty(\Omega_1)$, such that

$$0 \leq \psi(x) \leq 1 \qquad x \in \Omega_1 \tag{3-29}$$

$$\psi(x) = 1 \qquad x \in N_y \tag{3-30}$$

(See Sec. 1-3.) Clearly, $\psi\varphi = \varphi$ for $\varphi \in C_0^\infty(N_y)$. Now, for $\varphi \in C_0^\infty(N_y)$, $v \in S$, we have

$$\begin{aligned}(u, \bar{\varphi}\bar{P}(D)J_\varepsilon v) &= (u, \bar{P}(D)(\bar{\varphi}\psi J_\varepsilon v)) \\ &\quad - \sum_{|\mu|>0} \frac{(u, D^\mu \bar{\varphi} \bar{P}^{(\mu)}(D)(\psi J_\varepsilon v))}{\mu!} \\ &= (f, \bar{\varphi} J_\varepsilon v) - \sum_{j=1}^{N} (u, \varphi \bar{P}_j(D)(\bar{b}_j \psi J_\varepsilon v)) \\ &\quad - \sum_{j=0}^{N} \sum_{|\mu|>0} \frac{(uD^\mu\varphi, \bar{P}_j^{(\mu)}(D)(\bar{b}_j \psi J_\varepsilon v))}{\mu!}\end{aligned}$$

where we have used identity (1-102) of Sec. 1-7 as well as Eq. (3-23), and have set $P_0(\xi) = P(\xi)$, $b_0(\xi) \equiv 1$. Put $\psi_j = \psi b_j$. Since u is assumed to be in H_P^{s-b}, $P_j(D)(\varphi u) \in H^{s-b}$ for

$$\begin{aligned}(P(D)J_\varepsilon(\varphi u), v) &= (J_\varepsilon(\varphi f), v) \\ &\quad - \sum_{j=1}^{N} (J_\varepsilon[\psi_j P_j(D)(\varphi u)], v) \\ &\quad - \sum_{j=0}^{N} \sum_{|\mu|>0} \frac{(J_\varepsilon[\psi_j P_j^{(\mu)}(D)(uD^\mu\varphi)], v)}{\mu!}\end{aligned}$$

Since this is true for all $v \in S$, we have, by Eq. (2-29) of Sec. 2-3,

$$\begin{aligned}|P(D)J_\varepsilon(\varphi f)|_s &\leq |J_\varepsilon(\varphi f)|_s + \sum_{j=1}^{N} |J_\varepsilon[\psi_j P_j(D)(\varphi u)]|_s \\ &\quad + \sum_{j=0}^{N} \sum_{|\mu|>0} \frac{|J_\varepsilon[\psi_j P_j^{(\mu)}(D)(uD^\mu\varphi)]|_s}{\mu!}\end{aligned}$$

Now by inequalities (3-25), (3-24) and (3-27), for $1 \leq j \leq N$

$$\begin{aligned}|J_\varepsilon[\psi_j P_j(D)(\varphi u)]|_s &\leq |\psi_j J_\varepsilon[P_j(D)(\varphi u)]|_s + C|P_j(D)(\varphi u)|_{s-1} \\ &\leq \max_{\Omega_1} |\psi_j| \; |P_j(D)J_\varepsilon(\varphi u)|_s + C_1 |P_j(D)(\varphi u)|_{s-a} \\ &\leq C_0 \max_{\Omega_1} |\psi_j| \; |P(D)J_\varepsilon(\varphi u)|_s + C_2 |P(D)(\varphi u)|_{s-a}\end{aligned}$$

(It is obvious that we can always take $a \leq 1$. Actually, we can never have $a > 1$.) Applications of Lemmas 3-4 and 3-5 give, for $|\mu| > 0$ and $0 \leq j \leq N$

$$\begin{aligned}|J_\varepsilon[\psi_j P_j^{(\mu)}(D)(uD^\mu\varphi)]|_s &\leq |\psi_j J_\varepsilon P_j^{(\mu)}(D)(uD^\mu\varphi)|_s + C_3 |P_j^{(\mu)}(D)(uD^\mu\varphi)|_{s-1} \\ &\leq C_4 |P(D)(uD^\mu\varphi)|_{s-b}\end{aligned}$$

Now, by inequalities (3-28) and (3-29)

$$\sum_{j=1}^{N} |\psi_j(x)| \leq \frac{1}{2C_0} \qquad x \in E^n$$

Hence, we have

$$|P(D)J_\varepsilon(\varphi u)|_s \leq |J_\varepsilon(\varphi f)|_s + \tfrac{1}{2}|P(D)J_\varepsilon(\varphi u)|_s$$
$$+ (N+1)(C_2 + C_4) \sum_{|\mu| \leq m} |P(D)(uD^\mu \varphi)|_{s-b}$$

Subtracting $\frac{1}{2}|P(D)J_\varepsilon(\varphi u)|_s$ from both sides, we get

$$|J_\varepsilon(\varphi u)|_s^P \leq 2|J_\varepsilon(\varphi f)|_s + C_5 \sum_{|\mu| \leq m} |uD^\mu \varphi|_{s-b}^P \tag{3-31}$$

which implies inequality (3-16).

It remains to show that the functions $c_j(x)$ in Eq. (3-20) are in $C^\infty(\Omega)$. To do this, first note

Lemma 3-6 Let $P_1(\xi), \ldots, P_N(\xi)$ be linearly independent polynomials. Then there exist real vectors $\xi^{(1)}, \ldots, \xi^{(N)}$, such that the determinant

$$|P_j(\xi^{(k)})| \qquad 1 \leq j, k \leq N \tag{3-32}$$

does not vanish.

Assume this lemma for the moment. Then we have, by Eq. (3-19)

$$P(x, \xi^{(k)}) = \sum_{j=1}^{N} c_j(x) P_j(\xi^{(k)}) \qquad 1 \leq k \leq N \tag{3-33}$$

Solve these N equations by Kramer's rule. This gives

$$c_j(x) = \sum_{k=1}^{N} \beta_{jk} P(x, \xi^{(k)}) \qquad 1 \leq j \leq N \tag{3-34}$$

where β_{jk} is the cofactor of $P_j(\xi^{(k)})$ in (3-32) divided by the determinant (3-32). This shows that the functions $c_j(x)$ are as smooth as the coefficients of $P(x, \xi)$.

All that remains is to give the

PROOF of Lemma 3-6 Assume that we cannot choose the $\xi^{(k)}$ so that (3-32) does not vanish. Let $l \leq N$ be the smallest integer, such that the determinant

$$|P_j(\xi^{(k)})| \qquad 1 \leq j, k \leq l \tag{3-35}$$

vanishes for all choices of $\xi^{(1)}, \ldots, \xi^{(l)}$. Then, there are vectors $\xi^{(1)}, \ldots, \xi^{(l-1)}$, such that the determinant

$$|P_j(\xi^{(k)})| \neq 0 \qquad 1 \leq j, k < l \tag{3-36}$$

while
$$\begin{vmatrix} P_1(\xi^{(1)}) \cdots P_1(\xi^{(l-1)}) & P_1(\xi) \\ \vdots \quad \vdots\vdots\vdots \quad \vdots & \vdots \\ P_l(\xi^{(1)}) \cdots P_l(\xi^{(l-1)}) & P_l(\xi) \end{vmatrix} \tag{3-37}$$
vanishes identically in ξ. Expanding (3-37), we get
$$\sum_{j=1}^{l} \gamma_j P_j(\xi) \equiv 0$$
and we know that $\gamma_l \neq 0$, by (3-36). This would mean that the $P_j(\xi)$ are linearly dependent, contrary to assumption. Hence, no such integer l exists. This completes the proof.

3-3 VECTOR SPACES

A vector space is a collection of elements $u, v, w, \ldots$, satisfying (1-45) to (1-50) and (1-56) of Sec. 1-5. None of the other properties of Hilbert space are assumed.

Let V be a vector space. The elements $v_1, \ldots, v_n$ are called *linearly independent* if the only scalars $\alpha_1, \ldots, \alpha_n$ which satisfy
$$\alpha_1 v_1 + \cdots + \alpha_n v_n = 0 \tag{3-38}$$
are $\alpha_1 = \cdots = \alpha_n = 0$. Otherwise the v_k are said to be *linearly dependent*. The space V is said to be of dimension $n \geq 0$ (dim $V = n$) if

1. there are n linearly independent elements in v
2. every set of $n + 1$ elements of V are linearly dependent.

(We take $n = 0$ when there are no linearly independent vectors in V, i.e., when V consists of only the zero vector.) If there is no finite n satisfying 1 and 2, then we say that V is infinite dimensional. Now suppose V is of dimension n, and let $v_1, \ldots, v_n$ be any set of n linearly independent elements. Then every $v \in V$ can be expressed uniquely in the form
$$v = \alpha_1 v_1 + \cdots + \alpha_n v_n \tag{3-39}$$
To see this, note that the set $v, v_1, \ldots, v_n$ is linearly dependent. Hence, there are scalars $\beta, \beta_1, \ldots, \beta_n$ not all zero, such that
$$\beta v + \beta_1 v_1 + \cdots + \beta_n v_n = 0$$
Moreover, β cannot vanish, for that would mean that the v_k are dependent. We now divide by β and obtain the form (3-39). Note that α_k are unique. For if
$$v = \alpha_1' v_1 + \cdots + \alpha_n' v_n$$
then
$$(\alpha_1 - \alpha_1')v_1 + \cdots + (\alpha_n - \alpha_n')v_n = 0$$
showing that $\alpha_k' = \alpha_k$ for each k. In an n-dimensional vector space V, every set of n linearly independent vectors is called a basis for V.

Lemma 3-7 If V is a vector space containing linearly independent elements $v_1, \ldots, v_n$, such that every $v \in V$ can be expressed in the form of Eq. (3-39), then $\dim V = n$.

PROOF All that we need to show is that every set of $n + 1$ elements is linearly dependent. Let $u_1, \ldots, u_{n+1}$ be any $n + 1$ elements of V. Then there are scalars α_{jk}, such that

$$u_j = \sum_{k=1}^{n} \alpha_{jk} v_k \qquad 1 \leq j \leq n + 1 \tag{3-40}$$

Now from the theory of linear equations, we know that we can find scalars $\beta_1, \ldots, \beta_{n+1}$ not all zero, such that

$$\sum_{j=1}^{n+1} \beta_j \alpha_{jk} = 0 \qquad 1 \leq k \leq n \tag{3-41}$$

since this is a system of n equations in $n + 1$ unknowns. Thus,

$$\sum_{j=1}^{n+1} \beta_j u_j = \sum_{j=1}^{n+1} \sum_{k=1}^{n} \beta_j \alpha_{jk} v_k = 0$$

showing that the u_j are linearly dependent.

Just as in the case of a Hilbert space, a subset W of V is called a subspace of V if $au + \beta v \in W$ for all scalars α, β whenever u, v are in W. Note that a subspace of a vector space is itself a vector space.

Lemma 3-8 If V is a vector space and every set of n vectors in V are linearly dependent, then $\dim V < n$.

PROOF Let k be the largest integer, such that V has k linearly independent vectors (if there are none we take $k = 0$). We know that such an integer exists since every set of n or more vectors is linearly dependent. Moreover $k < n$. Clearly, $k = \dim V$, for every set of $k + 1$ vectors is linearly dependent. This completes the proof.

Corollary 3-9 If W is a subspace of a vector space V, then $\dim W \leq \dim V$.

PROOF If $\dim V = n$, and $w_1, \ldots, w_{n+1}$ are any $n + 1$ vectors in W, then they must be linearly dependent. Thus, $\dim W \leq n$ (Lemma 3-8).

Lemma 3-10 If a_μ are complex constants, such that

$$\sum_{|\mu| \leq m} a_\mu \xi^\mu \equiv 0 \qquad \xi \text{ real} \tag{3-42}$$

then all of the a_μ vanish.

PROOF We employ induction. We know that

$$\sum_{k=0}^{m} a_k t^k = 0 \qquad t \text{ real}$$

implies that the a_k all vanish. Now suppose the lemma is proved for $n-1$. The identity (3-42) can be put in the form

$$\sum_{k=0}^{m} Q_k(\xi')\xi_n^k = 0 \tag{3-43}$$

where $\xi' = (\xi_1, \ldots, \xi_{n-1})$ and the Q_k are polynomials in ξ'. From this we see that the $Q_k(\xi')$ vanish identically. By the induction hypothesis, their coefficients must all vanish.

Corollary 3-11 The monomials ξ^μ are linearly independent. Let π_m be the set of polynomials of degree $\leq m$. Thus, every polynomial $P(\xi)$ in π_m, is of the form

$$P(\xi) = \sum_{|\mu| \leq m} a_\mu \xi^\mu \tag{3-44}$$

Clearly π_m is a vector space. By Lemma 3-7 and Corollary 3-11, we see that π_m is finite dimensional.

Lemma 3-12 Let $P(\xi)$ be a polynomial of degree m and let $Q(\xi)$ be a polynomial weaker than $P(\xi)$. Then $Q(\xi)$ is of degree $\leq m$.

PROOF Suppose

$$Q(\xi) = \sum_{|\mu| \leq k} b_\mu \xi^\mu$$

is of degree $k > m$. Then not all of the b_μ vanish for $|\mu| = k$. Thus,

$$q(\xi) = \sum_{|\mu| = k} b_\mu \xi^\mu$$

does not vanish identically (Lemma 3-10). Hence, there is a real vector $\tilde{\xi}$, such that $q(\tilde{\xi}) \neq 0$. If we put $Q_1(\xi) = Q(\xi) - q(\xi)$, then we have, for $\lambda > 0$

$$\begin{aligned} |Q(\lambda\tilde{\xi})| &\geq |q(\lambda\tilde{\xi})| - |Q_1(\lambda\tilde{\xi})| \\ &= \lambda^k |q(\tilde{\xi})| - |Q_1(\lambda\tilde{\xi})| \end{aligned}$$

Let $j \leq m$ be the highest degree in λ among the polynomials $P^{(\mu)}(\lambda\tilde{\xi})$.

Then
$$M_\lambda = \frac{\sum |P^{(\mu)}(\lambda\tilde{\xi})|}{\lambda^j}$$

is bounded and bounded away from zero as $\lambda \to \infty$. Thus,

$$\frac{|Q(\lambda\xi)|}{\lambda^{k-1}M_\lambda} \geq \frac{\lambda|q(\xi)|}{M_\lambda} - \frac{|Q_1(\lambda\xi)|}{\lambda^{k-1}M_\lambda}$$

$$\to \infty \quad \text{as} \quad \lambda \to \infty$$

Since $j < k$, this shows that $Q(\xi)$ is not weaker than $P(\xi)$. This completes the proof.

Corollary 3-13 For a given polynomial $P(\xi)$, the set of all polynomials weaker than $P(\xi)$ is a finite dimensional vector space.

3-4 PROOF OF THE LEMMAS

Let us now turn to the proof of inequalities (3-24) to (3-26). To prove the first, we shall employ some simple inequalities.

Let s be any real number. Then, by the theorem of the mean

$$t_1^s - t_2^s = st_3^{s-1}(t_1 - t_2) \tag{3-45}$$

where t_3 is some value between t_1 and t_2. Now

$$1 + |\xi| \leq (1 + |\xi - \eta|)(1 + |\eta|) \tag{3-46}$$

$$1 + |\xi - \eta| \leq (1 + |\xi|)(1 + |\eta|) \tag{3-47}$$

From (3-46), it follows that for $\sigma > 0$

$$(1 + |\xi|)^\sigma \leq (1 + |\xi - \eta|)^\sigma(1 + |\eta|)^\sigma$$

From (3-47), it follows that for $\sigma < 0$

$$(1 + |\xi|)^\sigma \leq (1 + |\xi - \eta|)^\sigma(1 + |\eta|)^{-\sigma}$$

Hence, we have in general

$$(1 + |\xi|)^\sigma \leq (1 + |\xi - \eta|)^\sigma(1 + |\eta|)^{|\sigma|} \tag{3-48}$$

Set $t_1 = (1 + |\xi|)$, $t_2 = (1 + |\xi - \eta|)$ in Eq. (3-45). Then, we have

$$\begin{aligned} |(1 + |\xi|)^s - (1 + |\xi - \eta|)^s| &\leq |s| \, \big||\xi| - |\xi - \eta|\big| \cdot \max\,[(1 + |\xi|)^{s-1}, \\ &\qquad (1 + |\xi - \eta|)^{s-1}] \\ &\leq |s| \, |\eta|(1 + |\eta|)^{|s-1|}(1 + |\xi - \eta|)^{s-1} \end{aligned} \tag{3-49}$$

We are now ready to prove (3-24). It clearly suffices to prove it for $v \in S$. By Theorem 2-10 of Sec. 2-5

$$(2\pi)^n F(\psi v) = \int F\psi(\eta) Fv(\xi - \eta)\, d\eta$$

Hence, by (3-49)

$$(2\pi)^n(1 + |\xi|)^s |F(\psi v)|$$
$$= (1 + |\xi|)^s \left| \int F\psi(\eta)Fv(\xi - \eta)\, d\eta \right|$$
$$\leq \left| \int (1 + |\xi - \eta|)^s F\psi(\eta)Fv(\xi - \eta)\, d\eta \right|$$
$$+ |s| \int |\eta| (1 + |\eta|)^{|s-1|}(1 + |\xi - \eta|)^{s-1} |F\psi(\eta)Fv(\xi - \eta)|\, d\eta \tag{3-50}$$

Set
$$w(x) = G[(1 + |\xi|)^s Fv(\xi)]$$
where G is the inverse Fourier transform (see Sec. 2-5).

Then
$$\int (1 + |\xi - \eta|)^s F\psi(\eta)Fv(\xi - \eta)\, d\eta = (2\pi)^n F(\psi w) \tag{3-51}$$

If $\| \quad \|'$ denotes the usual $L^2(E^n)$ norm with respect to the ξ variables, then
$$(2\pi)^{n/2} \| w \| = \| Fw \|' = |v|_s \tag{3-52}$$

and hence
$$\| F(\psi w) \|' = (2\pi)^{n/2} \| \psi w \|$$
$$\leq \max_x |\psi(x)| (2\pi)^{n/2} \| w \| = \max_x |\psi(x)| \; |v|_s \tag{3-53}$$

By (3-50) and (3-51)

$$(2\pi)^n \| (1 + |\xi|)^s F(\psi v) \|' \leq (2\pi)^n \| F(\psi w) \|'$$
$$+ |s| \int |\eta| (1 + |\eta|)^{|s-1|} |F\psi(\eta)|$$
$$\times \| (1 + |\xi - \eta|)^{s-1} Fv(\xi - \eta) \|'\, d\eta$$

Thus, by (3-50) and (3-53)

$$|\psi v|_s \leq \max_x |\psi(x)| \; |v|_s + (2\pi)^{-n} |s| \; |v|_s \int |\eta| (1 + |\eta|)^{|s-1|} |F\psi(\eta)|\, d\eta \tag{3-54}$$

This gives (3-24).

To prove (3-25), we note that by Eq. (2-62) and Theorem 2-10 of Sec. 2-5

$$F[J_\varepsilon(\psi v) - \psi J_\varepsilon v] = \int F\psi(\eta)Fv(\xi - \eta)[Fj(\varepsilon\xi) - Fj(\varepsilon\xi - \varepsilon\eta)]\, d\eta$$

We shall show that there is a constant C, such that

$$|(1 + |\xi|)[Fj(\varepsilon\xi) - Fj(\varepsilon\eta - \varepsilon\xi)]| \leq C(1 + |\eta|) \tag{3-55}$$

Assuming this, we get by (3-48)

$$\| (1 + |\xi|^s F[J_\varepsilon(\psi v) - \psi J_\varepsilon v] \|'$$

$$\leq C \int (1 + |\eta|)^{|s-1|+1} |F\psi(\eta)| \; \| (1 + |\xi - \eta|)^{s-1} Fv(\xi - \eta) \|' \, d\eta$$

$$\leq C |v|_{s-1} \int (1 + |\eta|)^{|s-1|+1} |F\psi(\eta)| \, d\eta$$

and this gives (3-25). To prove (3-55), we note that $|Fj(\xi)| \leq 1$ for all ξ, and that

$$\varepsilon\xi_k[Fj(\varepsilon\xi) - Fj(\varepsilon\xi - \varepsilon\eta)] = - \int (D_k \, e^{-i\varepsilon(\xi,x)})(1 - e^{i\varepsilon(\eta,x)}) j(x) \, dx$$

$$= \int e^{-i\varepsilon(\xi,x)}[-\varepsilon\eta_k j(x) + (1 - e^{i\varepsilon(\eta,x)}) D_k j(x)] \, dx$$

Since
$$|1 - e^{i\varepsilon(\eta,x)}| \leq \varepsilon |\eta| \; |x|$$

(see inequality (2-66) of Sec. 2-5), we have

$$|\xi_k[Fj(\varepsilon\xi) - Fj(\varepsilon\eta - \varepsilon\xi)]| \leq C |\eta|$$

and since $|\xi| \leq \Sigma |\xi_k|$, (3-55) is proved.

In order to prove (3-26), we first prove

Lemma 3-14 Let $r = r(m, n)$ denote the number of distinct multi-indices μ of norm $|\mu| \leq m$ (i.e., the number of distinct derivatives of orders $\leq m$). Let $\theta_1, \ldots, \theta_r$ be any r vectors in E^n, such that $\det(\theta_j^\mu) \neq 0$. Then, for each multi-index μ of norm $|\mu| \leq m$, there are real numbers $\lambda_1, \ldots, \lambda_r$, such that

$$t^{|\mu|} Q^{(\mu)}(\xi) = \sum_{j=1}^{r} \lambda_j Q(\xi + t\theta_j) \tag{3-56}$$

holds for all polynomials $Q(\xi)$ of degree $\leq m$, all $\xi \in E^n$, and all real t.

Note that it follows from Lemma 3-6 and Corollary 3-11 that there always exist vectors $\theta_j \in E^n$, such that $\det(\theta_j^\mu) \neq 0$. Before proving Lemma 3-14, we give some consequences.

Theorem 3-15 If $Q(\xi)$ is weaker than $P(\xi)$ then there is a constant C, such that

$$\sum_\mu t^{|\mu|} |Q^{(\mu)}(\xi)| \leq C \sum_\mu t^{|\mu|} |P^{|\mu|}(\xi)| \tag{3-57}$$

holds for all $\xi \in E^n$ and $t \geq 1$.

Proof Suppose $P(\xi)$ is of order m. Let $\theta_1, \ldots, \theta_2$ be any $r(m, n)$ vectors in E^n satisfying the hypothesis of Lemma 3-14. If μ is any index of norm $|\mu| \leq m$, we can find $\lambda_1, \ldots, \lambda_r$ such that Eq. (3-56) holds. Thus,

$$t^{|\mu|}|Q^{(\mu)}(\xi)| \le \sum |\lambda_j| \, |Q(\xi + t\theta_j)|$$

$$\le C \sum_{\nu, j} |P^{(\nu)}(\xi + t\theta_j)| \tag{3-58}$$

Now, by Taylor's formula

$$P^{(\nu)}(\xi + t\theta_j) = \sum_\rho \frac{P^{(\nu+\rho)}(\xi)t^{|\rho|}\theta_j^\rho}{\rho!}$$

and hence
$$|P^{(\nu)}(\xi + t\theta_j)| \le \sum_\rho t^{|\rho|}|P^{(\nu+\rho)}(\xi)|$$

$$\le \sum_\sigma t^{|\sigma|}|P^{(\sigma)}(\xi)| \tag{3-59}$$

where we made use of the fact that $t \ge 1$. Combining inequalities (3-58) and (3-59), we obtain inequality (3-57).

Another consequence is

Theorem 3-16 If $Q(\xi)$ is weaker than $P(\xi)$, then there is a constant C, such that

$$\sum_\mu |Q^{(\mu)}(D)v|_{s+c|\mu|} \le C \sum_\mu |P^{(\mu)}(D)v|_{s+c|\mu|} \tag{3-60}$$

for all real $s, c > 0$ and $v \in S$.

PROOF By Theorem 3-15 there is a constant C_1, such that

$$\sum_\mu t^{2|\mu|}|Q^{(\mu)}(\xi)|^2 \le C_1 \sum_\mu t^{2|\mu|}|P^{(\mu)}(\xi)|^2 \tag{3-61}$$

for all $\xi \in E^n$ and $t \ge 1$. Take $t = (1 + |\xi|)^c$, multiply by $(1 + |\xi|)^{2s}|Fv|^2$, and integrate over E^n with respect to ξ. This gives an inequality equivalent to (3-60).

PROOF of Lemma 3-5 Clearly $b|\mu| \le a$ for $|\mu| \le m$. Hence, by inequality (2-84) of Sec. 2-8

$$\sum_{|\mu|>0} |P^{(\mu)}(D)\varphi|_{s+b|\mu|} \le |||\varphi|||_{s+a}$$

$$\le C(|P(D)\varphi|_s + |\varphi|_s) \qquad \varphi \in C_0^\infty(\Omega)$$

Hence by (3-60)

$$\sum_\mu |Q^{(\mu)}(D)\varphi|_{s+b|\mu|} \le C_1(|P(D)\varphi|_s + |\varphi|_s) \tag{3-62}$$

which is a stronger form of (3-26).

It remains only to give the

PROOF of Lemma 3-14 By Taylor's formula we have

$$Q(\xi + t\theta_j) = \sum \frac{Q^{(\mu)}(\xi)t^{|\mu|}\theta_j^\mu}{\mu!}$$

Thus, for any $\lambda_1, \ldots, \lambda_r$

$$\sum_{j=1}^{r} \lambda_j Q(\xi + t\theta_j) = \sum_{\mu} \left(\sum_{j=1}^{r} \lambda_j \theta_j^\mu \right) \frac{t^{|\mu|}Q^{(\mu)}(\xi)}{\mu!} \tag{3-63}$$

Since the matrix (θ_j^μ) is non-singular, for any multi-index v with $|v| \leq m$, we can solve the system

$$\sum_{j=1}^{r} \lambda_j \theta_j^\mu = \delta_{\mu v} v! \qquad |\mu| \leq m \tag{3-64}$$

for the λ_j, where $\delta_{\mu v} = 0$ for $\mu \neq v$ and $\delta_{vv} = 1$. Note that the solution does not depend on Q, t, or ξ. Substituting the solution of Eq. (3-64) into Eq. (3-63), we obtain Eq. (3-56). This completes the proof.

We can now prove inequality (2-78) of Sec. 2-7. We merely take $t = 1$ and $Q(\xi) = P(\xi)$ in Eq. (3-56). Thus, for each μ we have

$$|P^{(\mu)}(\xi)| \leq \max |\lambda_j| \sum_{j=1}^{r} |P(\xi + \theta_j)|$$

If we now sum over all μ, we get the desired inequality.

3-5 EXISTENCE

There is something missing. We have seen that every weak solution of a formally hypoelliptic equation is in $C^\infty(\Omega)$ provided that the right-hand side is in $C^\infty(\Omega)$. But we have not shown whether such an equation has a weak solution in the first place. To help remedy this situation, we shall prove

Theorem 3-17 Let $P(x, D)$ be a formally hypoelliptic operator in a domain Ω. Then each $y \in \Omega$ has a neighborhood N_y, such that

$$P(x, D)u = f \tag{3-65}$$

has a weak solution in N_y for each $f \in L^2(N_y)$.

Note that we do not conclude that Eq. (3-65) has a weak solution in the whole of Ω.

PROOF By Theorem 1-12 of Sec. 1-6, it suffices to find a neighborhood N_y of y, such that

$$\| \varphi \| \leq C \| P'(x, D)\varphi \| \qquad \varphi \in C_0^\infty(N_y) \tag{3-66}$$

We may assume that Ω is bounded. By Eq. (3-23)

$$P'(x, D)\varphi = \bar{P}(D)\varphi + \sum_{j=1}^{N} \bar{P}_j(D)(\bar{b}_j\varphi) \tag{3-67}$$

Moreover, since the $P_j(\xi)$ are weaker than $P(\xi)$, there is a constant C_0, such that

$$\| P_j(D)\varphi \| \leq C_0 \| P(D)\varphi \| \qquad \varphi \in C_0^\infty(\Omega) \qquad 1 \leq j \leq N \tag{3-68}$$

By Eq. (3-22), there is a neighborhood N_y of y, of the form

$$|x_k - y_k| < \delta \qquad 1 \leq k \leq n$$

such that
$$\sum_{j=1}^{N} |b_j(x)| < \frac{1}{3C_0} \qquad x \in N_y \tag{3-69}$$

By Eq. (3-67), we have

$$\begin{aligned} \| P'(x, D)\varphi \| &\geq \| \bar{P}(D)\varphi \| - \sum_{j=1}^{N} \| \bar{P}_j(D)(\bar{b}_j\varphi) \| \\ &\geq \| P(D)\varphi \| - C_0 \sum_{j=1}^{N} \| P(D)(\bar{b}_j\varphi) \| \end{aligned}$$

where we have used Eq. (1-104) of Sec. 1-7 and inequality (3-68). Thus, if $\varphi \in C_0^\infty(N_y)$, we have

$$\begin{aligned} \| P'(x, D)\varphi \| &\geq \| P(D)\varphi \| - C_0 \sum_{j=1}^{N} \sum_{|\mu| \leq m} \| (D^\mu \bar{b}_j) P^{(\mu)}(D)\varphi \| \\ &\geq \| P(D)\varphi \| - C_0 \sum_{j=1}^{N} \sum_{|\mu| \leq m} \operatorname*{lub}_{N_y} |D^\mu \bar{b}_j| \; \| P^{(\mu)}(D)\varphi \| \\ &\geq \| P(D)\varphi \| - C_0 \operatorname*{lub}_{N_y} \sum_{j=1}^{N} |b_j| \; \| P(D)\varphi \| \\ &\quad - C_0 \sum_{j=1}^{N} \sum_{|\mu| > 0} \operatorname*{lub}_{\Omega} |D^\mu \bar{b}_j| \; \| P^{(\mu)}(D)\varphi \| \\ &\geq \tfrac{2}{3} \| P(D)\varphi \| - K\delta \| P(D)\varphi \| \end{aligned}$$

where we have used Lemma 1-15 of Sec. 1-7 and (3-69). Now take $\delta < 1/3K$. This gives

$$\| P(D)\varphi \| \leq 3 \| P'(x, D)\varphi \| \qquad \varphi \in C_0^\infty(N_y) \tag{3-70}$$

Since $P(D)$ is a constant-coefficient operator, Theorem 1-13 of Sec. 1-7 says that there is a constant C, such that

$$\| \varphi \| \leq C \| P(D)\varphi \| \qquad \varphi \in C_0^\infty(\Omega) \tag{3-71}$$

Combining inequalities (3-70) and (3-71) we get (3-66). This completes the proof.

3-6 EXAMPLES

It is about time that we considered some specific examples of operators. First, let us consider constant-coefficient operators. Some well-known operators are

1 The Laplacian $\sum_{k=1}^{n} D_k^2$. This is clearly elliptic.

2 The heat operator $D_1 - i\sum_{k=2}^{n} D_k^2$. It is easily checked that this operator is hypoelliptic.

3 The wave operator $D_1^2 - \sum_{k=2}^{n} D_k^2$. This is not hypoelliptic, because its corresponding polynomial has real roots no matter how large $|\xi|$ is.

To help us recognize other operators, we can use

Theorem 3-18 If $P_1(\xi)$ and $P_2(\xi)$ are hypoelliptic, then the same is true of $P = P_1P_2$.

PROOF Since $P^{(j)} = P_1^{(j)}P_2 + P_1P_2^{(j)}$, we have

$$\left|\frac{P^{(j)}(\xi)}{P(\xi)}\right| \le \left|\frac{P_1^{(j)}(\xi)}{P_1(\xi)}\right| + \left|\frac{P_2^{(j)}(\xi)}{P_2(\xi)}\right| \to 0 \quad \text{as} \quad |\xi| \to \infty \qquad 1 \le j \le n$$

Thus, P is hypoelliptic.

From this we see that products of operators of the form

$$\left(\sum_{k=1}^{n} D_k^2\right)^r \left(D_1 \pm i\sum_{k=2}^{n} D_k^2\right)^s \left(D_1 \pm \sum_{k=2}^{n} D_k^4\right)^t \cdots$$

are all hypoelliptic.

PROBLEMS

3-1 Show that $w_k \to u$ in $H_P^s(\Omega)$ implies that $Q(D)w_k \to Q(D)u$ in H^s.

3-2 Prove Corollary 3-13.

3-3 Prove Eq. (3-45) and inequalities (3-46) and (3-47).

3-4 Show that the heat operator is hypoelliptic.

3-5 Determine the number $r(m, n)$ of distinct multi-indices μ of norm $|\mu| \le m$ in n dimensions. (Hint: first consider the multi-indices satisfying $|\mu| = m$.)

3-6 Show that one can always find $\beta_1, \ldots, \beta_{n+1}$, not all 0, such that Eq. (3-41) holds.

3-7 Prove the first assertion in the proof of Lemma 3-10.

CHAPTER

FOUR

THE CAUCHY PROBLEM

4-1 STATEMENT OF THE PROBLEM

In the previous chapters we investigated the question as to when

$$P(x, D)u = f \tag{4-1}$$

has solutions in a domain Ω. As we saw at the beginning of Chapter 1, solutions of Eq. (4-1) are, in general, not unique. It should not, therefore, come as any surprise that, in most applications, further restrictions are required of solutions of partial differential equations. These additional restrictions are usually in the form of *boundary conditions*, and the equation together with the boundary conditions is called a boundary value problem.

One of the oldest and more important boundary problems bears the name of Cauchy. To describe it for Eq. (4-1), we must distinguish one of the coordinate axes. To do this most easily, let us work in E^{n+1} instead of E^n. If we set

$$x = (x_1, \ldots, x_n) \qquad t = x_{n+1}$$

and denote points of E^{n+1} by (x, t), we have in place of $P(x, D)$ the operator

$$P(x, t, D_x, D_t) = \sum_{|\mu| + k \leq m} a_{\mu,k}(x, t) D_x^\mu D_t^k \tag{4-2}$$

where
$$D_x = (D_1, \ldots, D_n) \qquad D_t = D_{n+1}$$

and
$$D_x^\mu = D_1^{\mu_1} \cdots D_n^{\mu_n}$$

Let Ω_0 be a domain in E^{n+1} containing the origin, and let Ω be the intersection of Ω_0 with the half-space $t > 0$. The Cauchy problem for Eq. (4-2) in Ω is to find a function $u(x, t) \in C^m(\bar{\Omega})$, such that

$$P(x, t, D_x, D_t)u = f(x, t) \quad \text{in} \quad \Omega \tag{4-3}$$

and
$$D_t^k u(x, 0) = g_k(x) \quad \text{on} \quad \partial_0 \Omega \qquad 0 \leq k < m \tag{4-4}$$

where f is a given function on Ω, and the g_k are given functions on the intersection $\partial_0\Omega$ of Ω_0 and the *hyperplane* $t = 0$. We shall say that the Cauchy problem (4-3), (4-4) is "well posed" if it has a unique solution for each choice of $f, g_0, \ldots, g_{n-1}$ sufficiently smooth.

We must make one remark before we proceed. In order that Eqs. (4-3) and (4-4) not contradict each other, it is important that the coefficient of D_t^m in Eq. (4-2) should not vanish anywhere on $\partial_0\Omega$. For if it vanished at a point $(x, 0) \in \partial_0\Omega$, then Eq. (4-3) would give

$$\sum_{k=0}^{m-1} \sum_{|\mu| \leq m-k} a_{\mu,k}(x, 0) D_x^\mu g_k(x) = f(x, 0) \tag{4-5}$$

which would impose a restriction on the g_k at the point $(x, 0)$. Since we want a solution for all choices of the g_k (provided they are sufficiently differentiable), we want to avoid this situation. Hence, we shall assume that

$$a_{(0,\ldots,0),m}(x, 0) \neq 0 \quad \text{on} \quad \partial_0\Omega \tag{4-6}$$

We describe this by saying that the hyperplane $t = 0$ is not *characteristic* for the operator (4-2).

4-2 WEAK SOLUTIONS

As usual, let us begin by considering a simple case. Let us first assume that the operator (4-2) has constant coefficients. Secondly, let us take the g_k in (4-4) to vanish, and even let us assume that $f \in C_0^\infty(\Omega)$. Thus, our problem is of the form

$$P(D)u = f \quad \text{in} \quad \Omega \tag{4-7}$$

$$D_t^k u(x, 0) = 0 \quad \text{on} \quad \partial_0\Omega \qquad 0 \leq k < m \tag{4-8}$$

where $D = (D_x, D_t)$ is used to save time and labor.

We begin by making a simple observation. Since $f \in C_0^\infty(\Omega)$, we can extend it to vanish in $\Omega_0 - \Omega$ (the portion of Ω_0 contained in the half-space $t \leq 0$). This extended function f_1 is clearly in $C_0^\infty(\Omega_0)$. Now, suppose $u \in C^\infty(\bar{\Omega})$ is a solution of Eqs. (4-7) and (4-8). We observe that if we extend u to vanish in $\Omega_0 - \Omega$, then the extension u_1 is in $C^m(\bar{\Omega}_0)$. To see this, note that all derivatives $D_t^k u$ for $0 \leq k < m$ are continuous across the surface $\partial_0\Omega$. Moreover, since

$$P(D) = P(D_x, D_t) = \sum_{|\mu|+k \leq m} a_{\mu,k} D_x^\mu D_t^k \tag{4-9}$$

and $$a = a_{(0,\dots,0),m} \neq 0 \tag{4-10}$$

we see that $D_t^m u(x,0) = 0$ on $\partial_0\Omega$ (recall that f also vanishes there). Since both f_1 and u_1 vanish in $\Omega_0 - \Omega$, it follows that u_1 is a solution of

$$P(D)u_1 = f_1 \quad \text{in} \quad \Omega_0 \tag{4-11}$$

In particular, it is a weak solution, and hence

$$\int_{\Omega_0} u_1 \overline{\bar{P}(D)\varphi}\, dx\, dt = \int_{\Omega_0} f_1 \bar{\varphi}\, dx\, dt \qquad \varphi \in C_0^\infty(\Omega_0) \tag{4-12}$$

Since both u_1 and f_1 vanish in $\Omega_0 - \Omega$, this reduces to

$$(u, \bar{P}(D)\varphi) = (f, \varphi) \qquad \varphi \in C_0^\infty(\Omega_0) \tag{4-13}$$

where $$(v, w) = \int_\Omega v\bar{w}\, dx\, dt \tag{4-14}$$

The upshot of all this is that every solution of Eqs. (4-7), (4-8) satisfies Eq. (4-13). Now we prove the converse. If $u \in C^m(\bar{\Omega})$ satisfies Eq. (4-13), then it is a solution of Eqs. (4-7) and (4-8). To see this we must return to our integration by parts formula (see (1-26) of Sec. 1-3), since the functions $\varphi \in C_0^\infty(\Omega_0)$ are not required to vanish on $\partial_0\Omega$.

In our present notation, the integration by parts formula is

$$\int_\Omega D_k h\, dx\, dt = -i \int_{\partial\Omega} h\gamma_k\, d\sigma \qquad 1 \le k \le n+1 \tag{4-15}$$

where γ_k is the cosine of the angle between the x_k axis and the outward normal to $\partial\Omega$. Let $\partial_1\Omega$ denote that part of $\partial\Omega$ not contained in the hyperplane $t = 0$. If h vanishes near $\partial_1\Omega$, then Eq. (4-15) gives

$$\int_\Omega D_k h\, dx\, dt = 0 \qquad 1 \le k \le n \tag{4-16}$$

and $$\int_\Omega D_t h\, dx\, dt = -i \int_{\partial_0\Omega} h\, dx \tag{4-17}$$

In particular, if $w \in C^1(\bar{\Omega})$ and $\varphi \in C_0^\infty(\Omega_0)$, then

$$(D_k w, \varphi) = (w, D_k\varphi) \qquad 1 \le k \le n \tag{4-18}$$

and $$(D_t w, \varphi) = (w, D_t\varphi) - i \int_{\partial_0\Omega} w\bar{\varphi}\, dx \tag{4-19}$$

Repeated applications of Eq. (4-19) give

$$(D_t^k w, \varphi) = (w, D_t^k\varphi) - i \sum_{j=1}^{k} \int_{\partial_0\Omega} D_t^{k-j} w \overline{D_t^{j-1}\varphi}\, dx \tag{4-20}$$

for $w \in C^k(\bar{\Omega})$ and $\varphi \in C_0^\infty(\Omega_0)$. Let us now apply these formulas to the expression $(P(D)u, \varphi)$ for $u \in C^m(\bar{\Omega})$ and $\varphi \in C_0^\infty(\Omega_0)$. They give

$$
\begin{aligned}
(P(D)u, \varphi) &= \sum_{|\mu|+k\le m} a_{\mu,k}(D_x^\mu D_t^k u, \varphi) \\
&= \sum_{|\mu|+k\le m} a_{\mu,k}\left[(u, D_t^k D_x^\mu \varphi) - i\sum_{j=1}^{k}\int_{\partial_0\Omega} D_t^{k-j}u\overline{D_t^{j-1}D_x^\mu\varphi}\,dx\right] \\
&= (u, \bar{P}(D)\varphi) - i\sum_{j=1}^{m}\int_{\partial_0\Omega} N_j u\overline{D_t^{j-1}\varphi}\,dx
\end{aligned}
\tag{4-21}
$$

where
$$N_j u = \sum_{k=j}^{m}\sum_{|\mu|\le m-k} a_{\mu,k}D_x^\mu D_t^{k-j}u \qquad 1\le j\le m \tag{4-22}$$

Now suppose $u \in C^m(\bar{\Omega})$ satisfies Eq. (4-13) for all $\varphi \in C_0^\infty(\Omega_0)$. In particular, this holds for all $\varphi \in C_0^\infty(\Omega)$, and hence u is a solution of Eq. (4-7) (see Sec. 1-6). Thus, Eq. (4-21) yields

$$\sum_{j=1}^{m}\int_{\partial_0\Omega} N_j u\overline{D_t^{j-1}\varphi}\,dx = 0 \qquad \varphi\in C_0^\infty(\Omega_0) \tag{4-23}$$

We claim that Eq. (4-23) implies that

$$N_j u = 0 \quad \text{on} \quad \partial_0\Omega \qquad 1\le j\le m \tag{4-24}$$

Let us postpone the proof of this until the end of this section. We now note that Eq. (4-24) implies Eq. (4-8). In fact, by Eq. (4-22)

$$N_m u = au$$

$$N_{m-1}u = aD_t u + \sum_{|\mu|\le 1} a_{\mu,m-1}D_x^\mu u$$

and, in general, $N_{m-k}u = aD_t^k u +$ a linear expression involving only $u, D_t u, \ldots, D_t^{k-1}u$ and their derivatives with respect to $x_1, \ldots, x_n$. Since $a \ne 0$ by (4-10), $N_m u = 0$ on $\partial_0\Omega$ implies $u = 0$ on $\partial_0\Omega$ and, consequently, $N_{m-1}u = 0$ implies $D_t u = 0$, and so forth. Thus, we have proved

Theorem 4-1 A function $u \in C^m(\bar{\Omega})$ is a solution of Eqs. (4-7), (4-8) if and only if it satisfies Eq. (4-13).

We wish to remark that Theorem 4-1 holds for any $f \in C(\bar{\Omega})$. In fact, if u is a solution of Eq. (4-7), then by Eq. (4-21)

$$(u, \bar{P}(D)\varphi) - (f, \varphi) = i\sum_{j=1}^{m}\int_{\partial_0\Omega} N_j u\overline{D_t^{j-1}\varphi}\,dx \tag{4-25}$$

Now, an examination of Eq. (4-22) shows that $D_t^m u$ does not appear in any of the $N_j u$. Hence, if u satisfies Eq. (4-8), then $N_j u = 0$ on $\partial_0\Omega$ for each j, showing that Eq. (4-13) holds. Conversely, if Eq. (4-13) holds, then so does Eq. (4-23) and hence Eq. (4-8).

In view of Theorem 4-1 it appears natural to call $u \in L^2(\Omega)$ a weak solution of Eqs. (4-7), (4-8) if it satisfies Eq. (4-13). In Sec. 4-3 we shall study weak solutions of (4-7), (4-8).

It remains to show that Eq. (4-23) implies Eq. (4-24). This follows from

Lemma 4-2 If $w_1(x), \ldots, w_m(x)$ are continuous functions on $\partial_0\Omega$, and

$$\sum_{j=1}^{m} \int_{\partial_0\Omega} w_j \overline{D_t^{j-1}\varphi}\, dx = 0 \qquad \varphi \in C_0^\infty(\Omega_0) \tag{4-26}$$

then the w_j vanish identically.

PROOF Suppose that for some j and some $x_0 \in \partial_0\Omega$ we have $w_j(x_0) \neq 0$. We may assume that $\operatorname{Re} w_j(x_0) > 0$. By continuity there is a neighborhood N of x_0, such that $\operatorname{Re} w_j(x) > 0$ for $x \in N$. Now, there exist positive constants r, b such that the cylinder

$$|x - x_0| < 2r \qquad |t| < 2b$$

is contained in Ω_0 and $|x - x_0| < 2r$ is contained in N. Let $\psi(x)$ be a function in $C_0^\infty(\partial_0\Omega)$, such that

$$\begin{aligned} \psi(x) &= 1 \qquad |x - x_0| \leq r \\ \psi(x) &= 0 \qquad |x - x_0| \geq 2r \\ 0 \leq \psi(x) &\leq 1 \qquad x \in \partial_0\Omega \end{aligned}$$

and let $\rho(t)$ be a function in $C_0^\infty(E^1)$, satisfying

$$\begin{aligned} \rho(t) &= 1 \qquad |t| \leq b \\ \rho(t) &= 0 \qquad |t| \geq 2b \end{aligned}$$

(for the construction of such functions see Sec. 1-3). Set

$$\varphi(x, t) = (it)^{j-1}\psi(x)\rho(t)$$

Clearly, $\varphi \in C_0^\infty(\Omega_0)$. Moreover, $D_t^k\varphi(x, 0) = 0$ for $k \neq j - 1$, and

$$D_t^{j-1}\varphi(x, 0) = (j - 1)!\,\psi(x)$$

showing that this function is positive in $|x - x_0| < r$, nonnegative in N, and vanishing outside N. Hence, we must have

$$\operatorname{Re} \int_{\partial_0\Omega} w_j(x) D_t^{j-1}\varphi(x, 0)\, dx > 0 \tag{4-27}$$

But this is impossible since the left-hand side of (4-27) equals the real part of the left-hand side of (4-26) for this choice of φ. This completes the proof.

4-3 HYPERBOLIC EQUATIONS

Let us examine weak solutions of Eqs. (4-7), (4-8) a bit more clearly. First of all, let us note that the reasoning of Sec. 1-6 gives immediately

Theorem 4-3 For any $f \in L^2(\Omega)$, a necessary and sufficient condition that Eqs. (4-7), (4-8) have a weak solution, is that

$$|(f, \varphi)| \leq C \| \bar{P}(D)\varphi \| \qquad \varphi \in C_0^\infty(\Omega_0) \tag{4-28}$$

In addition, we have

Theorem 4-4 A necessary and sufficient condition that Eqs. (4-7), (4-8) have a weak solution for all $f \in L^2(\Omega)$, is that

$$\| \varphi \| \leq C \| \bar{P}(D)\varphi \| \qquad \varphi \in C_0^\infty(\Omega_0) \tag{4-29}$$

PROOF The sufficiency is obvious, since (4-29) implies, for each $f \in L^2(\Omega)$,

$$|(f, \varphi)| \leq \| f \| \; \| \varphi \| \leq \| f \| \, C \, \| \bar{P}(D)\varphi \|$$

by the Schwarz inequality (1-62). To prove the necessity, suppose inequality (4-28) holds for each $f \in L^2(\Omega)$. If (4-29) did not hold, there would be a sequence $\{\varphi_k\}$ of functions in $C_0^\infty(\Omega_0)$, for which

$$\| \varphi_k \| \to \infty \qquad \| \bar{P}(D)\varphi_k \| = 1 \qquad k \to \infty \tag{4-30}$$

Set $$F_k(f) = (f, \varphi_k) \qquad f \in L^2(\Omega) \qquad k = 1, 2, \ldots$$

Then, for each k, F_k is a bounded linear functional on $L^2(\Omega)$ and, for each $f \in L^2(\Omega)$, we have by (4-28)

$$|F_k(f)| \leq |(f, \varphi_k)| \leq C \| \bar{P}(D)\varphi_k \| = C$$

so that $$\operatorname*{lub}_k |F_k(f)| < \infty$$

By the Banach–Steinhaus theorem (Theorem 1-10 of Sec. 1-5), there is a constant C, such that

$$|F_k(f)| \leq C \| f \| \qquad f \in L^2(\Omega) \qquad k = 1, 2, \ldots$$

Hence $$|(f, \varphi_k)| \leq C \| f \| \qquad f \in L^2(\Omega) \qquad k = 1, 2, \ldots$$

Taking $f = \varphi_k$, we get

$$\| \varphi_k \| \leq C$$

which contradicts (4-30). Thus, (4-29) holds, and the proof is complete.

In particular, we know that for any domain $\Omega_0 \subset E^{n+1}$, Eqs. (4-7), (4-8) have a weak solution for each $f \in L^2(\Omega)$ if and only if inequality (4-29) holds. A natural question to ask is what operators $P(D)$ satisfy (4-29)? In this connection, we have

Theorem 4-5 If Ω contains a slab of the form $0 < t < 2b$, $b > 0$, and $P(D)$ satisfies (4-29), then there is a constant K, such that

$$P(\xi, \tau) = 0 \quad \text{for } \xi \text{ real, implies} \quad |\operatorname{Im} \tau| \leq K \tag{4-31}$$

PROOF Let (ξ, τ) be such that ξ is real and $P(\xi, \tau) = 0$. Let $\psi(x) \neq 0$ be a function in $C_0^\infty(E^n)$, and let $\rho(t)$ be a function in $C_0^\infty(E^1)$, which satisfies

$$\rho(t) = 1 \qquad |t| \leq b$$

$$\rho(t) = 0 \qquad |t| \geq 2b$$

and

$$0 \leq \rho(t) \leq 1 \qquad |t| \leq 2b$$

Set

$$\varphi_\varepsilon(x, t) = \psi(\varepsilon x)\rho(t) \exp(i(x, \xi) + it\tau) \tag{4-32}$$

We can take Ω_0 to contain the slab $|t| \leq 2b$. In this case, $\varphi_\varepsilon \in C_0^\infty(\Omega_0)$ and hence inequality (4-29) is assumed to hold for it. Now

$$\| \varphi_\varepsilon(x, t) \|^2 = \int |\psi(\varepsilon x)|^2 \, dx \int_0^{2b} \exp(2t \operatorname{Im} \tau) |\rho(t)|^2 \, dt$$

$$\geq \varepsilon^{-n} \| \psi(x) \|^2 \int_0^b \exp(2t \operatorname{Im} \tau) \, dt$$

On the other hand,

$$\overline{P}(D)\varphi_\varepsilon = \sum_{|\mu|+k>0} \frac{1}{\mu!k!} \overline{P}^{(\mu,k)}(D) \exp(i(x, \xi) + it\tau) D_x^\mu \psi(\varepsilon x) D_t^k \rho(t)$$

$$= \exp(i(x, \xi) + it\bar{\tau}) \sum_{|\mu|+k>0} \frac{1}{\mu!k!} \overline{P}^{(\mu,k)}(\xi, \bar{\tau}) D_x^\mu \psi(\varepsilon x) D_t^k \rho(t)$$

where

$$P^{(\mu,k)}(\xi, \tau) = \frac{\partial^{|\mu|+k} P(\xi, \tau)}{\partial \xi^\mu \, \partial \tau^k}$$

Thus

$$|\overline{P}(D)\varphi_\varepsilon| \leq \exp(t \operatorname{Im} \tau) \sum_{|\mu|+k>0} \frac{1}{\mu!k!} |P^{(\mu,k)}(\xi, \tau)| \; |D_x^\mu \psi(\varepsilon x) D_t^k \rho(t)|$$

and consequently

$$\| \overline{P}(D)\varphi_\varepsilon \|^2 \leq C \sum_{|\mu|+k>0} |P^{(\mu,k)}(\xi, \tau)|^2 \; \| D_x^\mu \psi(\varepsilon x) \|^2 \int_0^{2b} \exp(2t \operatorname{Im} \tau) |D_t^k \rho(t)|^2 \, dt$$

$$\leq \varepsilon^{-n} C_1 \sum_{k>0} |P^{(0,k)}(\xi, \tau)|^2 \; \| \psi(x) \|^2 \int_b^{2b} \exp(2t \operatorname{Im} \tau) \, dt$$

$$+ \varepsilon^{-n} C_2 \sum_{|\mu|>0} \sum_{k \geq 0} |P^{(\mu,k)}(\xi, \tau)|^{2|\mu|} \; \| D_x^\mu \psi(x) \|^2 \int_0^{2b} \exp(2t \operatorname{Im} \tau) \, dt$$

Substituting into (4-29), we have

$$\int_0^b \exp(2t \operatorname{Im} \tau) \, dt \leq C_3 \sum_{k>0} |P^{(0,k)}(\xi, \tau)|^2 \int_b^{2b} \exp(2t \operatorname{Im} \tau) \, dt$$
$$+ C_4 \sum_{|\mu|>0} \varepsilon^{2|\mu|} \sum_{k \geq 0} |P^{(\mu,k)}(\xi, \tau)|^2 \int_b^{2b} \exp(2t \operatorname{Im} \tau) \, dt \tag{4-33}$$

Let $\varepsilon \to 0$. If $\operatorname{Im} \tau \leq 0$ this yields

$$1 - \exp(2b \operatorname{Im} \tau) \leq C_5(1 + |\xi| + |\tau|)^{2m}(\exp(2b \operatorname{Im} \tau) - \exp(4b \operatorname{Im} \tau))$$

showing that

$$1 \leq C_6(1 + |\xi| + |\tau|)^{2m} \exp(2b \operatorname{Im} \tau)$$

Thus

$$|\operatorname{Im} \tau| \leq \frac{1}{2b}[\ln C_6 + 2m \ln(1 + |\xi| + |\tau|)] \tag{4-34}$$

Now, τ depends algebraically on ξ and, hence, $\operatorname{Im} \tau$ depends algebraically on ξ and $\operatorname{Re} \tau$. Hence, if $|\operatorname{Im} \tau|$ were not bounded as $|\xi|^2 + (\operatorname{Re} \tau)^2 \to \infty$, it would tend to infinity like a power of $|\xi|^2 + (\operatorname{Re} \tau)^2$. This possibility is eliminated by (4-34). Thus, (4-31) is proved for $\operatorname{Im} \tau \leq 0$ (see Sec. 2-6). To prove it for $\operatorname{Im} \tau > 0$, note that we can write $P(\xi, \tau)$ in the form

$$P(\xi, \tau) = a\tau^m + a_1(\xi)\tau^{m-1} + \cdots + a_m(\xi) \tag{4-35}$$

where $a_j(\xi)$ is a polynomial in ξ of degree $\leq j$. Now, by the fundamental theorem of algebra, for each ξ there are m roots $\tau_1, \ldots, \tau_m$ of $P(\xi, \tau) = 0$. Furthermore

$$\sum_1^m \tau_j = \frac{-a_1(\xi)}{a} \tag{4-36}$$

so that

$$\sum_1^m \operatorname{Im} \tau_j = -\operatorname{Im}\left[\frac{a_1(\xi)}{a}\right]$$

Now, $\operatorname{Im}[a_1(\xi)/a]$ is a polynomial in ξ of degree ≤ 1. Moreover, we have just shown that

$$\sum_1^m \operatorname{Im} \tau_j \geq -C_7 \tag{4-37}$$

This shows that

$$\operatorname{Im}\left[\frac{a_1(\xi)}{a}\right] \leq C_7$$

Now the only polynomial of degree ≤ 1 which is bounded above by a constant, is itself a constant. Thus,

$$\sum_1^m \operatorname{Im} \tau_j = C_8$$

Thus, for each k

$$\operatorname{Im} \tau_k = C_8 - \sum_{j \neq k} \operatorname{Im} \tau_j \leq C_8 + (m-1)K$$

and the proof of (4-31) is complete.

An operator $P(D)$ satisfying (4-31) is called *hyperbolic* with respect to the t axis. When no ambiguity will arise, we shall merely call it hyperbolic. In Sec. 4-4 we shall study some properties of such operators.

4-4 PROPERTIES OF HYPERBOLIC OPERATORS

From the definition, we see immediately

Lemma 4-6 The operator $P(D)$ is hyperbolic if and only if $\overline{P}(D)$ is.

Another simple consequence is

Theorem 4-7 Let

$$p(D) = \sum_{|\mu|+k=m} a_{\mu,k} D_x^\mu D_t^k \tag{4-38}$$

denote the principal part of $P(D)$ (see Sec. 2-4). If $P(D)$ is hyperbolic, then for real ξ, $p(\xi, \tau)$ has only real roots τ. Hence $p(D)$ is hyperbolic.

PROOF The last statement follows from the definition of hyperbolicity (Sec. 4-3). To show that $p(\xi, \tau)$ has only real roots, suppose that ξ is real and that $p(\xi, \tau) = 0$ with $\operatorname{Im} \tau = c > 0$. Now

$$\frac{P(\lambda\xi, \lambda z)}{\lambda^m} = p(\xi, z) + \frac{Q(\lambda\xi, \lambda z)}{\lambda^m} \tag{4-39}$$

where Q is a polynomial of degree $<m$. Let Γ be a circle in the upper half-plane with center τ and radius $\leq c/2$, and such that $p(\xi, z)$ has no roots on Γ. Thus,

$$|p(\xi, z)| \geq \delta > 0 \tag{4-40}$$

for z on Γ. Moreover, there is a constant M, such that

$$|Q(\lambda\xi, \lambda z)| \leq M\lambda^{m-1} \tag{4-41}$$

for z on Γ. Hence, for λ sufficiently large

$$\frac{|Q(\lambda\xi, \lambda z)|}{\lambda^m} < |p(\xi, z)|$$

for z on Γ. By Rouché's theorem and Eq. (4-39) we see that $P(\lambda\xi, \lambda z)/\lambda^m$ has at least one root inside Γ. Call it σ. By the definition of hyperbolicity, we have

$$|\operatorname{Im} \lambda\sigma| \leq K$$

or

$$|\operatorname{Im} \sigma| \leq \frac{K}{\lambda}$$

On the other hand, since σ is inside Γ, we must have

$$\operatorname{Im} \sigma \geq \frac{c}{2} \tag{4-42}$$

This gives $c \leq 2K/\lambda$. Letting $\lambda \to \infty$, we obtain a contradiction, showing that we cannot have $\operatorname{Im} \tau > 0$. Similar reasoning shows that we cannot have $\operatorname{Im} \tau < 0$. This completes the proof.

Corollary 4-8 A homogeneous polynomial $P(D)$ is hyperbolic if and only if Eq. (4-10) holds, and the equation $P(\xi, \tau)$ has only real roots when ξ is real.

A useful criterion is given by

Theorem 4-9 If the principal part $p(D)$ of $P(D)$ is hyperbolic and $P(D) - p(D)$ is weaker than $p(D)$, then $P(D)$ is hyperbolic.

Before proving Theorem 4-9, let me give an important application. A homogeneous operator $p(D)$ of order m is called *totally hyperbolic* if $p(\xi, \tau)$ has m simple real roots τ for each real $\xi \neq 0$. The importance of this concept stems from

Theorem 4-10 If $p(D)$ is totally hyperbolic and of order m, and $Q(D)$ is any operator of order $<m$, then $P(D) = p(D) + Q(D)$ is a hyperbolic operator.

Proof Consider the polynomials

$$p^{(k)}(\xi, \tau) = \frac{\partial p(\xi, \tau)}{\partial \xi_k} \qquad 1 \leq k \leq n$$

$$p^{(n+1)}(\xi, \tau) = \frac{\partial p(\xi, \tau)}{\partial \tau}$$

By hypothesis

$$p(\lambda\xi, \lambda\tau) = \lambda^m p(\xi, \tau) \qquad \lambda > 0 \tag{4-43}$$

Differentiating both sides of Eq. (4-43) with respect to λ and setting $\lambda = 1$, we obtain

$$\sum_1^n \xi_k p^{(k)}(\xi, \tau) + \tau p^{(n+1)}(\xi, \tau) = m p(\xi, \tau) \tag{4-44}$$

Now

$$\sum_1^{n+1} |p^{(k)}(\xi, \tau)| \neq 0 \qquad \xi, \tau \text{ real} \qquad |\xi| + |\tau| \neq 0 \tag{4-45}$$

For otherwise there would be real $(\xi, \tau) \neq 0$ for which each $p^{(k)}(\xi, \tau) = 0$. This would imply $p(\xi, \tau) = 0$ by Eq. (4-44). But this would mean that τ is at least a double root of $p(\xi, \tau) = 0$, contrary to assumption.

Now by (4-45), the expression

$$\frac{1}{\sum_1^{n+1} |p^{(k)}(\xi, \tau)|} \tag{4-46}$$

is continuous on the set $|\xi| + |\tau| = 1$, ξ, τ real, and hence has a maximum M on this set. Furthermore, the denominator in (4-46) is homogeneous of degree $m - 1$. Thus, if (ξ, τ) is any real vector with $|\xi| + |\tau| = \lambda \neq 0$, then

$$\frac{(|\xi| + |\tau|)^{m-1}}{\sum_1^{n+1} |p^{(k)}(\xi, \tau)|} = \frac{1}{\sum_1^{n+1} \left| p^{(k)}\left(\frac{\xi}{\lambda}, \frac{\tau}{\lambda}\right) \right|} \leq M \tag{4-47}$$

This implies that there is another constant M', such that

$$(|\xi| + |\tau| + 1)^{m-1} \leq M'\left(\sum_1^{n+1} |p^{(k)}(\xi, \tau)| + 1\right) \tag{4-48}$$

which says that every operator of order $<m$ is weaker than $p(D)$. We now apply Theorem 4-9.

The proof of Theorem 4-9 will be based on the following lemmas. In them we revert to the notation of Chapters 1 to 3.

Lemma 4-11 Let Q be a polynomial of degree m and suppose that, for complex vectors ξ and ζ, we have

$$Q(\xi + z\zeta) \neq 0 \quad \text{when} \quad |z| < 1, \quad z \text{ complex} \tag{4-49}$$

Then

$$|Q(\xi + \zeta)| \leq 2^m |Q(\xi)| \tag{4-50}$$

PROOF Set $g(z) = Q(\xi + z\zeta)$. If $z_1, \ldots, z_m$ are the complex roots of $g(z)$, then by hypothesis $|z_j| \geq 1$ for j. On the other hand

$$\frac{Q(\xi + \zeta)}{Q(\xi)} = \frac{g(1)}{g(0)} = \prod_1^m \frac{(1 - z_j)}{z_j}$$

Hence

$$\left| \frac{Q(\xi + \zeta)}{Q(\xi)} \right| \leq \prod_1^m \left(\frac{1}{|z_j|} + 1 \right) \leq 2^m$$

This is precisely inequality (4-50).

Lemma 4-12 If the polynomial $Q(\xi)$ is weaker than a polynomial $P(\xi)$, then for each complex ζ the polynomial $Q(\xi + \zeta)$ in ξ is weaker than the polynomial $P(\xi + \zeta)$.

PROOF By the Taylor expansion

$$Q(\xi + \zeta) = \sum \frac{Q^{(\mu)}(\xi)\zeta^\mu}{\mu!}$$

Hence $$|Q(\xi + \zeta)| \le |\zeta|^k \sum |Q^{(\mu)}(\xi)|$$

where k is the degree of $Q(\xi)$. By Theorem 3-15 of Sec. 3-4, this gives

$$|Q(\xi + \zeta)| \le C|\zeta|^k \sum |P^{(\nu)}(\xi)| \tag{4-51}$$

Now $$P^{(\nu)}(\xi) = P^{(\nu)}(\xi + \zeta - \zeta) = \sum \frac{P^{(\nu+\mu)}(\xi + \zeta)(-\zeta)^\mu}{\mu!}$$

and hence $$|P^{(\nu)}(\xi)| \le |\zeta|^m \sum |P^{(\mu)}(\xi + \zeta)| \tag{4-52}$$

where m is the degree of $P(\xi)$. Combining (4-51) and (4-52), we get

$$|Q(\xi + \zeta)| \le C|\zeta|^{k+m} \sum |P^{(\mu)}(\xi + \zeta)| \tag{4-53}$$

which shows that $Q(\xi + \zeta)$ is weaker than $P(\xi + \zeta)$. This completes the proof.

Lemma 4-13 Suppose

$$Q(\xi) = \sum_0^m Q_j(\xi)$$

where each $Q_j(\xi)$ is a homogeneous polynomial of degree j. If $p(\xi)$ is a homogeneous polynomial of degree m, and $Q(\xi)$ is weaker then $p(\xi)$, then each $Q_j(\xi)$ is weaker than $p(\xi)$.

PROOF Let $\lambda_0, \ldots, \lambda_m$ be $m + 1$ different real numbers. Then

$$\sum_{j=0}^m \lambda_k^j Q_j(\xi) = Q(\lambda_k \xi) \qquad 0 \le k \le m \tag{4-54}$$

Since the matrix (λ_k^j) is nonsingular, we can solve the system (4-54) for the $Q_j(\xi)$ in terms of the $Q(\lambda_k \xi)$. Thus

$$Q_j(\xi) = \sum_{k=0}^m \alpha_{jk} Q(\lambda_k \xi) \qquad 0 \le j \le m$$

Hence $$|Q_j(\xi)| \le C \sum_k |Q(\lambda_k \xi)|$$

$$\le C' \sum_{\mu,k} |p^{(\mu)}(\lambda_k \xi)| \le C'' \sum_\mu |p^{(\mu)}(\xi)|$$

since the $p^{(\mu)}(\xi)$ are homogeneous. This shows that each $Q_j(\xi)$ is weaker than $p(\xi)$.

We are now ready for the

PROOF of Theorem 4-9 We claim that it suffices to prove that there is a $\delta > 0$, such that

$$\xi, \eta \text{ real} \quad \zeta, \tau \text{ complex} \quad |\zeta| + |\tau| < \delta \tag{4-55}$$

imply that $$p(\xi + \zeta, \eta + i + \tau) \neq 0$$

For suppose (4-55) holds. Then, by Lemma 4-11

$$|p(\xi + \zeta, \eta + i + \tau)| \leq 2^m |p(\xi, \eta + i)| \qquad |\zeta| + |\tau| < \delta \tag{4-56}$$

Now, by Lemmas 3-6 and 3-14 of Chapter 3 there is a finite set of real vectors $(\zeta^{(k)}, \tau^{(k)})$, satisfying $|\zeta^{(k)}| + |\tau^{(k)}| < \delta$, such that each derivative of $p(\xi, \eta + i)$ can be expressed in the form

$$p^{(\mu)}(\xi, \eta + i) = \sum \lambda_k p(\xi + \zeta^{(k)}, \eta + i + \tau^{(k)})$$

Thus, by (4-56)

$$|p^{(\mu)}(\xi, \eta + i)| \leq C |p(\xi, \eta + i)| \qquad \xi, \eta \text{ real} \tag{4-57}$$

Now $$P(\xi, \tau) = \sum_0^m p_k(\xi, \tau)$$

where p_k is homogeneous of degree k, $p_m = p$. By hypothesis and Lemma 4-13, each p_k is weaker than p. Hence, by Lemma 4-12 each $p_k(\xi, \eta + i)$ is weaker than $p(\xi, \eta + i)$. By (4-57) this gives

$$|p_k(\xi, \eta + i)| \leq C |p(\xi, \eta + i)| \qquad 0 \leq k < m \tag{4-58}$$

Then, if λ is real

$$\begin{aligned} |p_k(\xi, \eta + \lambda i)| &= |\lambda|^k \left| p_k\left(\frac{\xi}{\lambda}, \frac{\eta}{\lambda} + i\right) \right| \\ &\leq |\lambda|^k C \left| p\left(\frac{\xi}{\lambda}, \frac{\eta}{\lambda} + i\right) \right| \\ &\leq |\lambda|^{k-m} C |p(\xi, \eta + \lambda i)| \end{aligned}$$

In other words

$$|P(\xi, \tau) - p(\xi, \tau)| \leq \frac{mC |p(\xi, \tau)|}{|\text{Im}\, \tau|} \qquad |\text{Im}\, \tau| \geq 1$$

In particular

$$\begin{aligned} |P(\xi, \tau)| &\geq |p(\xi, \tau)| - |P(\xi, \tau) - p(\xi, \tau)| \\ &\geq |p(\xi, \tau)| \left(1 - \frac{mC}{|\text{Im}\, \tau|}\right) \end{aligned} \tag{4-59}$$

Since $p(\xi, \tau)$ is hyperbolic, it does not vanish for $\operatorname{Im} \tau \neq 0$. Inequality (4-59) now shows that $P(\xi, \tau)$ cannot vanish for $|\operatorname{Im} \tau| > \max[1, mC]$. Then P is hyperbolic.

It remains, therefore, only to prove (4-55). In proving it I shall make use of the important

Lemma 4-14 Let

$$P(\xi, z) = \sum_{k=0}^{m} a_k(\xi) z^k \tag{4-60}$$

be a polynomial in z with coefficients $a_k(\xi)$ continuous at $\xi = \xi_0$. Assume that $a_m(\xi_0) \neq 0$ and let $z_1(\xi), \ldots, z_m(\xi)$ be the z roots of

$$P(\xi, z) = 0 \tag{4-61}$$

for given ξ. Then, by suitably ordering these roots, we have

$$z_j(\xi) \to z_j(\xi_0) \quad \text{as} \quad \xi \to \xi_0 \qquad 1 \le j \le m \tag{4-62}$$

Assuming this lemma for the moment, let us proceed to establish (4-55). First note that by absorbing $\operatorname{Re} \zeta$ into ξ and $\operatorname{Re} \tau$ into η, (4-55) is implied by

$$p(\xi + i\theta, \eta + i + i\lambda) \neq 0 \quad \text{for} \quad \xi, \theta, \eta, \lambda \text{ real} \qquad |\theta| + |\lambda| < \delta \tag{4-63}$$

In other words we do not want $z = i$ to be a root of

$$p\left[\xi + z\theta, \eta + \frac{i}{2} + z\left(\lambda + \frac{1}{2}\right)\right] = 0 \tag{4-64}$$

Since

$$p(0, \tfrac{1}{2} + z) = a(\tfrac{1}{2} + z)^m$$

this polynomial in z has only the root $z = -\frac{1}{2}$. By Lemma 4-14, there is a $\delta > 0$ so small that the polynomial in z

$$p(\theta, \tfrac{1}{2} + \lambda + z) \tag{4-65}$$

has only negative roots for $|\theta| + |\lambda| < \delta$ (note that the coefficient of z^m in (4-65) is $p(0, 1) = a \neq 0$). Now, the hyperbolicity of p tells us that

$$p(\xi + z\theta, \eta + \tau + z(\lambda + \tfrac{1}{2})) = 0 \qquad \xi, \theta, \eta, \lambda \text{ real} \qquad |\theta| + |\lambda| < \delta \tag{4-66}$$

has no real z roots for $\operatorname{Im} \tau > 0$. Moreover, the coefficient of z^m in Eq. (4-66) is $p(\theta, \lambda + \frac{1}{2}) \neq 0$ for $|\theta| + |\lambda| < \delta$, since (4-65) has only negative roots there. Hence, the roots of Eq. (4-66) depend continuously on τ, and cannot be real if $\operatorname{Im} \tau > 0$. If one can show that they all have negative imaginary parts for some τ satisfying $\operatorname{Im} \tau > 0$, it will then follow that they all have negative imaginary parts for any τ satisfying $\operatorname{Im} \tau > 0$. In particular, it will follow that $z = i$ cannot be a root of Eq. (4-66) for $\tau = i/2$. This is precisely what we want.

To show that there is a τ satisfying $\operatorname{Im} \tau > 0$, such that all roots of Eq. (4-66) have negative imaginary parts, set $\tau = i\rho$ and let z be any root of Eq. (4-66).

If we put $\sigma = z/\rho$, then σ is a root of

$$p\left(\frac{\xi}{\rho} + \sigma\theta, \frac{\eta}{\rho} + i + \sigma(\lambda + \tfrac{1}{2})\right) = 0$$

Now, this equation tends to

$$p(\sigma\theta, i + \sigma(\lambda + \tfrac{1}{2})) = 0 \tag{4-67}$$

as $\rho \to \infty$, and hence σ tends to a root of Eq. (4-67) as $\rho \to \infty$ (Lemma 4-14). But any σ satisfying (4-67), satisfies

$$p\left(\theta, \lambda + \tfrac{1}{2} + \frac{i}{\sigma}\right) = 0$$

Hence i/σ must be negative, since (4-65) has only negative roots. This shows that $\text{Im } z \to -\infty$ as $\rho \to \infty$, when $\tau = i\rho$. Since z was any root of Eq. (4-66), it follows that they all have negative imaginary parts for $\tau = i\rho$ and ρ sufficiently large. This completes the proof.

It remains to give the

PROOF of Lemma 4-14 Let $P(\xi_0, z(\xi_0)) = 0$. Let $\varepsilon > 0$ be any number, such that $P(\xi_0, z) \neq 0$ for $|z - z(\xi_0)| = \varepsilon$ (since there are at most $m - 1$ other roots, this is true for all $\varepsilon > 0$ except at most $m - 1$ values). Let Γ be a circle with center $z(\xi_0)$ and radius ε. Since $P(\xi_0, z) \neq 0$ on Γ, there is a $\delta > 0$, such that $|P(\xi_0, z)| \geq \delta$ on Γ. Set

$$M = \max_{z \in \Gamma} |z|$$

and take ξ so close to ξ_0, that

$$\sum_0^m |a_k(\xi) - a_k(\xi_0)| M^k < \frac{\delta}{2}$$

This means that for such ξ and for $z \in \Gamma$

$$|P(\xi, z) - P(\xi_0, z)| \leq \frac{\delta}{2} < |P(\xi_0, z)|$$

Hence, by Rouché's theorem, the polynomial

$$P(\xi, z) = P(\xi_0, z) + [P(\xi, z) - P(\xi_0, z)]$$

has as many roots (counting multiplicities) inside Γ as the polynomial $P(\xi_0, z)$. Since they each have exactly m roots, the lemma is proved.

An operator $P(D)$ will be called *totally hyperbolic* if its principal part (i.e., the terms of highest order) is totally hyperbolic. Theorem 4-10 asserts that every totally hyperbolic operator is hyperbolic.

4-5 ORDINARY DIFFERENTIAL EQUATIONS

We would now like to examine a problem in ordinary differential equations. Let $P(\tau)$ be a polynomial of the form

$$P(\tau) = \sum_0^m a_k \tau^k \tag{4-68}$$

where the a_k are complex constants. We assume that $a_m \neq 0$, and for convenience let us put

$$a_m = 1 \tag{4-69}$$

Consider the following problem

$$P(D_t)u(t) = f(t) \qquad t \geq 0 \tag{4-70}$$

$$D_t^k u(D) = g_k \qquad 0 \leq k < m \tag{4-71}$$

where $f(t)$ is a continuous function in $t \geq 0$ and the g_k are given constants. We want to solve Eqs. (4-70), (4-71) for all f and g_k. To this end, let us first observe that all one needs to know is how to solve

$$P(D_t)u(t) = 0 \qquad t \geq 0 \tag{4-72}$$

$$D_t^k u(0) = 0 \qquad 0 \leq k < m - 1 \tag{4-73}$$

$$D_t^{m-1} u(0) = 1 \tag{4-74}$$

For suppose one can solve Eqs. (4-72) to (4-74). Let $w(t)$ be a solution. Set $u_0(t) = g_0 w(t)$, and inductively define

$$u_j(t) = \left[g_j - \sum_{i=1}^{j} D_t^{m+j-i} u_{i-1}(0) \right] w(t) \qquad 1 \leq j < m \tag{4-75}$$

Set

$$u(t) = \sum_{j=1}^{m} D_t^{m-j} u_{j-1}(t) \tag{4-76}$$

Clearly $u(t)$ is a solution of Eq. (4-72). Moreover

$$D_t^k u(0) = \sum_{j=1}^{m} D_t^{m+k-j} u_{j-1}(0) = \sum_{j=1}^{k+1} D_t^{m+k-j} u_{j-1}(0)$$

$$= D_t^{m-1} u_k(0) + \sum_{j=1}^{k} D_t^{m+k-j} u_{j-1}(0) = g_k$$

by Eqs. (4-73), (4-74) and (4-75). Thus, Eq. (4-76) is a solution of (4-71), (4-72). To obtain a solution of (4-70), (4-71) all we need to do is add to Eq. (4-76) a solution of

$$P(D_t)v(t) = f(t) \qquad t \geq 0 \tag{4-77}$$

and

$$D_t^k v(0) = 0 \qquad 0 \leq k < m \tag{4-78}$$

Set
$$v(t) = i \int_0^t f(s)w(t-s)\,ds \tag{4-79}$$

Differentiating, we get

$$D_t v(t) = f(t)w(0) + i \int_0^t f(s)D_t w(t-s)\,ds \tag{4-80}$$

and, by Eq. (4-73)

$$D_t^k v(t) = i \int_0^t f(s)D_t^k w(t-s)\,ds \qquad 0 \le k < m \tag{4-81}$$

In particular, we see from Eq. (4-81) that v satisfies Eq. (4-78). Differentiating Eq. (4-81) for $k = m - 1$, we get

$$D_t^m v(t) = f(t)D_t^{m-1} w(0) + i \int_0^t f(s)D_t^m w(t-s)\,ds$$

Hence
$$P(D_t)v(t) = f(t) + i \int_0^t f(s)P(D_t)w(t-s)\,ds = f(t)$$

and $v(t)$ is a solution of Eqs. (4-77), (4-78). Thus, if we can find a solution $w(t)$ of Eqs. (4-72) to (4-74), we will have solved Eqs. (4-70), (4-71). Again we can exhibit a solution very easily. In fact, set

$$w(t) = \frac{1}{2\pi i} \oint_\Gamma \frac{e^{itz}}{P(z)}\,dt \tag{4-82}$$

where Γ is any simple closed curve in the complex z plane, containing the roots of $P(z)$ in its interior. Clearly, the function given by Eq. (4-82) is infinitely differentiable, and

$$D_t^k w(t) = \frac{1}{2\pi i} \oint_\Gamma \frac{z^k\, e^{itz}}{P(z)}\,dt \tag{4-83}$$

In particular

$$P(D_t)w(t) = \frac{1}{2\pi i} \oint_\Gamma \frac{P(z)\, e^{itz}}{P(z)}\,dz = \frac{1}{2\pi i} \oint_\Gamma e^{itz}\,dz = 0 \tag{4-84}$$

Moreover, by taking Γ to be a circle of radius R about the origin with R sufficiently large, we obtain

$$D_t^k w(0) = \frac{1}{2\pi i} \int_0^{2\pi} \frac{R^k\, e^{ik\theta} \cdot i\, e^{i\theta}\, R\, d\theta}{R^m\, e^{im\theta} + Q(R\, e^{i\theta})}$$

where $Q(z) = P(z) - z^m$ is a polynomial of degree $< m$. Letting $R \to \infty$, we get 0 for $0 \le k < m - 1$, and 1 for $k = m - 1$. This shows that Eq. (4-82) is a solution of Eqs. (4-72) to (4-74).

In conclusion, we want to derive some useful inequalities for the functions

$v(t)$ and $w(t)$ given above. Suppose the roots of $P(z)$ are contained in the strip $|\operatorname{Im} z| \le K$. If $\tau_1, \ldots, \tau_m$ denote these roots, surround each τ_j with a square with sides equal to 2, parallel to the real and imaginary axes and center τ_j. Let G be the union of these squares and let Γ be the boundary of G. Although Γ may not be a simple closed curve, it can be made so by using cross-cuts. Moreover the length of Γ is at most $8m$. By the way Γ was chosen, we have

$$|z - \tau_j| \ge 1 \qquad z \in \Gamma \tag{4-85}$$

Set

$$P^{(k)}(z) = \frac{d^k P(z)}{dz^k} \tag{4-86}$$

Then

$$\frac{P^{(k)}(z)}{P(z)} = \sum \frac{1}{(z - \tau_{j_1}) \cdots (z - \tau_{j_k})}$$

Thus, we have by Eq. (4-83)

$$|P^{(k)}(D_t)w(t)| \le \frac{4mm!}{\pi(m-k)!} e^{t(K+1)} \qquad 0 < k \le m \tag{4-87}$$

Now

$$P^{(j)}(z) = \sum_{k=j}^{m} \frac{k!}{(k-j)!} a_k z^{k-j} \tag{4-88}$$

This enables us to estimate $D_t^k w$ for each k. If we agree to set $a_{m-j} = 0$ for $j > m$, a simple induction gives

$$|D_t^k w(t)| \le C\left(1 + \sum_1^k |a_{m-j}|\right)^k e^{t(K+1)} \qquad k = 0, 1, 2, \ldots \tag{4-89}$$

Where the constant C depends only on k. To see this, note that it is surely true for $k = 0$. We merely use the fact that $P^{(m)}(z) = m!$, and apply (4-87). Now assume that (4-89) holds for all $k < j$. We shall prove it for $k = j$. If $j \le m$, we have, by Eq. (4-88)

$$\begin{aligned} P^{(m-j)}(z) &= \sum_{k=m-j}^{m} \frac{k!}{(k+j-m)!} a_k z^{k+j-m} \\ &= \frac{m!}{j!} z^j + \sum_{m-j}^{m-1} \frac{k!}{(k+j-m)!} a_k z^{k+j-m} \end{aligned}$$

Thus, by (4-87) and the induction hypothesis

$$\begin{aligned} |D_t^j w(t)| &\le C\left(e^{t(K+1)} + \sum_{m-j}^{m-1} |a_k| \, |D_t^{k+j-m} w|\right) \\ &\le C'\left[1 + \sum_{k=m-j}^{m-1} |a_k|\left(1 + \sum_{i=1}^{k+j-m} |a_{m-i}|\right)^{k+j-m}\right] e^{t(K+1)} \\ &\le C'\left[1 + \sum_{k=m-j}^{m-1} |a_k|\left(1 + \sum_{i=1}^{j} |a_{m-i}|\right)^{j-1}\right] e^{t(K+1)} \\ &\le C'\left(1 + \sum_{i=1}^{j} |a_{m-i}|\right)^j e^{t(K+1)} \end{aligned} \tag{4-90}$$

which is precisely what we wanted to show. If $j > m$, we use the fact that

$$z^{j-m}P(z) = z^j + \sum_{k=0}^{m-1} a_k z^{k+j-m}$$

This gives (4-90) and, consequently, (4-89) if we agree to take $a_{m-i} = 0$ for $i > m$. Thus, inequality (4-89) holds for all k.

We can now estimate $v(t)$ given by Eq. (4-79). By Eq. (4-81) we have for $k < m$

$$|D_t^k v(t)| \leq C\left(1 + \sum_1^k |a_{m-j}|\right)^k \int_0^t |f(s)| \, e^{(t-s)(K+1)} \, ds$$

$$\leq C'\left(1 + \sum_1^k |a_{m-j}|\right)^k e^{t(K+1)} \sum_{0 \leq s \leq t} |f(s)| \tag{4-91}$$

If $k \geq m$, we have

$$D_t^k v(t) = \sum_{j=m}^{k} D_t^{k-j} f(t) D_t^{j-1} w(0) + i \int_0^t f(s) D_t^k w(t-s) \, ds \tag{4-92}$$

The same reasoning now gives

$$|D_t^k v(t)| \leq C\left(1 + \sum_1^m |a_{m-j}|\right)^k e^{t(K+1)} \sum_{i=0}^{k-m} \max_{0 \leq s \leq t} |D_t^k f(s)| \tag{4-93}$$

4-6 EXISTENCE OF SOLUTIONS

We are now ready to prove an existence theorem. In this section we assume that Ω is the slab $0 < t < b$. We refer to the definition of $S(\Omega)$ given in Sec. 2-3.

Theorem 4-15 Suppose $P(D)$ is hyperbolic, and let f be any function in $S(\Omega)$. Then there exists a function $u \in S(\Omega)$ which satisfies Eqs. (4-7), (4-8).

PROOF Let F denote the Fourier transform with respect to the variables $x_1, \ldots, x_n$ (see Sec. 2-3 and 2-5). Suppose Eqs. (4-7), (4-8) have a solution $u \in S(\Omega)$. Then $u(x, t)$ is in $S(E^n)$ for each fixed t. Setting

$$h(\xi, t) = Fu(\xi, t)$$

we have

$$P(\xi, D_t)h(\xi, t) = Ff(\xi, t) \qquad t \geq 0 \qquad \xi \in E^n \tag{4-94}$$

$$D_t^k h(\xi, 0) = 0 \qquad 0 \leq k < m \qquad \xi \in E^n \tag{4-95}$$

For each fixed ξ this is an ordinary differential problem of the type studied in Sec. 4-5. Note that the coefficient of D_t^m in $P(D)$ is a nonvanishing constant independent of ξ. We may take it to be 1. By the results of Sec. 4-5, a solution of Eqs. (4-94), (4-95) is given by

$$h(\xi, t) = i \int_0^t Ff(\xi, t) w(\xi, t-s) \, ds \tag{4-96}$$

where
$$w(\xi, t) = \frac{1}{2\pi i} \oint_{\Gamma} \frac{e^{itz}}{P(\xi, z)}\, dz \tag{4-97}$$

and Γ is a simple closed curve in the complex plane containing the roots of $P(\xi, z)$ in its interior. If we can show that $h(\xi, z)$ is in $S(\Omega)$, it will follow that the same is true of $u(x, t)$. Furthermore, an application of the inverse Fourier transform to Eqs. (4-94), (4-95) will show that u is a solution of Eqs. (4-7), (4-8).

It remains, therefore, to show that $h \in S(\Omega)$. We take Γ to be the curve described in Sec. 4-5. Since

$$D_t^k w(\xi, t) = \frac{1}{2\pi i} \oint_{\Gamma} \frac{z^k\, e^{itz}\, dz}{P(\xi, z)}$$

by Eq. (4-83), we have by Eq. (4-89)

$$|\, D_t^k w(\xi, t) \,| \le C(1 + |\,\xi\,|)^{2k}\, e^{t(K+1)}$$

Here we have made use of the fact that the coefficient of z^{m-j} in $P(\xi, z)$ is a polynomial in ξ of degree $\le j$. Hence, it is bounded by a constant times $(1 + |\,\xi\,|)^j$. Now, it is a simple matter to prove

$$D_\xi^\mu \left[\frac{1}{P(\xi, z)} \right] = \frac{Q_\mu(\xi, z)}{P(\xi, z)^{|\mu|+1}} \tag{4-98}$$

where Q_μ is a polynomial of degree $\le (m-1)\,|\,\mu\,|$. Thus, each derivative $D_\xi^\mu D_t^k w(\xi, t)$ is the sum of terms of the form

$$\text{constant } \xi^\nu \oint_{\Gamma} \frac{z^{j+k}\, e^{itz}\, dz}{P(\xi, z)^{|\mu|+1}}$$

where $|\,\nu\,| + j \le (m-1)\,|\,\mu\,|$. Now, $P(\xi, z)^r$ satisfies all of the hypotheses of $P(\xi, t)$. Hence, we may apply Eq. (4-89) to the above expression as well. This gives

$$|\, D_\xi^\mu D_t^k w(\xi, t) \,| \le C(1 + |\,\xi\,|)^{2(m-1)|\mu| + 2k}\, e^{t(K+1)} \tag{4-99}$$

Since $f \in S(\Omega)$, the same is true of Ff. Hence, for each j, k, μ, there is a constant C, such that

$$|\, D_\xi^\mu D_t^k Ff(\xi, t) \,| \le \frac{C}{(1 + |\,\xi\,|)^j}$$

By Eq. (4-92)

$$D_\xi^\mu D_t^k h(\xi, t) = \sum_{j=m}^{k} D_\xi^\mu [D_t^{k-j} Ff(\xi, t) D_t^{j-1} w(\xi, 0)]$$
$$+ i D_\xi^\mu \int_0^t Ff(\xi, t) D_t^k w(\xi, t - s)\, ds$$

Applying the last two inequalities, we get

$$|D_\xi^\mu D_t^k h(\xi,t)| \leq \frac{C}{(1+|\xi|)^j}$$

for each j, k, and μ. This shows that $h \in S(\Omega)$, and the proof is complete.

In contrast to the above, we now prove a theorem concerning weak solutions.

Theorem 4-16 If $P(D)$ is hyperbolic, then Eqs. (4-7), (4-8) have a weak solution for each $f \in L^2(\Omega)$.

Our proof of Theorem 4-16 makes use of the fundamental

Lemma 4-17 For any $\Omega, C_0^\infty(\Omega)$ is dense in $L^2(\Omega)$.

Let us postpone the proof of Lemma 4-17 a moment and use it to give the

PROOF of Theorem 4-16 Let f be any function in $L^2(\Omega)$. Then there exists a sequence $\{f_k\}$ of functions in $C_0^\infty(\Omega)$ converging to f in $L^2(\Omega)$. By Theorem 4-15, for each f_k, there is a solution u_k of (4-7), (4-8) satisfying

$$\| u_k \| \leq C \| f_k \| \qquad k = 1, 2, \dots \tag{4-100}$$

In particular, u_k is a weak solution, and satisfies

$$(u_k, \bar{P}(D)\varphi) = (f_k, \varphi) \qquad \varphi \in C_0^\infty(\Omega_0) \tag{4-101}$$

Since the f_k converge to f in $L^2(\Omega)$, the norms $\| f_k \|$ are bounded, and by (4-100) so are the norms $\| u_k \|$. By Theorem 1-8 of Sec. 1-5, there is a subsequence (we denote the subsequence also by $\{u_k\}$) converging weakly to a function $u \in L^2(\Omega)$. Letting $k \to \infty$ in Eq. (4-101), gives

$$(u, \bar{P}(D)\varphi) = (f, \varphi) \qquad \varphi \in C_0^\infty(\Omega_0) \tag{4-102}$$

showing that u is a weak solution of (4-7), (4-8). This completes the proof.

We can now give a converse of Theorem 4-5.

Theorem 4-18 If $P(D)$ is hyperbolic, then inequality (4-29) holds as well as

$$\| \varphi \| \leq C \| P(D)\varphi \| \qquad \varphi \in C_0^\infty(\Omega_0) \tag{4-103}$$

where the constant C depends only on $P(D)$ and b.

PROOF By Theorem 4-16, the Eqs. (4-7), (4-8) have a weak solution for each $f \in L^2(\Omega)$. Hence (4-29) holds by Theorem 4-4. Inequality (4-103) follows from the same reasoning applied to $\bar{P}(D)$, which is also hyperbolic by Lemma 4-6.

Let me now give the

PROOF of Lemma 4-17 Let Ω be any domain in E^n and let u be any function in $L^2(\Omega)$. Define u to be zero outside Ω. Thus $u \in L^2(E^n)$. For any $\varepsilon > 0$ we can find a function $w \in C^\infty(E^n) \cap L^2(E^n)$, such that

$$\| u - w \| < \frac{\varepsilon}{2} \tag{4-104}$$

(Theorem 2-3 of Sec. 2-2; we can take $w = J_\delta u$ for δ sufficiently small.) Now there exists a bounded domain Ω_1 with $\overline{\Omega}_1 \subset \Omega$, such that

$$\int_{\Omega - \Omega_1} |w|^2 \, dx < \frac{\varepsilon^2}{4} \tag{4-105}$$

Moreover, there is a $\psi \in C_0^\infty(\Omega)$, such that

$$0 \leq \psi(x) \leq 1 \qquad x \in \Omega \tag{4-106}$$

and $\psi(x) = 1$ for $x \in \Omega_1$ (see (1-36), (1-37) of Sec. 1-3). Now $\varphi = \psi w \in C_0^\infty(\Omega)$ and

$$\begin{aligned} \| u - \varphi \| &\leq \| u - w \| + \| w - \varphi \| \\ &< \frac{\varepsilon}{2} + \left[\int_{\Omega - \Omega_1} |(1 - \psi)w|^2 \, dx \right]^{1/2} < \varepsilon \end{aligned}$$

and the proof is complete.

Now let us show how to solve the problem

$$P(D)u(x,t) = f(x,t) \text{ in } \Omega \tag{4-107}$$

$$D_t^k u(x,0) = g_k(x) \text{ on } \partial_0\Omega \qquad 0 \leq k < m \tag{4-108}$$

Theorem 4-19 Suppose $P(D)$ is hyperbolic. Then for each $f \in S(E^{n+1})$ and each set of functions $g_0, \ldots, g_{m-1} \in S(E^n)$, there is a solution $u \in C^\infty(\overline{\Omega}) \cap L^2(\Omega)$ of Eqs. (4-107), (4-108) satisfying Eq. (4-94).

PROOF Theorem 4-19 follows from Theorem 4-15 and the simple observation that one can find a $w \in S(E^{n+1})$, such that

$$D_t^k w(x,0) = g_k(x) \text{ on } \partial_0\Omega \qquad 0 \leq k < m \tag{4-109}$$

Once we have w, we merely solve the problem

$$P(D)v = f - P(D)w \text{ in } \Omega \tag{4-110}$$

$$D_t^k v(x,0) = 0 \text{ on } \partial_0\Omega \qquad 0 \leq k < m \tag{4-111}$$

This can be done by Theorem 4-15. It is then a simple matter to verify that $u = v + w$ is a solution of Eqs. (4-107), (4-108).

To find a w satisfying Eq. (4-109), let $\rho(t)$ be a function in $C_0^\infty(E^1)$ which equals one in a neighborhood of $t = 0$. Then set

$$w(x,t) = \rho(t) \sum_{k=0}^{m-1} \frac{1}{k!} t^k g_k(x) \tag{4-112}$$

This function has the desired properties.

4-7 UNIQUENESS

Now that we know how to solve the Cauchy problem (4-107), (4-108) for smooth functions $f, g_0, \ldots, g_{m-1}$, we should ask whether or not the solution is unique. That the solution is always unique follows from a general theorem due to Holmgren (1901). However, we shall not prove this theorem in its generality. Instead we shall prove the following special case.

Theorem 4-20 Let $P(\xi, \tau)$ be a polynomial of degree m, such that the coefficient of τ^m does not vanish. Let Ω be the slab $0 < t < b$, where $b > 0$. Let $u \in C^\infty(\bar{\Omega})$ be a solution of

$$P(D)u = 0 \text{ in } \Omega \tag{4-113}$$

$$D_t^k u(x, 0) = 0 \qquad 0 \le k < m \tag{4-114}$$

If $P^{(\mu,k)}(D)u \in L^2(\Omega)$ for each μ, k, then $u \equiv 0$ in Ω.

Note that this theorem shows that the solution given by Theorem 4-15 is unique.

Theorem 4-20 can be proved by using

Lemma 4-21 There is a constant C depending only on $P(D)$ and b, such that

$$\int_\Omega e^{-\lambda t} |v|^2 \, dx \, dt \le C \int_\Omega e^{-\lambda t} |P(D)v|^2 \, dx \, dt \tag{4-115}$$

holds for all real λ, and all $v \in C^\infty(\bar{\Omega})$ which vanish near $t = b$ satisfy

$$D_t^k v(x, 0) = 0 \qquad 0 \le k < m \tag{4-116}$$

and such that $P^{(\mu,k)}(D)v \in L^2(\Omega)$ for each μ, k.

The proof of Lemma 4-21 requires a bit of preparation. Let us first show how it can be used to give the

PROOF of Theorem 4-20 Let $\rho(t)$ be a function in $C^\infty(E^1)$, satisfying

$$\begin{aligned} \rho(t) &= 1 \qquad 0 \le t \le a + \varepsilon \\ &= 0 \qquad a + 2\varepsilon \le t \le b \\ 0 \le \rho(t) &\le 1 \qquad 0 \le t < \infty \end{aligned}$$

where a and ε are any positive numbers such that $a + 2\varepsilon < b$. Set $v = u\rho$, where $u \in C^\infty(\bar{\Omega}) \cap L^2(\Omega)$ and satisfies Eqs. (4-113) and (4-114). Then v satisfies the hypotheses of Lemma 4-21. Thus (4-115) holds, and consequently

$$\int_0^a \int_{E^n} e^{-\lambda t} |u|^2 \, dx \, dt \le C \int_{a+\epsilon}^b \int_{E^n} e^{-\lambda t} |P(D)v|^2 \, dx \, dt \tag{4-117}$$

holds for all real λ. This implies for $\lambda > 0$

$$\int_0^a \int_{E^n} |u|^2 \, dx \, dt \le e^{-\lambda\varepsilon} C \int_{a+\varepsilon}^b \int_{E^n} |P(D)v|^2 \, dx \, dt \tag{4-118}$$

Note that $$P(D)v = \sum P^{(0,k)}(D)uD_t^k \frac{\rho}{k!} \in L^2(\Omega)$$

Letting $\lambda \to \infty$, we see that $u \equiv 0$ for $0 \le t \le a$. Since a can be taken to be any number $< b$, the proof is complete.

The following theorem is an important tool in the proof of Lemma 4-21. In it we revert to our previous notation.

Theorem 4-22 Let Ω be a domain contained in the slab $|x_k - a_k| < M/2$ in E^n. Then

$$\int_\Omega e^{(\xi,x)} |P^{(k)}(D)\varphi|^2 \, dx \le mM \int_\Omega e^{(\xi,x)} |P(D)\varphi|^2 \, dx \tag{4-119}$$

for all real vectors ξ, all $\varphi \in C_0^\infty(\Omega)$ and all polynomials $P(\xi)$ of degree $\le m$.

PROOF If

$$P(\xi) = \sum_{|\mu| \le m} a_\mu \xi^\mu$$

define, for ζ a complex vector

$$P(D + \zeta) = \sum_{|\mu| \le m} a_\mu (D + \zeta)^\mu$$

Then
$$\begin{aligned} P(D)(\varphi\, e^{i(\zeta,x)}) &= \sum \frac{P^{(\mu)}(D)\, e^{i(\zeta,x)}\, D^\mu \varphi}{\mu!} \\ &= e^{i(\zeta,x)} \sum \frac{P^{(\mu)}(\zeta) D^\mu \varphi}{\mu!} \\ &= e^{i(\zeta,x)}\, P(D + \zeta)\varphi \end{aligned} \tag{4-120}$$

Now for any complex ζ, $P(D - \zeta)$ is a constant-coefficient operator. Hence, by Lemma 1-15 of Sec. 1-7

$$\int |P^{(k)}(D - \zeta)(\varphi\, e^{i(\zeta,x)})|^2 \, dx \le mM \int |P(D - \zeta)(\varphi\, e^{i(\zeta,x)})|^2 \, dx \qquad \varphi \in C_0^\infty(\Omega)$$

Thus, by Eq. (4-120)

$$\int e^{-2(\mathrm{Im}\,\zeta,x)} |P^{(k)}(D)\varphi|^2 \, dx \leq mM \int e^{-2(\mathrm{Im}\,\zeta,x)} |P(D)\varphi|^2 \, dx$$

Since ζ was any complex vector, we can take $\mathrm{Im}\,\zeta = -\xi/2$. This completes the proof.

By repeated applications of Theorem 4-22, we have

Corollary 4-23

$$\int e^{(\xi,x)} |\varphi|^2 \, dx \leq m!\, M^m \int e^{(\xi,x)} |P(D)\varphi|^2 \, dx \tag{4-121}$$

for all $\varphi \in C_0^\infty(\Omega)$ and all real vectors ξ.

We also have

Corollary 4-24 The inequality

$$\int e^{\lambda x_k} |v|^2 \, dx \leq m!\, M^m \int e^{\lambda x_k} |P(D)v|^2 \, dx \tag{4-122}$$

holds for all real λ and all $v \in C^\infty(\overline{\Omega})$ which vanish for $|x|$ large, and such that $D^\mu v = 0$ on $\partial\Omega$ for $|\mu| < m$.

Proof Let Ω_1 be the slab $|x_k - a_k| < (M + \delta)/2$, where $\delta > 0$. Define v to be zero outside Ω. Then $J_\varepsilon v \in C_0^\infty(\Omega_1)$ for $\varepsilon < \delta/2$. Moreover

$$\begin{aligned} P(D)J_\varepsilon v &= \int v(y)P(D_x)j_\varepsilon(y - x) \, dy \\ &= \int v(y)P(-D_y)j_\varepsilon(y - x) \, dy \\ &= \int P(D_y)v(y)j_\varepsilon(y - x) \, dy \\ &= J_\varepsilon P(D)v \end{aligned} \tag{4-123}$$

(The integration by parts produced no boundary terms because all derivatives of v up to order $m - 1$ vanish on $\partial\Omega$.) Applying Corollary 4-23 to $J_\varepsilon v$, we get, by Eq. (4-123)

$$\int e^{\lambda x_k} |J_\varepsilon v|^2 \, dx \leq m!\,(M + \delta)^m \int e^{\lambda x_k} |J_\varepsilon P(D)v|^2 \, dx$$

Now the function $e^{\lambda x_k}$ is bounded in Ω_1. Hence

$$\int e^{\lambda x_k} |J_\varepsilon v - v|^2 \, dx \le K \| J_\varepsilon v - v \|^2$$

$$\to 0 \quad \text{as} \quad \varepsilon \to 0$$

A similar statement holds for $J_\varepsilon P(D)v$. Hence

$$\int e^{\lambda x_k} |v|^2 \, dx \le m!\,(M+\delta)^m \int e^{\lambda x_k} |P(D)v|^2 \, dx$$

Since this is true for any $\delta > 0$, we finally obtain inequality (4-122). This completes the proof.

And, lastly, we have

Corollary 4-25 Inequality (4-122) holds for all $v \in C^\infty(\bar{\Omega})$, such that $P^{(\mu)}(D)v \in L^2(\Omega)$ for all μ, and $D^\mu v = 0$ on $\partial\Omega$ for $|\mu| < m$.

PROOF Let $\psi(x)$ be a function in $C_0^\infty(E^n)$, such that

$$\begin{aligned} \psi(x) &= 1 \quad \text{for} \quad |x| \le 1 \\ &= 0 \quad \text{for} \quad |x| \ge 2 \\ 0 \le \psi(x) &\le 1 \quad \text{for} \quad x \in E^n \end{aligned}$$

Set

$$\psi_R(x) = \psi\left(\frac{x}{R}\right)$$

Then there are constants C_μ, such that

$$|D^\mu \psi_R| \le \frac{C_\mu}{R^{|\mu|}}$$

Now for each R, the function $v\psi_R$ satisfies the hypotheses of Corollary 4-24.

Moreover

$$P(D)(v\psi_R) = \sum \frac{P^{(\mu)}(D)vD^\mu\psi_R}{\mu!}$$

Hence

$$\int e^{\lambda x_k} |v\psi_R|^2 \, dx \le m!\,M^m \left[\sum \frac{C_\mu \left(\int e^{\lambda x_k} |P^{(\mu)}(D)v|^2 \, dx \right)^{1/2}}{\mu!\,R^{|\mu|}} \right]^2$$

Letting $R \to \infty$ we get (4-122). The proof is complete.

It now only remains to note that Lemma 4-21 is a special case of Corollary 4-25.

PROBLEMS

4-1 If $u \in C^m(\bar{\Omega})$ and $D_t^k u = 0$ for $t = 0$ and $0 \leq k < m$, show that the function which equals u in Ω and vanishes in $\Omega_0 - \Omega$ is in $H^m(\Omega_0)$ when Ω_0 is bounded.

4-2 Prove that $N_{m-k}u$ is of the form claimed in Sec. 4-2.

4-3 In the proof of Theorem 4-5, what modification must be made when Ω_0 does not contain a slab of the form $|t| \leq 2b$.

4-4 Give the details of the proof of inequality (4-34).

4-5 Show that the only polynomial of degree ≤ 1 which is bounded above by a constant is itself a constant.

4-6 Prove inequality (4-41).

4-7 Prove inequality (4-48).

4-8 If Ω is the slab $0 < t < b$ and $v \in S(\Omega)$, show that $fv \in S(\Omega)$.

4-9 Prove Eq. (4-98).

4-10 If $P(\xi, \tau)$ is a homogeneous polynomial and $\tau_1(\xi), \ldots, \tau_m(\xi)$ are its roots, show that

$$\tau_k(\lambda\xi) = \lambda\tau_k(\xi) \qquad \lambda > 0 \qquad \xi \in E^n \qquad 1 \leq k \leq m$$

4-11 Fill in the details in the proof of Corollary 4-25.

CHAPTER

FIVE

PROPERTIES OF SOLUTIONS

5-1 EXISTENCE OF STRONG SOLUTIONS

If we make use of the Holmgren uniqueness theorem (see Theorem 4-20 of Sec. 4-7) we can improve Theorem 4-16 of Sec. 4-6. Let Ω be the slab $0 < t < b$, and suppose we wish to solve

$$P(D)u = f \quad \text{in } \Omega \tag{5-1}$$

$$D_t^k u(x,0) = 0 \qquad 0 \le k < m \tag{5-2}$$

where $P(D)$ is a hyperbolic operator of order m. If f is any given function in $L^2(\Omega)$, we know that there is a sequence $\{\varphi_j\}$ of functions in $C_0^\infty(\Omega)$, such that $\varphi_j \to f$ in $L^2(\Omega)$ (Lemma 4-17 of Sec. 4-6). Moreover, by Theorem 4-15 of Sec. 4-6 there is a $u_j \in C^\infty(\overline{\Omega})$ which is a solution of Eq. (5-2) and $P(D)u_j = \varphi_j$ with $D_x^\mu D_t^k u_j \in L^2(\Omega)$ for each j and k. Thus, $u_j - u_k$ is a solution of $P(D)u = \varphi_j - \varphi_k$ and Eq. (5-2) for each j, k. By Theorem 4-20 of Sec. 4-7, the solution is unique. Hence, we may apply Eq. (4-94) of Sec. 4-6 to conclude

$$\| u_j - u_k \| \le C \| \varphi_j - \varphi_k \| \to 0 \quad \text{as} \quad j, k \to \infty$$

Hence, there is a function $u \in L^2(\Omega)$, such that

$$u_k \to u \qquad P(D)u_k \to f \qquad \text{in } L^2(\Omega) \tag{5-3}$$

We call u a *strong solution* of Eqs. (5-1), (5-2). We have proved

Theorem 5-1 If $P(D)$ is hyperbolic, then Eqs. (5-1), (5-2) have a strong solution for each $f \in L^2(\Omega)$.

Clearly, every strong solution is a weak solution as well. Since the requirements on a strong solution are more than those on a weak solution, one would expect that a weak solution need not necessarily be a strong solution. Indeed, this is true in general. However, we shall show that, in our particular case, every weak solution is a strong solution as well. We shall do this by proving

Theorem 5-2 If $P(D)$ is hyperbolic, then a weak solution of Eqs. (5-1), (5-2) is unique.

Corollary 5-3 If $P(D)$ is hyperbolic, then every weak solution is a strong solution.

The corollary follows easily from the theorems. In fact, let u_1 be a weak solution of Eqs. (5-1), (5-2). By Theorem 5-1, there is a strong solution u_2, which is also a weak solution. Since there is only one weak solution (Theorem 5-2), we have $u_1 = u_2$. Hence, u_1 is a strong solution.

It remains, therefore, only to prove Theorem 5-2. Let Ω_0 be the strip $-1 < t < b$. We must show that if $u \in L^2(\Omega)$, and

$$(u, \bar{P}(D)\varphi) = 0 \qquad \varphi \in C_0^\infty(\Omega_0) \tag{5-4}$$

then $u \equiv 0$ in Ω. Since $\bar{P}(D)$ is hyperbolic if and only if $P(D)$ is, we can replace Eq. (5-4) by

$$(u, P(D)\varphi) = 0 \qquad \varphi \in C_0^\infty(\Omega_0) \tag{5-5}$$

We are going to use

Theorem 5-4 Suppose $P(D)$ is hyperbolic, and let g be any function in $C_0^\infty(\Omega)$. Then there are functions v, w in $S(\Omega_0)$, such that v vanishes near $t = 0$, w vanishes near $t = b$, and

$$P(D)v = P(D)w = g \tag{5-6}$$

Let us postpone the proof of Theorem 5-4 for the moment and finish the

PROOF of Theorem 5-2 Let g be any function in $C_0^\infty(\Omega)$, and let w be the function given by Theorem 5-4. Let $\psi(x)$ be a function in $C_0^\infty(E^n)$, such that $0 \le \psi \le 1$, $\psi(x) = 1$ for $|x| < 1$ and $\psi(x) = 0$ for $|x| > 2$ (see Sec. 1-3). Let $\rho(t)$ be a function in $C_0^\infty(E^1)$, such that $\rho(t) = 1$ for $0 < t < b$ and $\rho(t) = 0$ for $t < -\frac{1}{2}$. Set

$$\varphi_R(x, t) = w(x, t)\rho(t)\psi(x/R) \qquad R > 0 \tag{5-7}$$

Clearly $\varphi_R \in C_0^\infty(\Omega_0)$ for each $R > 0$. Moreover, since $\rho(t) = 1$ in Ω, we have, by Eq. (1-102) of Sec. 1-7

$$P(D)\varphi_R = \sum \frac{P^{(\mu,0)}(D)wD_x^\mu\psi(x/R)}{\mu!}$$

$$= \sum \frac{P^{(\mu,0)}(D)w\psi_\mu(x/R)}{\mu!\,R^{|\mu|}} \tag{5-8}$$

in Ω, where $\psi_\mu(x) = D_x^\mu\psi(x)$. Thus

$$\| P(D)(\varphi_R - w) \| \leq \| [1 - \psi(x/R)]P(D)w \| + M \sum_{|\mu|\neq 0} \frac{\| P^{(\mu,0)}(D)w \|}{\mu!\,R^{|\mu|}}$$

where
$$M = \max_{|\mu|\leq m} \max_x |\psi_\mu(x)|$$

Let $\varepsilon > 0$ be given, and take R so large that

$$\int_0^b \int_{|x|>R} |P(D)w|^2\,dx\,dt < \frac{\varepsilon^2}{4}$$

and
$$M \sum_{|\mu|\neq 0} \| P^{(\mu,0)}(D)w \| < \frac{\varepsilon R}{2}$$

Since $\psi(x/R) = 1$ for $|x| < R$, this gives

$$\| P(D)(\varphi_R - w) \| < \varepsilon$$

This shows that $P(D)\varphi_R \to P(D)w = g$, as $R \to \infty$. Now suppose that $u \in L^2(\Omega)$ satisfies Eq. (5-5). In particular, we have

$$(u, P(D)\varphi_R) = 0$$

for each R. Taking the limit as $R \to \infty$, we see that $(u, g) = 0$. Since g was an arbitrary function in $C_0^\infty(\Omega)$, we have

$$(u, \varphi) = 0 \qquad \varphi \in C_0^\infty(\Omega) \tag{5-9}$$

But $C_0^\infty(\Omega)$ is dense in $L^2(\Omega)$ (Lemma 4-17 of Sec. 4-6). Hence, there is a sequence $\{\varphi_k\} \in C_0^\infty$ converging to u. Thus, (5-9) implies $\| u \| = 0$ and consequently that $u = 0$. This completes the proof of Theorem 5-2.

We now turn to the

PROOF of Theorem 5-4 Let $\delta > 0$ be such that $g = 0$ for $0 < t < \delta$, and let Ω_δ denote the slab $\delta < t < b$. By Theorem 4-15 of Sec. 4-6, there is a function $v \in S(\Omega_\delta)$, such that

$$P(D)v = g \quad \text{in } \Omega_\delta \tag{5-10}$$

$$D_t^k v(x, \delta) = 0 \qquad 0 \leq k < m \tag{5-11}$$

Since the coefficient of D_t^m in $P(D)$ does not vanish, and since $g(x, \delta) = 0$, we see that $D_t^m v(x,\delta) = 0$. Now, both v and g are in $C^\infty(\bar{\Omega}_\delta)$. Thus, we can differentiate both sides of Eq. (5-10) with respect to t. Since $D_t^k g(x,\delta) = 0$ for each k, we see that $D_t^k v(x,\delta) = 0$ for each k. If we define $v(x,t)$ to vanish for $0 < t < \delta$, we see that $v \in S(\Omega)$.

To construct w, note that the operator $P(D_x, -D_t)$ is hyperbolic when $P(D_x, D_t)$ is. If we let Ω_- denote the slab $-b < t < 0$, then, by what we have just shown, there is a function $h \in S(\Omega_-)$, vanishing near $t = -b$, and such that

$$P(D_x, -D_t)h(x,t) = g(x, -t) \qquad -b < t < 0$$

If we now set $w(x,t) = h(x, -t)$, we see that w has the desired properties. This completes the proof.

5-2 PROPERTIES OF STRONG SOLUTIONS

In this section we discuss strong solutions of Eqs. (5-1), (5-2), and give some of their properties. It will be convenient to introduce a family of scalar products and norms. As in Sec. 5-1, we let Ω denote the slab $0 < t < b$. For r a nonnegative integer and s real, we set

$$(u, v)_{r,s} = \sum_{k=0}^{r} \int_0^b (D_t^k u, D_t^k v)_{r-k+s}\, dt \tag{5-12}$$

where $(\ ,\ )_s$ denotes the scalar product given by Eq. (2-25) of Sec. 2-3 with respect to the x variables. The corresponding norm is given by

$$|u|_{r,s}^2 = \sum_{k=0}^{r} \int_0^b |D_t^k u|_{r-k+s}^2\, dt \tag{5-13}$$

We let $H^{r,s} = H^{r,s}(\Omega)$ denote the completion of $S(\Omega)$ with respect to the norm (5-13). Clearly, $H^{r,s}$ is a Hilbert space for each r, s and $H^{0,0} = L^2(\Omega)$. We put $H^r = H^{r,0}$.

If $u \in L^2(\Omega)$ is a strong solution of Eqs. (5-1), (5-2), then there is a sequence $\{u_j\}$ of functions in $C^\infty(\bar{\Omega}) \cap H^m$, satisfying

$$D_t^k u_j(x, 0) = 0 \qquad 0 \le k < m \tag{5-14}$$

and, such that

$$u_j \to u \qquad P(D)u_j \to f \quad \text{in } L^2(\Omega) \tag{5-15}$$

We can say more about the solution. For example, we have

Theorem 5-5 Assume that $P(D)$ is hyperbolic. If $\{u_j\}$ is a sequence of functions in $C^\infty(\bar{\Omega}) \cap H^{m,s}$ satisfying Eq. (5-14), and such that $P(D)u_j$ converges in $H^{0,s}$, then $P^{(0,k)}(D)u_j$ converges in $H^{0,s}$ for each k.

For the notation, see Sec. 4-3. The proof of Theorem 5-5 depends on the inequality

$$\sum |P^{(0,k)}(D)u|_{0,s} \leq C|P(D)u|_{0,s} \tag{5-16}$$

holding for all $u \in C^m(\bar{\Omega}) \cap H^{m,s}$ satisfying Eq. (5-2). We shall prove this inequality in Sec. 5-4. A slight variation can be described as follows.

Let $\tau_1(\xi), \ldots, \tau_m(\xi)$ be the roots of $P(\xi, \tau) = 0$.

If $J = (j_1, \ldots, j_k)$ is any subset of the integers $(1, 2, \ldots, m)$, we form the polynomial in τ

$$P_J(\xi, \tau) = (\tau - \tau_{j_1}) \cdots (\tau - \tau_{j_m}) \tag{5-17}$$

Note that it need not be a polynomial in ξ. However, for $u(x,t) \in C^\infty(\Omega)$ we can define $P_j(D)u$, by setting

$$P_J(D)u = F^{-1}P_j(\xi, D_t)Fu \tag{5-18}$$

where F is the Fourier transform with respect to the variables x. If $P_j(\xi, \tau)$ happens to be a polynomial, this agrees with the usual definition by Eqs. (2-22) and (2-23) of Sec. 2-3. With this notion, we have

Theorem 5-6 Under the hypotheses of Theorem 5-5, for each subset J of $(1, \ldots, m)$ the functions $P_J(D)u_j$ converge in $H^{0,s}$.

This theorem follows from the inequality

$$\sum_J |P_J(D)u|_{0,s} \leq C|P(D)u|_{0,s} \tag{5-19}$$

holding for functions $u \in C^\infty(\bar{\Omega}) \cap H^{m,s}$ and satisfying Eq. (5-2). This inequality will also be proved in Sec. 5-4.

Let A be an operator defined on $C^\infty(\Omega)$, and let u be a function in $H^{r,s}$. We shall say that A *applies strongly* to u if there is an $f \in H^{r,s}$, such that

1. If $\{v_k\}$ is a sequence of functions in $C^\infty(\bar{\Omega}) \cap H^{r,s}$, such that $v_k \to u$, $Av_k \to g$ in $H^{0,s}$, then $g = f$.
2. There exists such a sequence $\{v_k\}$ for which $v_k \to u$, $Av_k \to f$ in $H^{r,s}$.

Using this definition we have

Corollary 5-7 If $P(D)$ is hyperbolic and $u \in H^{0,s}$ is a strong solution of Eqs. (5-1), (5-2), then $P^{(0,k)}(D)$ and $P_J(D)$ apply strongly to u for each k and J.

PROOF The fact that requirement 2 is satisfied follows from Theorems 5-5 and 5-6. Requirement 1 follows from the fact that u is also a weak solution. In fact, let $Q(D)$ represent any of the operators mentioned, and suppose

$\{v_k\} \subset C^\infty(\bar{\Omega}) \cap H^{0,s}$ is such that $v_k \to u$ and $Q(D)v_k \to g$ in $H^{0,s}$. If ψ is any function in $C^\infty(\bar{\Omega}) \cap H^{0,-s}$ vanishing near $t = 0, b$, then

$$(v_k, \bar{Q}(D)\psi) = (Q(D)v_k, \psi) \tag{5-20}$$

This follows from the fact that ψ and each v_k are the limits in $H^{0,s}$ of functions which are in $S(\Omega)$. Taking the limit as $k \to \infty$, we get

$$(u, \bar{Q}(D)\psi) = (g, \psi)$$

for all such ψ. Now, if there is another sequence $\{u_k\} \subset C^\infty(\bar{\Omega}) \cap H^{0,s}$, such that $u_k \to u$, $Q(D)u_k \to f$ in $H^{0,s}$, we have

$$(f - g, \psi) = 0 \tag{5-21}$$

for all such ψ. Let w be any function in $C^\infty(\bar{\Omega}) \cap H^{0,s}$, and set

$$w_0 = F^{-1}(1 + |\xi|)^{-s}Fw$$

where F is the Fourier transform with respect to the x variables. Then $w_0 \in L^2(\Omega)$. Consequently, there is a sequence $\{\varphi_k\}$ of functions in $C_0^\infty(\Omega)$ converging to w_0 in $L^2(\Omega)$ (Lemma 4-17 of Sec. 4-6). Then $\varphi_{ok} = F^{-1}(1 + |\xi|)^s F\varphi_k$ is in $C^\infty(\bar{\Omega}) \cap H^{0,-s}$ and vanishes near $t = 0, b$. Thus

$$(f - g, \varphi_{ok}) = 0 \qquad k = 1, 2, \ldots$$

But $$|\varphi_{ok} - w|_{0,-s} = |\varphi_k - w_0|_{0,0} \to 0 \quad \text{as} \quad k \to \infty$$

This implies

$$(f - g, w) = 0 \tag{5-22}$$

Thus, Eq. (5-22) holds for all $w \in C^\infty(\bar{\Omega}) \cap H^{0,-s}$. Next, let h be any function in $C^\infty(\bar{\Omega}) \cap H^{0,s}$, and set

$$w = F^{-1}(1 + |\xi|)^{2s}Fh$$

Then, $w \in C^\infty(\bar{\Omega}) \cap H^{0,-s}$. Moreover

$$(f - g, h)_{0,s} = (f - g, w)_{0,0} = 0$$

by Eq. (5-22). Thus

$$(f - g, h)_{0,s} = 0 \tag{5-23}$$

for all $h \in C^\infty(\bar{\Omega}) \cap H^{0,s}$. But there exists a sequence $\{h_k\}$ of such functions converging to $f - g$ in $H^{0,s}$. Hence $g = f$, and the proof is complete.

5-3 ESTIMATES IN ONE DIMENSION

In this section we shall derive some very useful inequalities which will be employed in the proofs of Theorems 5-5 and 5-6, and used for other purposes later on. To begin we consider functions of a single variable t.

Lemma 5-8 If $u(t)$ has a continuous derivative in $[a, b]$ and τ is a complex number, then

$$|u(b)|^2 = |u(a)|^2 - 2 \operatorname{Im} \tau \int_a^b |u|^2 \, dt + 2 \operatorname{Re} i \int_a^b \bar{u}(D_t - \tau)u \, dt \tag{5-24}$$

PROOF We have

$$\operatorname{Re} i \int_a^b \bar{u}(D_t - \tau)u \, dt = \operatorname{Re} \int_a^b \bar{u} \frac{du}{dt} \, dt - (\operatorname{Re} i\tau) \int_a^b |u|^2 \, dt$$

$$= \tfrac{1}{2}[|u(b)|^2 - |u(a)|^2] + \operatorname{Im} \tau \int_a^b |u|^2 \, dt$$

This gives Eq. (5-24).

Lemma 5-9 Suppose $r(t)$ and $\rho(t)$ are integrable functions in $[a, b]$, and

$$r(t) \le c \int_a^t r(s) \, ds + \rho(t) \qquad a \le t \le b \tag{5-25}$$

where $c \ge 0$. Then

$$r(t) \le c \int_a^t e^{c(t-s)} \rho(s) \, ds + \rho(t) \qquad a \le t \le b \tag{5-26}$$

If $\rho(t)$ is nondecreasing, then $r(t) \le e^{c(t-a)}\rho(t)$.

PROOF Set

$$v(t) = e^{-ct} \int_a^t r(s) \, ds$$

Then
$$v'(t) = -cv(t) + e^{-ct} r(t)$$

Thus, inequality (5-25) says that

$$v'(t) \le e^{-ct}\rho(t)$$

Integrating from a to t gives

$$v(t) \le \int_a^t e^{-cs} \rho(s) \, ds$$

Substituting into (5-25) we obtain (5-26). The second statement follows immediately.

Corollary 5-10 If

$$\rho(t) = \alpha + \int_a^t \sigma(s) \, ds$$

in (5-25), then

$$r(t) \le \alpha\, e^{c(t-a)} + \int_a^t e^{c(t-s)}\,\sigma(s)\,ds \tag{5-27}$$

PROOF We have

$$c\,e^{ct}\int_a^t\int_a^s e^{-cs}\sigma(\lambda)\,d\lambda\,ds = c\,e^{ct}\int_a^t \sigma(\lambda)\int_\lambda^t e^{-cs}\,ds\,d\lambda$$
$$= e^{ct}\int_a^t (e^{-c\lambda} - e^{-ct})\sigma(\lambda)\,d\lambda$$

Theorem 5-11 If $u(t)$ has a continuous derivative in $[a, b]$ and $\operatorname{Im}\tau < 0$, then

$$|u(b)|^2 \le e^{2|\operatorname{Im}\tau|(b-a)}|u(a)|^2 + 2\operatorname{Re} i\int_a^b e^{2|\operatorname{Im}\tau|(b-t)}\,\bar{u}(D_t - \tau)u\,dt \tag{5-28}$$

If $\operatorname{Im}\tau \ge 0$, we have

$$|u(b)|^2 \le |u(a)|^2 + 2\operatorname{Re} i\int_a^b \bar{u}(D_t - \tau)u\,dt \tag{5-29}$$

PROOF Inequality (5-28) follows from (5-27), while (5-29) follows from Eq. (5-24).

Corollary 5-12 If $\operatorname{Im}\tau < 0$, then

$$|u(b)|^2 \le e^{(2|\operatorname{Im}\tau|+1)(b-a)}|u(a)|^2 + \int_a^b e^{(2|\operatorname{Im}\tau|+1)(b-t)}|(D_t - \tau)u|^2\,dt \tag{5-30}$$

If $\operatorname{Im}\tau \ge 0$, then

$$|u(b)|^2 \le e^{b-a}|u(a)|^2 + \int_a^b e^{b-t}|(D_t - \tau)u|^2\,dt \tag{5-31}$$

PROOF By Eq. (5-24) we have

$$|u(t)|^2 \le |u(a)|^2 + c\int_a^t |u|^2\,dt + \int_a^t |(D_t - \tau)u|^2\,dt$$

where $c = 2|\operatorname{Im}\tau| + 1$, or $c = 1$, depending on whether $\operatorname{Im}\tau < 0$ or $\operatorname{Im}\tau \ge 0$.

Next, let $P(\tau)$ be a polynomial of degree m. If $\tau_1, \dots, \tau_m$ are its complex roots,

$$P(\tau) = a_0 \prod_{k=1}^m (\tau - \tau_k) \tag{5-32}$$

where the constant $a_0 \neq 0$. Set

$$P_k(\tau) = \frac{P(\tau)}{(\tau - \tau_k)} \qquad 1 \le k \le m \tag{5-33}$$

$$P'(\tau) = \frac{dP(\tau)}{d\tau} \tag{5-34}$$

We have

Theorem 5-13 If $u(t)$ has continuous derivatives up to order m in $[a, b]$, then

$$|P_k(D_t)u(b)|^2 = 2 \operatorname{Re} i \int_a^b P(D_t)u\overline{P_k(D_t)u}\, dt - 2 \operatorname{Im} \tau_k \int_a^b |P_k(D_t)u|^2\, dt + |P_k(D_t)u(a)|^2 \qquad 1 \le k \le m \tag{5-35}$$

and

$$\sum_1^m |P_k(D_t)u(b)|^2 = 2 \operatorname{Re} i \int_a^b P(D_t)u\overline{P'(D_t)u}\, dt - 2\sum_1^m \operatorname{Im} \tau_k \int_a^b |P_k(D)u|^2\, dt + \sum_1^m |P_k(D_t)u(a)|^2 \tag{5-36}$$

PROOF These identities follow immediately from Eq. (5-24). Note that $P'(\tau) = \Sigma\, P_k(\tau)$.

Theorem 5-14 If $u(t)$ has continuous derivatives up to order m in $[a, b]$, then

$$|P_k(D_t)u(b)|^2 \le e^{2|\operatorname{Im} \tau_k|(b-a)} |P_k(D_t)u(a)|^2 + 2 \operatorname{Re} i \int_a^b e^{2|\operatorname{Im} \tau_k|(b-t)} P(D_t)u\overline{P_k(D_t)u}\, dt \qquad 1 \le k \le m \tag{5-37}$$

If

$$-\operatorname{Im} \tau_k \le K > 0 \qquad 1 \le k \le m \tag{5-38}$$

then

$$|P_k(D_t)u(b)|^2 \le \int_a^b e^{(2K+1)(b-t)} |P(D_t)u|^2\, dt + e^{(2K+1)(b-a)} |P_k(D_t)u(a)|^2 \qquad 1 \le k \le m \tag{5-39}$$

PROOF Apply Theorem 5-11 and Corollary 5-12.

Corollary 5-15 Under the same hypotheses

$$|P'(D_t)u(b)|^2 \le m \int_a^b e^{(2K+1)(b-t)} |P(D_t)u|^2 \, dt \tag{5-40}$$

Next, suppose the roots $\tau_1, \ldots, \tau_m$ are continuous functions $\tau_k(\xi)$ on E^n, which are distinct and homogeneous, of degree 1. By this we mean

$$\tau_j(\xi) \neq \tau_k(\xi) \qquad 1 \le j < k \le m \qquad \xi \in E^n \tag{5-41}$$

and

$$\tau_k(\lambda\xi) = \lambda\tau_k(\xi) \qquad \lambda > 0 \qquad \xi \in E^n \tag{5-42}$$

Theorem 5-16 Under the above hypotheses

$$\sum_0^{m-1} |\xi|^{2(m-k-1)} |D_t^k u(t)|^2 \le C \sum_0^{m-1} |\xi|^{2(m-k-1)} |D_t^k u(a)|^2 + C \operatorname{Re} i \sum_1^m \int_a^t e^{-2|\operatorname{Im} \tau_k|\lambda} P(D_t)u(\lambda) P_k(D_t)u(\lambda) \, d\lambda \qquad a \le t \le b \tag{5-43}$$

and

$$\sum_0^{m-1} |\xi|^{2(m-k-1)} |D_t^k u(t)|^2 \le \sum_0^{m-1} |\xi|^{2(m-k-1)} |D_t^k u(a)|^2 + C \int_a^t |P(D_t)u(\lambda)|^2 \, d\lambda \qquad a \le t \le b \tag{5-44}$$

for all $u \in C^m[a, b]$.

In proving this theorem we shall make use of

Lemma 5-17 Suppose the numbers $\tau_1, \ldots, \tau_m$ are distinct, and that the polynomials $P_k(\tau)$ given by Eq. (5-33) are of the form

$$P_j(\tau) = \sum_0^{m-1} a_{jk}\tau^k \qquad 1 \le j \le m \tag{5-45}$$

Suppose $w_0, \ldots, w_{m-1}$ are complex numbers, such that

$$\sum_0^m a_{jk} w_k = 0 \qquad 1 \le j \le m \tag{5-46}$$

Then $w_0 = \cdots = w_{m-1} = 0$.

Proof Let $b_0, \ldots, b_{m-1}$ be any numbers, and set

$$R(\tau) = \sum_0^{m-1} b_k \tau^k \tag{5-47}$$

By partial fractions

$$\frac{R(\tau)}{P(\tau)} = \sum_{1}^{m} \frac{c_k}{\tau - \tau_k}$$

where $c_k = R(\tau_k)/P_k(\tau_k)$. Thus

$$R(\tau) = \sum c_j P_j(\tau) \tag{5-48}$$

From Eqs. (5-45), (5-47), and (5-48), we see that

$$b_k = \sum_{0}^{m-1} c_j a_{jk} \qquad 0 \le k < m$$

Hence

$$\sum b_k w_k = \sum c_j a_{jk} w_k = 0$$

by Eq. (5-46). Since the b_k were arbitrary, we see that all the w_k vanish.

PROOF of Theorem 5-16 Set

$$w_k = \frac{D_t^k u(t)}{|\xi|^k} \qquad 0 \le k < m \tag{5-49}$$

and consider the expression

$$G(\xi, w) = \frac{|\xi|^{2m-2} \sum |w_k|^2}{\sum_j \left| \sum_k a_{jk}(\xi) |\xi|^k w_k \right|^2} \tag{5-50}$$

where the $a_{jk}(\xi)$ are the coefficients of $P_j(\tau)$. By Eq. (5-42), $a_{jk}(\xi)$ is homogeneous of degree $m - k - 1$. By Lemma 5-17, the denominator in $G(\xi, w)$ does not vanish unless the w_k all vanish. Thus, $G(\xi, w)$ is a continuous function on the set $|\xi| = |w_0|^2 + \cdots + |w_{m-1}|^2 = 1$ (note that the components of ξ are real and the w_k are complex). Since this set is closed and bounded, it is compact. Since $G(\xi, w)$ is continuous on this compact set, it is bounded there. Thus

$$G(\xi, w) \le M \tag{5-51}$$

on this set. Next, notice that both the numerator and denominator of Eq. (5-50) are homogeneous in ξ of degree $2m - 2$, and in the w_k of degree 2. It follows, therefore, that (5-51) holds for all ξ and w as long as $\xi \ne 0$ and not all of the w_k vanish. In view of Eq. (5-49), this gives

$$\sum_k |\xi|^{2(m-k-1)} |D_t u(t)|^2 \le M \sum_j |P_j(D_t) u(t)|^2 \tag{5-52}$$

Since $a_{jk}(\xi)$ is homogeneous of degree $m - k - 1$, the opposite inequality is immediate. If we now apply them to inequalities (5-37) and (5-39), we obtain (5-43) and (5-44), respectively. This completes the proof.

Corollary 5-18 Under the hypotheses of Theorem 5-16

$$\sum_0^{m-1} (1+|\xi|)^{2(m-k-1)}|D_t^k u(t)|^2 \le C \sum_0^{m-1} |\xi|^{2(m-k-1)}|D_t^k u(a)|^2 + C \operatorname{Re} i \sum_1^m \int_a^t e^{N(t-\lambda)-2|\operatorname{Im}\tau_k|\lambda} \times P(D_t)u(\lambda)\overline{P_k(D_t)u(\lambda)}\, d\lambda \qquad a \le t \le b \tag{5-53}$$

and

$$\sum_0^{m-1} (1+|\xi|)^{2(m-k-1)}|D_t^k u(t)|^2 \le C \sum_0^{m-1} (1+|\xi|)^{2(m-k-1)}|D_t^k u(a)|^2 + C\int_a^t |P(D_t)u(\lambda)|^2\, d\lambda \qquad a < t \le b \tag{5-54}$$

for all $u \in C^m[a,b]$.

Proof We apply inequality (2-45) of Sec. 2-4 to inequality (5-43). This gives

$$\sum_0^{m-1} (1+|\xi|)^{2(m-k-1)}|D_t^k u(t)|^2 \le C \sum_0^{m-1} |\xi|^{2(m-k-1)}|D_t^k u(a)|^2 + C' \sum_0^{m-2} |D_t^k u(t)|^2 + C \operatorname{Re} i \int_a^t \sum_1^m e^{-2|\operatorname{Im}\tau_k|\lambda} P(D_t)u(\lambda)\overline{P_k(D_t)u(\lambda)}\, d\lambda \tag{5-55}$$

Now
$$|D_t^k u(t)| \le \int_a^t |D_t^{k+1}u(\lambda)|\, d\lambda$$

Hence
$$\sum_0^{m-2} |D_t^k u(t)|^2 \le (b-a)\int_a^t \sum_1^{m-1} |D_t^k u(\lambda)|^2\, d\lambda$$

Thus, if we let $r(t)$ denote the left-hand side of (5-53), we have

$$r(t) \le N\int_a^t r(\lambda)\, d\lambda + C\sum_0^{m-1} |\xi|^{2(m-k-1)}|D_t^k u(a)|^2 + C \operatorname{Re} i \int_a^t \sum_1^m e^{-2|\operatorname{Im}\tau_k|\lambda} P(D_t)u(\lambda)P_k(D_t)u(\lambda)\, d\lambda$$

If we now apply Corollary 5-10, we get

$$r(t) \le C\, e^{N(t-a)} \sum_0^{m-1} |\xi|^{2(m-k-1)}|D_t^k u(a)|^2 + C \operatorname{Re} i \int_a^t \sum_1^m e^{N(t-\lambda)-2|\operatorname{Im}\tau_k|\lambda} P(D_t)u(\lambda)\overline{P_k(D_t)u(\lambda)}\, d\lambda$$

This gives (5-53). To prove (5-54), note that

$$\left| \int_a^t e^{-Nt-2|\operatorname{Im}\tau_j|\lambda} P(D_t)u(\lambda)P_j(D_t)u(\lambda)\, d\lambda \right|$$

$$\leq \int_a^t |P(D_t)u(\lambda)|^2\, d\lambda + K \sum_0^{m-1} |\xi|^{2(m-k-1)} \int_a^t |D_t^k u(\lambda)|^2\, d\lambda \qquad (5\text{-}56)$$

Thus
$$r(t) \leq CK \int_a^t r(\lambda)\, d\lambda + C \int_a^t |P(D_t)u(\lambda)|^2\, d\lambda$$
$$+ C \sum_0^{m-1} |\xi|^{2(m-k-1)} |D_t^k u(a)|^2$$

Another application of Corollary 5-10 gives inequality (5-54). This completes the proof.

5-4 ESTIMATES IN $n + 1$ DIMENSIONS

We now return to the situation in Sec. 5-2. In particular, Ω is the slab $0 < t < b$ in E^{n+1}. As a modification of Eq. (5-12), we write

$$(u, v)_{r,s}^{(a)} = \sum_{k=0}^{r} \int (1 + |\xi|)^{2(r-k+s)} D_t^k Fu(\xi, a) D_t^k Fv(\xi, a)\, d\xi \qquad 0 \leq a \leq b \qquad (5\text{-}57)$$

and
$$(u, v)_{r,s}^{(a,c)} = \int_a^c (u, v)_{r,s}^{(t)}\, dt \qquad 0 \leq a < c \leq b \qquad (5\text{-}58)$$

The corresponding norms are denoted by $|u|_{r,s}^{(a)}$ and $|u|_{r,s}^{(a,c)}$, respectively. In the notation of Sec. 5-2, we have

$$(u, v)_{r,s} = (u, v)_{r,s}^{(0,b)} \qquad |u|_{r,s} = |u|_{r,s}^{(0,b)} \qquad (5\text{-}59)$$

Note that

$$(u, v)_{r-1,s+1}^{(a)} + (D_t^r u, D_t^r v)_{0,s}^{(a)} = (u, v)_{r,s}^{(a)} \qquad (5\text{-}60)$$

with a similar statement for the scalar product (5-58). In particular, we have

$$|u|_{r-1,s+1}^{(a)} \leq |u|_{r,s}^{(a)} \qquad (5\text{-}61)$$

which means that $H^{r,s} \subset H^{r-1,s+1}$. We begin with

Theorem 5-19 If $P(D)$ is hyperbolic, then

$$|P_k(D)u|_{0,s}^{(t)} \leq C |P_k(D)u|_{0,s}^{(0)} + C |P(D)u|_{0,s}^{(0,t)} \qquad 0 < t \leq b \qquad 1 \leq k \leq m \qquad (5\text{-}62)$$

holds for all $u \in C^m(\overline{\Omega})$, where the constant C depends only on b and the hyperbolicity constant K in (4-31) of Sec. 4-3.

PROOF By (5-39), we have

$$|P_k(\xi, D_t)Fu(\xi, t)|^2 \le C \int_0^t |P(\xi, D_t)Fu(\xi, \lambda)|^2 \, d\lambda + C|P_k(\xi, D_t)Fu(\xi, 0)|^2 \qquad (5\text{-}63)$$

If we multiply both sides by $(1 + |\xi|^2)^{2s}$ and integrate with respect to ξ, we obtain (5-62).

We can now give the proofs of Theorems 5-5 and 5-6. We first give the

PROOF of Theorem 5-6 Let J_k be the set $(1, \ldots, m)$ with k missing. Then P_{J_k} is just P_k given by Eq. (5-33). (Of course it now depends on ξ as well). By (5-62), we have

$$|P_k(D)u|_{0,s}^{(t)} \le C|P(D)u|_{0,s}^{(0,b)} \qquad 0 < t \le b \qquad (5\text{-}64)$$

for all $u \in C^m(\bar{\Omega}) \cap H^{m,s}$ satisfying Eq. (5-2). If we square both sides and integrate with respect to t, we obtain (5-19) for $J = J_k$. Next, we note that P_k has all of the properties of P (except that it may not be a polynomial in ξ). Moreover, every J containing $m - 2$ numbers is obtained from some J_k by removing a number. Hence, each P_J for such a J is obtained from some P_k in the same way that each P_k was obtained from P. Thus, (5-19) can be obtained by repeated applications of (5-64). This completes the proof.

PROOF of Theorem 5-5 We merely note that $P^{(0,k)}(D)$ is the sum of all $P_J(D)$, with J containing exactly k numbers. Thus, (5-16) follows from (5-19).

Recall that a homogeneous operator $p(D)$ is called totally hyperbolic if $p(\xi, \tau)$ has m simple real roots for each $\xi \ne 0$ (see Sec. 4-4). Using the results of Sec. 5-3, we have

Theorem 5-20 If $p(D)$ is a totally hyperbolic operator of order m, then there are constants C, N, such that

$$[|u|_{m-1,s}^{(t)}]^2 \le C \operatorname{Re} i(gp(D)u, p^{(0,1)}(D)u)_{0,s}^{(0,t)} + [|u|_{m-1,s}^{(0)}]^2 \qquad 0 < t \le b \qquad (5\text{-}65)$$

holds for all $u \in C^m(\bar{\Omega})$, where

$$g = e^{N(t-\lambda)} \qquad (5\text{-}66)$$

PROOF We employ inequality (5-53). Since the roots of $p(\xi, \tau)$ are distinct and real, Corollary 5-18 applies. (Note that the roots are continuous, by Lemma 4-14 of Sec. 4-4.) Since $p^{(0,1)} = \Sigma\, p_k$, we have

$$\sum_0^{m-1} (1 + |\xi|)^{2(m-k-1)} |D_t^k Fu(\xi, t)|^2 \le C \sum_0^{m-1} |\xi|^{2(m-k-1)} |D_t^k Fu(\xi, 0)|^2$$

$$+ C \operatorname{Re} i \int_0^t e^{N(t-\lambda)} p(\xi, D_t)Fu(\xi, \lambda)\overline{p^{(0,1)}(\xi, D_t)Fu(\xi, \lambda)}\, d\lambda \qquad (5\text{-}67)$$

If we multiply by $(1 + |\xi|)^{2s}$ and integrate with respect to ξ, we obtain (5-65).

Theorem 5-21 Let $p(D)$ be a totally hyperbolic operator of order m, and let $Q(D)$ be any operator of order $<m$. Set $P(D) = p(D) + Q(D)$. Then $P(D)$ is hyperbolic, and there are constants C, N, such that

$$[|u|_{m-1,s}^{(t)}]^2 \le C \operatorname{Re} i(gP(D)u, p^{(0,1)}(D)u)_{0,s}^{(0,t)} + [|u|_{m-1,s}^{(0)}]^2 \qquad 0 < t \le b \qquad (5\text{-}68)$$

for all $u \in C^m(\bar{\Omega})$, where g is of the form (5-66).

Proof That $P(D)$ is hyperbolic has already been proved (Theorem 4-10 of Sec. 4-4). To prove the inequality, we employ (5-65). Since both $Q(D)$ and $p^{(0,1)}(D)$ are of order $<m$, there is a constant M, such that

$$|(gQ(D)u, p^{(0,1)}(D)u)_{0,s}^{(0,t)}| \le M[|u|_{m-1,s}^{(0,t)}]^2$$

$$= M\int_0^t [|u|_{m-1,s}^{(0,\lambda)}]^2\, d\lambda \qquad (5\text{-}69)$$

Thus, if we set $r(t) = (|u|_{m-1,s}^{(t)})^2$, we have, by (5-65)

$$r(t) \le CM \int_0^t r(\lambda)\, d\lambda + \operatorname{Re} i(gP(D)u, p^{(0,1)}(D)u)_{0,s}^{(0,t)} + Cr(0) \qquad 0 < t \le b$$

If we now apply Corollary 5-10, we obtain (5-68).

Corollary 5-22 Under the same hypotheses

$$|u|_{m-1,s}^{(t)} \le C(|P(D)u|_{0,s}^{(0,t)} + |u|_{m-1,s}^{(0)}) \qquad 0 < t \le b \qquad u \in C^m(\bar{\Omega}) \qquad (5\text{-}70)$$

Proof Note that for any $\varepsilon > 0$ there is a constant C', such that

$$|(gP(D)u, p^{(0,1)}(D)u)_{0,s}^{(0,t)}| \le C'[|P(D)u|_{0,s}^{(0,t)}]^2 + \varepsilon[|u|_{m-1,s}^{(0,t)}]^2 \qquad (5\text{-}71)$$

If we combine this with (5-68), we obtain (5-70).

Corollary 5-23 Under the same hypotheses

$$|u|_{m,s-1}^{(0,t)} \le C(|P(D)u|_{0,s}^{(0,t)} + |u|_{m-1,s}^{(0)}) \qquad 0 < t \le b \qquad u \in C^m(\bar{\Omega}) \qquad (5\text{-}72)$$

Proof For $0 \le \lambda \le t$ we have, by (5-70)

$$|u|_{m-1,s}^{(\lambda)} \le C(|P(D)u|_{0,s}^{(0,\lambda)} + |u|_{m-1,s}^{(0)})$$

$$\le C(|P(D)u|_{0,s}^{(0,t)} + |u|_{m-1,s}^{(0)})$$

Integrating with respect to λ, we get

$$|u|_{m-1,s}^{(0,t)} \le C(|P(D)u|_{0,s}^{(0,t)} + |u|_{m-1,s}^{(0)}) \tag{5-73}$$

Since the coefficient of D_t^m in $P(D)$ does not vanish, we have also

$$|D_t^m u|_{0,s-1}^{(0,t)} \le C(|P(D)u|_{0,s-1}^{(0,t)} + |u|_{m-1,s}^{(0,t)}) \tag{5-74}$$

Since $$[|u|_{m,s-1}^{(0,t)}]^2 = [|u|_{m-1,s}^{(0,t)}]^2 + [|D_t^m u|_{0,s-1}^{(0,t)}]^2 \tag{5-75}$$

(cf. Eq. (5-60)), inequality (5-72) follows from (5-73) and (5-74).

Theorem 5-24 For each integer $k \ge 0$ there is a constant C independent of s, such that

$$|u|_{m+k,s-1}^{(0,t)} \le C(|P(D)u|_{k,s}^{(0,t)} + |u|_{m-1,s+k}^{(0)}) \qquad 0 < t \le b \qquad u \in C^{m+k}(\bar{\Omega}) \tag{5-76}$$

Proof We employ induction. Assume that (5-76) holds for k. Then (5-61) gives

$$\begin{aligned} |u|_{m+k,s}^{(0,t)} &\le C(|P(D)u|_{k,s+1}^{(0t)} + |u|_{m-1,s+k+1}^{(0)}) \\ &\le C(|P(D)u|_{k+1,s}^{(0,t)} + |u|_{m-1,s+k+1}^{(0)}) \end{aligned} \tag{5-77}$$

Moreover, there is a constant K independent of s, such that

$$\begin{aligned} |D_t^{m+k}u|_{0,s}^{(0,t)} &\le |D_t^k P(D)u|_{0,s}^{(0,t)} + K|u|_{m+k-1,s+1}^{(0,t)} \\ &\le |P(D)u|_{k,s}^{(0,t)} + K|u|_{m+k,s}^{(0,t)} \end{aligned} \tag{5-78}$$

If we now apply (5-75) to (5-77) and (5-78), we obtain (5-76) for $k + 1$. By Corollary 5-23 it is true for $k = 0$. Hence, it is true for all k.

5-5 EXISTENCE THEOREMS

We now show how the estimates of the preceding sections can be used to give various existence theorems for the Cauchy problem for hyperbolic equations. First, we have

Theorem 5-25 Let f be a function in $S(\Omega)$ and let $g_0(x), \ldots, g_{m-1}(x)$ be functions in $S(E^n)$. If $P(D)$ is an hyperbolic operator of order m, then there is a function $u \in S(\Omega)$, such that

$$P(D)u = f \qquad \text{in } \Omega \tag{5-79}$$

and $$D_t^k u(x,0) = g_k(x) \qquad x \in E^n \tag{5-80}$$

If $P(D)$ is totally hyperbolic, then

$$|u|_{m-1,s}^{(t)} \le C\left(|f|_{0,s}^{(0,t)} + \sum_0^{m-1} |g_k|_{0,m-k+s-1}^{(0)}\right) \qquad 0 < t \le b \tag{5-81}$$

where the constant C depends only on $P(D)$ and b.

PROOF We follow the proof of Theorem 4-15 of Sec. 4-6 closely. By Sec. 4-5, for each $\xi \in E^n$ there is a solution $h(\xi, t)$ of

$$P(\xi, D_t)h(\xi, t) = Ff(\xi, t) \qquad t \geq 0 \tag{5-82}$$

$$D_t^k h(\xi, 0) = Fg_k(\xi) \qquad 0 \leq k < m \tag{5-83}$$

From the formulas for $h(\xi, t)$ given in Sec. 4-5, we see that $h(\xi, t)$ is in $S(\Omega)$. Thus, $u(x, t) = F^{-1}h$ is in $S(\Omega)$. Applying inverse Fourier transforms to Eqs. (5-82), (5-83), we see that u is a solution of Eqs. (5-79), (5-80). Inequality (5-81) follows from Corollary 5-22.

Theorem 5-26 Under the same hypotheses, for each integer $r \geq 0$

$$|u|_{m+r,s-1}^{(0,t)} \leq C\left(|f|_{r,s}^{(0,t)} + \sum_0^{m-1} |g_k|_{0,m+r-k+s-1}^{(0)}\right) \qquad 0 < t \leq b \tag{5-84}$$

where the constant C depends only on $P(D)$, b and r.

Corollary 5-27 Under the hypotheses of Theorem 5-25, if the support of f is bounded away from $t = 0$, then there is a function $u \in S(\Omega)$ having this property, and such that $P(D)u = f$. Likewise, if the support of f is bounded away from $t = b$, there exists a function $v \in S(\Omega)$, such that $P(D)v = f$, and the support of v is bounded away from $t = b$.

The proof of Corollary 5-27 is similar to that of Theorem 5-4 and is omitted.

Let r be an integer ≥ 0. If f is a function in $H^{r,s}$, we shall say that a function $u \in H^{r,s}$ is a strong solution of Eqs. (5-1), (5-2) if there is a sequence $\{u_j\}$ of functions in $S(\Omega)$ satisfying Eq. (5-14), and such that

$$u_j \to u \qquad P(D)u_j \to f \qquad \text{in } H^{r,s} \tag{5-85}$$

With this terminology, we have

Theorem 5-28 If $P(D)$ is totally hyperbolic and $f \in H^{r,s}$, then there is a strong solution $u \in H^{m+r,s-1}$ of Eqs. (5-1), (5-2), satisfying

$$|u|_{m+r,s-1} \leq C|f|_{r,s} \tag{5-86}$$

where the constant C depends only on $P(D)$, b, and r.

PROOF Since $f \in H^{r,s}$, there is a sequence $\{f_j\}$ of functions in $C^\infty(\overline{\Omega})$ which are in $S(E^n)$ uniformly in t, and converge to f in $H^{r,s}$ (this is merely the definition of $H^{r,s}$). Moreover, there exist functions $u_j \in C^\infty(\overline{\Omega})$ which are in $S(E^n)$ uniformly in t, satisfy Eq. (5-14), and

$$P(D)u_j = f_j \qquad \text{in } \Omega \qquad j = 1, 2, \ldots \tag{5-87}$$

(Theorem 5-25). By inequality (5-84)

$$|u_j - u_k|_{m+r,s-1} \leq C|f_j - f_k|_{r,s} \to 0 \quad \text{as} \quad j, k \to \infty$$

Thus, there is a function $u \in H^{m+r,s-1} \subset H^{r,s}$, such that $u_j \to u$ in $H^{m+r,s-1}$. In particular (5-85) holds. Thus, u is a strong solution of Eqs. (5-1), (5-2), and the proof is complete.

In proving our next theorem we shall make use of

Lemma 5-29 The set of functions in $S(\Omega)$ which vanish near $t = 0, b$ are dense in $H^{0,s}$ for each real s.

PROOF It suffices to show that, for each $\varepsilon > 0$ and each $v \in S(\Omega)$, there is another function $w \in S(\Omega)$ which vanishes near $t = 0, b$, such that

$$|v - w|_{0,s} < \varepsilon \tag{5-88}$$

Let $\delta > 0$ be given, and let $\rho(t)$ be a function in $C_0^\infty[0, b]$, such that

$$0 \le \rho(t) \le 1 \qquad 0 \le t \le b$$

$$\rho(t) = 1 \qquad \delta \le t \le b - \delta$$

Then $v\rho$ is in $S(\Omega)$ and vanishes near $t = 0, b$. Moreover

$$\begin{aligned} |v\rho - v|_{r,s}^2 &= \int_0^b (\rho - 1)^2 \int (1 + |\xi|)^{2s} |Fv|^2 \, d\xi \, dt \\ &\le \int_0^\delta + \int_{b-\delta}^b \left[\int (1 + |\xi|)^{2s} |Fv|^2 \, d\xi \right] dt \\ &\to 0 \quad \text{as} \quad \delta \to 0 \end{aligned}$$

Thus, we can take $w = v\rho$ for δ sufficiently small. The proof is complete.

We now let S_a denote the set of functions in $S(\Omega)$ which vanish in the slab $a \le t \le b$.

Theorem 5-30 If $P(D)$ is totally hyperbolic, there is a constant C_0 depending only on $P(D)$ and b, such that

$$|v|_{0,s}^{(a)} \le C_0 \operatorname*{lub}_{\psi \in S_a} \frac{|(P(D)v, \psi)_{0,s}|}{|\psi|_{m-1,s}} \tag{5-89}$$

holds for all v in $S(\Omega)$ which vanish near $t = 0$.

PROOF Let v be such a function, and let g be the function of the form (5-66) for which (5-68) holds. By the preceding lemma, there is a sequence $\{\varphi_k\}$ of functions in S_a, such that

$$|\varphi_k - gv|_{0,s+1}^{(0,a)} \to 0 \quad \text{as} \quad k \to \infty \tag{5-90}$$

Since $p(\xi, \tau)$ has m distinct real roots for each $\xi, p^{(0,1)}(\xi, \tau)$ must have $m - 1$ distinct real roots. Thus, the operator $p^{(0,1)}(D)$ is totally hyperbolic of order

$m-1$. Thus, by Corollary 5-27 there is a $\psi_k \in S_a$, such that

$$p^{(0,1)}(D)\psi_k = \varphi_k \qquad k = 1, 2, \ldots \tag{5-91}$$

For the same reason there is an $h \in S(\Omega)$ which vanishes near $t = 0$ and satisfies

$$p^{(0,1)}(D)h = v \tag{5-92}$$

Now
$$\begin{aligned}(P(D)v, \psi_k)_{0,s} &= (P(D)p^{(0,1)}(D)h, \psi_k)_{0,s} \\ &= (P(D)h, p^{(0,1)}(D)\psi_k)_{0,s} \to (P(D)h, gv)_{0,s}^{(0,a)} \\ &= (gP(D)h, p^{(0,1)}(D)h)_{0,s}^{(0,a)} \qquad \text{as} \quad k \to \infty\end{aligned} \tag{5-93}$$

We were able to integrate by parts and avoid boundary terms, because h vanishes near $t = 0$ and each ψ_k vanishes near $t = b$. Note that the coefficients of $p^{(0,1)}(D)$ are real. By (5-68), we have

$$\liminf |(P(D)v, \psi_k)_{0,s}| \geq \frac{[|h|_{m-1,s}^{(0,a)}]^2}{C} \tag{5-94}$$

Set $w_k = \psi_k - gh$. Then

$$p^{(0,1)}(D)w_k = \varphi_k - gv - \sum_{j>0} \frac{p^{(0,j+1)}(D)hD_t^j g}{j!}$$

Consequently
$$|p^{(0,1)}(D)w_k|_{0,s+1}^{(0,a)} \leq |\varphi_k - gv|_{0,s+1}^{(0,a)} + C\sum_{j>1} |p^{(0,j)}(D)h|_{0,s+1}^{(0,a)} \tag{5-95}$$

By Corollary 5-23 (employed backwards)

$$|w_k|_{m-1,s}^{(0,a)} \leq C(|p^{(0,1)}(D)w_k|_{0,s+1}^{(0,a)} + |gh|_{m-2,s+1}^{(a)}) \tag{5-96}$$

since ψ_k vanishes for $t = a$. It follows from (5-90), (5-95), and (5-96) that there is a constant C, such that

$$\limsup |\psi_k|_{m-1,s} \leq C(|h|_{m-2,s+1}^{(0,a)} + |h|_{m-2,s+1}^{(a)}) \tag{5-97}$$

Let $\rho(a)$ denote the right-hand side of (5-89). Clearly $\rho(a)$ is a nondecreasing function of a. Moreover, it follows from (5-94) and (5-97), that

$$[|h|_{m-1,s}^{(a)}]^2 \leq C\rho(a)(|h|_{m-1,s}^{(0,a)} + |h|_{m-1,s}^{(a)}) \qquad 0 < a \leq b$$

(we have used (5-61)). If we set

$$r(t) = |h|_{m-1,s}^{(t)} \tag{5-98}$$

this becomes

$$\begin{aligned}r(t)^2 &\leq C\rho(t)\left[\left(\int_0^t r(\lambda)^2\, d\lambda\right)^{1/2} + r(t)\right] \\ &\leq \tfrac{1}{2}C^2\rho(t)^2 + \tfrac{1}{2}\int_0^t r(\lambda)^2\, d\lambda + \tfrac{1}{2}C^2\rho(t)^2 + \tfrac{1}{2}r(t)^2 \qquad 0 < t \leq b\end{aligned}$$

and, consequently

$$r(t)^2 \le 2C^2\rho(t)^2 + \int_0^t r(\lambda)^2\,d\lambda \qquad 0 < t \le b$$

If we now apply the second statement in Lemma 5-9, we get

$$r(t)^2 \le 2C^2\, e^t\, \rho(t)^2 \qquad 0 < t \le b \tag{5-99}$$

This is precisely what we want to show. The proof is complete.

Theorem 5-31 If $P(D)$ is totally hyperbolic, then

$$|v|_{0,s}^{(a)} \le C_0 \operatorname*{lub}_{\psi \in S_a} \frac{|(v, \bar{P}(D)\psi)_{0,s}|}{|\psi|_{m-1,s}} \tag{5-100}$$

holds for all $v \in H^{0,s}$, where C_0 is the constant in (5-89).

PROOF Let $\rho(t)$ be a nonnegative function in $C_0^\infty[-1, 1]$, which satisfies

$$\rho(t) > 0 \qquad |t| < \tfrac{1}{2}$$

and

$$\int \rho(t)\,dt = 1$$

Let $j(x)$ be the function given by Eq. (1-35) of Sec. 1-3, satisfying

$$\int j(x)\,dx = 1$$

For $v \in H^{r,s}$ define v to be zero outside Ω, and set

$$\tilde{J}_\varepsilon v = \frac{1}{\varepsilon^{n+1}} \int\int \rho\left(\frac{\tau - t + 2\varepsilon}{\varepsilon}\right) j\left(\frac{y - x}{\varepsilon}\right) v(y, \tau)\,dy\,d\tau \tag{5-101}$$

and

$$\hat{J}_\varepsilon v = \frac{1}{\varepsilon^{n+1}} \int\int \rho\left(\frac{t - \tau + 2\varepsilon}{\varepsilon}\right) j\left(\frac{x - y}{\varepsilon}\right) v(y, \tau)\,dy\,d\tau \tag{5-102}$$

for $\varepsilon > 0$. It is easy to check that both $\tilde{J}_\varepsilon v$ and $\hat{J}_\varepsilon v$ are in $C^\infty(\bar{\Omega}) \cap H^{r,s}$ for each $\varepsilon > 0$, that $\tilde{J}_\varepsilon v$ vanishes near $t = 0$ and that $\hat{J}_\varepsilon v$ vanishes near $t = b$. Moreover, one has

$$|\tilde{J}_\varepsilon v|_{r,s} \le |v|_{r,s} \qquad |\hat{J}_\varepsilon v|_{r,s} \le |v|_{r,s} \tag{5-103}$$

$$|v - \tilde{J}_\varepsilon v|_{r,s}^{(a)} \to 0 \qquad |v - \hat{J}_\varepsilon v|_{r,s}^{(a)} \to 0 \tag{5-104}$$

The proofs of these facts are similar to those of (2-63) and (2-64) of Sec. 2-5. We also have

$$(\tilde{J}_\varepsilon v, w)_{0,s} = (v, \hat{J}_\varepsilon w)_{0,s} \qquad v, w \in H^{0,s} \tag{5-105}$$

which is easily checked. Now, if $v \in H^{0,s}$, then $\tilde{J}_\varepsilon v$ is in $S(\Omega)$ and vanishes near $t = 0$ for each $\varepsilon > 0$. Consequently, we may apply Theorem 5-30 to obtain

$$|\tilde{J}_\varepsilon v|_{0,s}^{(a)} \leq C_0 \operatorname*{lub}_{\psi \in S_a} \frac{|(P(D)\check{J}_\varepsilon v, \psi)_{0,s}|}{|\psi|_{m-1,s}}$$

Now $\hat{J}_\varepsilon \psi$ is in S_a whenever ψ is. Thus, by (5-103) and (5-105)

$$|\tilde{J}_\varepsilon v|_{0,s}^{(a)} \leq C_0 \operatorname*{lub}_{\psi \in S_a} \frac{|(v, \bar{P}(D)\hat{J}_\varepsilon \psi)_{0,s}|}{|\hat{J}_\varepsilon \psi|_{m-1,s}}$$

$$\leq C_0 \operatorname*{lub}_{\psi \in S_a} \frac{|(v, \bar{P}(D)\psi)_{0,s}|}{|\psi|_{m-1,s}}$$

Inequality (5-100) now follows from (5-104). This completes the proof.

The material of Secs. 5-3 to 5-5 is due to Leray (1954) (cf. Gårding, 1956).

5-6 PROPERLY HYPERBOLIC OPERATORS

Let $p(D)$ be a hyperbolic operator which is homogeneous, of order m. Thus, $p(\xi, \tau)$ has only real roots for each $\xi \in E^n$. We know from elementary calculus that each derivative of $p(\xi, \tau)$ with respect to τ must have real roots. We set

$$p^{(0,k)}(\xi, \tau) = a \prod_{j=1}^{m-k} (\tau - \tau_{kj}(\xi)) \qquad k = 0, 1, \ldots \tag{5-106}$$

where the $\tau_{kj}(\xi)$ are continuous, real-valued functions, and a is a constant $\neq 0$. In particular, each of the operators $p^{(0,k)}(D)$ is hyperbolic. Let J be a subset of the integers $(1, \ldots, m)$. We set

$$p_{k,J}(\xi, \tau) = a \prod_{j \in J} (\tau - \tau_{kj}(\xi)) \qquad k = 0, 1, \ldots \tag{5-107}$$

if each integer in J is $\leq m - k$. Otherwise we set $p_{k,J}(\xi, \tau) = 0$. Let $Q(D)$ be an operator of order $< m$, and set $P(D) = p(D) + Q(D)$. We shall call $P(D)$ properly hyperbolic if there are bounded functions $c_{k,J}(\xi)$, such that

$$Q(\xi, \tau) = \sum_{k,J} c_{k,J}(\xi) p_{k,J}(\xi, \tau) \tag{5-108}$$

The reason for this definition is

Theorem 5-32 There is a constant C depending only on $p(D)$ and b, such that

$$\sum_{k,J} |p_{k,J}(D)u|_{0,s}^{(t)} \leq C\, |p(D)u|_{0,s}^{(0,t)} \qquad 0 < t \leq b \tag{5-109}$$

hold for all $u \in C^m(\bar{\Omega}) \cap H^{0,s}$, satisfying Eq. (5-2).

This theorem follows from Theorem 5-19 and the fact that the operators $p^{(0,k)}(D)$ are hyperbolic. From Theorem 5-32, we have

Theorem 5-33 If $P(D)$ is properly hyperbolic, then there is a constant C depending only on $P(D)$ and b, such that

$$\sum_{k,J} |p_{k,J}(D)u|_{0,s}^{(t)} \leq C|P(D)u|_{0,s}^{(0,t)} \qquad 0 < t \leq b \tag{5-110}$$

holds for all $u \in C^m(\bar{\Omega}) \cap H^{0,s}$, satisfying Eq. (5-2).

PROOF By Eq. (5-108)

$$|Q(D)u|_{0,s}^{(0,t)} \leq C \sum |p_{k,J}(D)u|_{0,s}^{(0,t)} \qquad 0 < t \leq b \tag{5-111}$$

Thus, if $r(t)$ represents the left-hand side of (5-109), we have by (5-111) and Theorem 5-32

$$r(t)^2 \leq C[|p(D)u|_{0,s}^{(0,t)}]^2 + C\int_0^t r(\lambda)^2\, d\lambda$$

The desired inequality now follows from Corollary 5-10.

Corollary 5-34 If $P(D)$ is properly hyperbolic, then

$$\sum_{k,J} |p_{k,J}(D)u|_{0,s} \leq C|P(D)u|_{0,s} \tag{5-112}$$

holds for all $u \in C^m(\bar{\Omega}) \cap H^{0,s}$, satisfying Eq. (5-2).

Note that properly hyperbolic operators are hyperbolic. One way of seeing this is to observe that Eq. (5-108) implies that Q is weaker than p. Thus, $P(D)$ is hyperbolic, by Theorem 4-9 of Sec. 4-4. Another way is to notice that (5-112) implies

$$\|\varphi\| \leq C\|P(D)\varphi\|$$

for all φ having compact support in the slab $0 < t < 2b$. Hyperbolicity now follows from Theorem 4-5 of Sec. 4-3 (applied backwards).

The results of this section are due to Peyser (1963).

5-7 EXAMPLES

We now give some examples of operators and discuss their hyperbolicity.

1 The wave operator $D_t^2 - \Sigma D_k^2$. Here $p(\xi, \tau) = \tau^2 - \Sigma \xi_k^2$. For $\xi \neq 0$, this polynomial has two distinct real roots. Hence it is totally hyperbolic.

2 The operator $(D_t - D_1 + D_2)(D_t + D_1 - D_2) + \alpha D_t$. Here $p(\xi, \tau) = (\tau - \xi_1 + \xi_2)(\tau + \xi_1 - \xi_2)$, and $Q(\xi, \tau) = \alpha\tau$. The operator is not totally hyperbolic because 0 is a double root when $\xi_1 = \xi_2$. However, it is properly hyperbolic. For, we have

$$2\tau = (\tau - \xi_1 + \xi_2) + (\tau + \xi_1 - \xi_2)$$

showing that Eq. (5-108) holds.

3 Same $p(\xi, \tau)$ as in **2**, and $Q(\xi, \tau) = \alpha(\xi_1 - \xi_2)$. Again it is properly hyperbolic. For, we have

$$2(\xi_1 - \xi_2) = -(\tau - \xi_1 + \xi_2) + (\tau + \xi_1 - \xi_2)$$

4 Same $p(\xi, \tau)$, and $Q(\xi, \tau) = \alpha\xi_1$. This operator is not hyperbolic. The roots of $P(\xi, \tau) = p(\xi, \tau) + Q(\xi, \tau)$ are

$$\tau = \pm\sqrt{(\xi_1 - \xi_2)^2 - 4\alpha\xi_1} \tag{5-113}$$

If we take $\xi_2 = \xi_1$, and ξ_1 large with the appropriate sign, we can always make the imaginary parts of these roots as large as desired.

5 $p(\xi, \tau) = (\tau^2 - \xi_1^2 - \xi_2^2)(\tau^2 - \xi_1^2 - 2\xi_2^2)$ and $Q(\xi)$ is a polynomial in ξ_1, ξ_2 of degree <4 with real coefficients. Clearly, $p(D)$ is not totally hyperbolic. However, it is hyperbolic. To see this, note that

$$P(\xi, \tau) = \tau^4 - (\tau_1^2 + \tau_2^2)\tau^2 + \tau_1^2\tau_2^2 + Q \tag{5-114}$$

where $\tau_1^2 = \xi_1^2 + \xi_2^2$ and $\tau_2^2 = \xi_1^2 + 2\xi_2^2$. Thus, the roots of $P(\xi, \tau)$ satisfy

$$2\tau^2 = 2\xi_1^2 + 3\xi_2^2 \pm \sqrt{\xi_2^4 - 4Q} \tag{5-115}$$

As $|\xi| \to \infty$ this expression becomes positive. Hence, the roots of $P(\xi, \tau)$ are real for $|\xi|$ large. Thus, their imaginary parts are bounded for all ξ. Now suppose Q is homogeneous, of degree three. Let us see under what conditions $P(D)$ will be properly hyperbolic. The only polynomials among the $p_{k,J}$ which are of degree three are

$$(\tau - \tau_1)(\tau^2 - \tau_2^2)$$
$$(\tau + \tau_1)(\tau^2 - \tau_2^2)$$
$$(\tau^2 - \tau_1^2)(\tau - \tau_2)$$
$$(\tau^2 - \tau_1^2)(\tau + \tau_2)$$

Now, suppose we have

$$Q(\xi) = c_1(\xi)(\tau - \tau_1)(\tau^2 - \tau_2^2) + c_2(\xi)(\tau + \tau_2)(\tau^2 - \tau_2^2)$$
$$+ c_3(\xi)(\tau^2 - \tau_1^2)(\tau - \tau_2) + c_4(\xi)(\tau^2 - \tau_1^2)(\tau + \tau_2)$$

Setting $\tau = -\tau_1, \tau_1, -\tau_2$, and τ_2, respectively, we find that

$$c_1 = \frac{Q(\xi)}{2\tau_1\xi_2^2} \qquad c_2 = \frac{-Q(\xi)}{2\tau_1\xi_2^2}$$

$$c_3 = \frac{-Q(\xi)}{2\tau_2\xi_2^2} \qquad c_4 = \frac{Q(\xi)}{2\tau_2\xi_2^2}$$

These functions will remain bounded if and only if Q is of the form

$$Q(\xi) = \beta_1\xi_1\xi_2^2 + \beta_2\xi_2^3 \tag{5-116}$$

Thus, $P(D)$ is properly hyperbolic if and only if Q is of this form.

PROBLEMS

5-1 Show that a strong solution is a weak solution.

5-2 Verify that the function $h(x, -t)$ has the properties ascribed to it in the proof of Theorem 5-4.

5-3 Show that Eq. (5-12) is a scalar product and that $H^{r,s}$ is a Hilbert space.

5-4 Prove Eqs. (5-20), (5-21), and (5-22).

5-5 Prove Corollary 5-15.

5-6 Show that the functions $a_{jk}(\xi)$, described in the proof of Theorem 5-16, are homogeneous of degree $m - k - 1$.

5-7 Prove inequality (5-56).

5-8 Prove (5-69).

5-9 Prove inequality (5-71).

5-10 Prove inequality (5-74).

5-11 Let $p(\tau)$ be a polynomial of degree m having m distinct real roots. Show that its derivative $p'(\tau)$ has $m-1$ distinct real roots.

5-12 Prove inequality (5-78).

5-13 Prove the second statement in Lemma 5-9.

5-14 Prove (5-103), (5-104), and (5-105).

5-15 Prove that Eq. (5-115) becomes positive as $|\xi| \to \infty$.

5-16 Let $p(\xi, \tau)$ be given as in Example **5** of Sec. 5-7. If Q is of degree <3, when is $P(D)$ properly hyperbolic? Give examples.

5-17 Prove $P(D)\hat{J}_\varepsilon v = \hat{J}_\varepsilon P(D)v$ for $v \in H^{0,s}$ and vanishing near $t = b$, where $P(D)$ is any operator with constant coefficients, and $\hat{J}_\varepsilon$ is given by Eq. (5-102). Is this true for $\tilde{J}_\varepsilon$?

CHAPTER

SIX

BOUNDARY VALUE PROBLEMS IN A HALF-SPACE (ELLIPTIC)

6-1 INTRODUCTION

We saw in Chapter 4 that we could solve the Cauchy problem

$$P(D_t)u(x,t) = f(x,t) \qquad 0 \le t \le b < \infty \tag{6-1}$$

$$D_t^k u(x,0) = g_k(x) \qquad 0 \le k < m \tag{6-2}$$

for any functions $g_k \in S(E^n)$ and $f \in S(\Omega)$, where $P(D)$ is a hyperbolic operator of order m and Ω is the slab $0 < t < b$ (Theorem 4-15 of Sec. 4-6). Moreover, it was shown that the solution $u(x,t)$ is in $S(\Omega)$. We may wonder whether or not this is true if we take $b = \infty$, i.e., if we wish to solve Eqs. (6-1), (6-2) in a half-space.

The method of Chapter 4 shows that we can solve Eqs. (6-1), (6-2) in a half-space Ω. But the solution will not necessarily be in $L^2(\Omega)$. This can be seen from a simple example. Consider the Cauchy problem in two dimensions.

$$u_{tt} - u_{xx} = 0 \qquad t > 0 \qquad -\infty < x < \infty \tag{6-3}$$

$$u(x,0) = g(x) \qquad u_t(x,0) = 0 \qquad -\infty < x < \infty \tag{6-4}$$

where $g(x)$ is in S. The solution of Eqs. (6-3), (6-4) is

$$u(x,t) = g(x+t) + g(x-t) \tag{6-5}$$

If $g > 0$, this is clearly not in $L^2(\Omega)$, where Ω is the half-plane $t > 0$. In fact

$$\iint\limits_{\Omega} |u(x,t)|^2 \, dx \, dt \ge \int_0^\infty \int_{-\infty}^\infty g(x+t)^2 \, dx \, dt = \infty$$

The question now arises as to whether there is an integer r, such that

$$P(D)u(x,t) = f(x,t) \qquad t > 0 \tag{6-6}$$

$$D_t^k u(x,0) = g_k(x) \qquad 0 \le k < r \tag{6-7}$$

has a solution in $L^2(\Omega)$, where Ω is the half-space $t > 0$. At this point we do not know if hyperbolicity is still an asset. Therefore, let us phrase the question as follows. For which constant-coefficient operator $P(D)$ do there exist integers r, such that Eqs. (6-6), (6-7) have a solution in $L^2(\Omega)$ for all choices of $f \in S(E^{n+1})$ and $g_k \in S(E^n)$? Clearly, if such a solution exists for some value of r, then it will exist for any smaller value. Of course, we are interested in obtaining the largest value. Another way of looking at this is to ask when Eqs. (6-6), (6-7) have at most one solution in $L^2(\Omega)$, i.e., when is the solution unique. For if there is only one solution of Eqs. (6-6), (6-7), then $g_r(x)$ is already determined and cannot be chosen arbitrarily.

The present chapter is devoted to the study of Eqs. (6-6), (6-7) with specific reference to existence and uniqueness of solutions in $L^2(\Omega)$. As usual we have no particular idea of attack. We resort, therefore, to the consideration of special cases. In Sec. 4-5 we found it useful to study ordinary differential operators. We try this again in Secs. 6-2 to 6-6. If G is the interval (a, b), we let $S(a, b)$ denote the set $S(G)$ (see Sec. 2-3).

6-2 THE PROBLEM IN A HALF-LINE

We now consider the case when $P(D)$ is an operator in t alone, i.e., when it is an ordinary differential operator. In this case, Eqs. (6-6), (6-7) reduce to

$$P(D_t)u(t) = f(t) \qquad t > 0 \tag{6-8}$$

$$D_t^k u(0) = g_k \qquad 0 \le k < r \tag{6-9}$$

where $f \in S(E^1)$ and the g_k are constants. We are looking for solutions in $L^2(0, \infty)$. We assume that $P(z)$ is of degree m, and that the coefficient of z^m is 1.

Let us first consider the case $f \equiv 0$. The general solution of

$$P(D_t)u = 0 \qquad t > 0 \tag{6-10}$$

is

$$u = \sum_{k=1}^{m} c_k e^{i\tau_k t} \tag{6-11}$$

where the τ_k are the roots of $P(z) = 0$. (If there are multiple roots, then the c_k become polynomials in t. This modification does not affect the present discussion.) Now, in order for u to be in $L^2(0, \infty)$, the coefficient c_k must vanish for any k, such that $\operatorname{Im} \tau_k \le 0$. For, otherwise, u would be bounded away from 0 as $t \to \infty$ and, therefore, could not be in $L^2(0, \infty)$. Now, Eq. (6-9) imposes r conditions on the coefficients. Therefore, it seems reasonable to state

Theorem 6-1 If there are r roots (counting multiplicities $\tau_1, \ldots, \tau_r$ of $P(z) = 0$, such that

$$\operatorname{Im} \tau_k > 0 \qquad 1 \leq k \leq r \tag{6-12}$$

then Eqs. (6-9), (6-10) has a solution in $S(0, \infty)$ for each choice of the constants g_k.

PROOF We expect to find a solution of the form

$$u(t) = \sum_{k=1}^{r} c_k \, e^{i\tau_k t} \tag{6-13}$$

Now, if Γ is any simple closed curve enclosing τ_k, we know that

$$e^{i\tau_k t} = \frac{1}{2\pi i} \oint_\Gamma \frac{e^{ixt} \, dz}{z - \tau_k} \tag{6-14}$$

Thus, if Γ encloses $\tau_1, \ldots, \tau_r$, Eq. (6-13) can be put in the form

$$u(t) = \frac{1}{2\pi i} \oint_\Gamma \frac{Q(z) \, e^{izt} \, dz}{P_+(z)} \tag{6-15}$$

where
$$P_+(z) = (z - \tau_1) \cdots (z - \tau_r) \tag{6-16}$$

In analogy with Eq. (4-82) of Sec. 4-5, let us consider the function

$$w_r(t) = \frac{1}{2\pi i} \oint_\Gamma \frac{e^{izt} \, dt}{P_+(z)} \tag{6-17}$$

Differentiating under the integral sign, we get

$$D_t^k w_r(t) = \frac{1}{2\pi i} \oint_\Gamma \frac{z^k \, e^{izt} \, dz}{P_+(z)} \qquad k = 0, 1, 2, \ldots \tag{6-18}$$

which shows that w_r is a solution of Eq. (6-10). Now, if we take Γ to be a circle of radius R about the origin, with R large enough so that Γ encloses $\tau_1 \ldots, \tau_r$, we have by Eq. (6-18)

$$D_t^k w_r(0) = \frac{1}{2\pi i} \int_0^{2\pi} \frac{R^k \, e^{ik\theta} \cdot i \, e^{i\theta} \, R \, d\theta}{R^r \, e^{ir\theta} + Q(R \, e^{i\theta})} \tag{6-19}$$

where $Q(z)$ is a polynomial of degree $< r$. If $k < r - 1$, this tends to 0 as $R \to \infty$. If $k = r - 1$, it tends to 1. Thus, $w_r(t)$ is a solution of

$$P(D_t)u = 0 \qquad t > 0 \tag{6-20}$$

$$D_t^k u(0) = 0 \qquad 0 \leq k < r - 1 \tag{6-21}$$

$$D_t^{r-1} u(0) = 1 \tag{6-22}$$

We must show that $w_r \in S(0, \infty)$. Let $\delta > 0$ be such that

$$\operatorname{Im} \tau_k > \delta \qquad 1 \leq k \leq r \tag{6-23}$$

Since w_r is of the form (6-13), for each k there is a constant C_k, such that

$$|D_t^k w_r(t)| \leq C_k(1+t)^{r+k-1} e^{-\delta t} \qquad k = 0, 1, \ldots \tag{6-24}$$

This shows that $w_r \in S(0, \infty)$. Now, we can solve Eqs. (6-9), (6-10) as in Sec. 4-5. In fact, we set $u_0(t) = g_0 w_r(t)$ and inductively define

$$u_j(t) = \left[g_j - \sum_{i=1}^{j} D_r^{r+j-1} u_{j-1}(0) \right] w_r(t) \qquad 1 \leq j < r \tag{6-25}$$

Then, the function

$$u(t) = \sum_{j=1}^{r} D_t^{r-j} u_{j-1}(t) \tag{6-26}$$

is a solution of Eqs. (6-9), (6-10). The verification is the same as that for Eq. (4-76) of Sec. 4-5. Note that inequality (6-24) assures us that the solution $u(t)$ given by Eq. (6-26) is in $S(0, \infty)$. The proof is complete.

We now turn to the original system (6-8), (6-9). In order to solve it, all we need do is add a solution of

$$P(D_t)u(t) = f(t) \qquad t > 0 \tag{6-27}$$

$$D_t^k u(0) = 0 \qquad 0 \leq k < r \tag{6-28}$$

to Eq. (6-26). In analogy to the development of Sec. 4-5, we might suggest

$$v(t) = i \int_0^t f(s) w_r(t-s)\, ds \tag{6-29}$$

as a candidate. If we follow the reasoning given there, we find

$$D_t^k v(t) = i \int_0^t f(s) D_t^k w_r(t-s)\, ds \qquad 0 \leq k < r \tag{6-30}$$

$$D_t^k v(t) = \sum_{j=r}^{k} D_t^{k-j} f(t) D_t^{j-1} w(0) + i \int_0^t f(s) D_t^k w(t-s)\, ds \qquad k \geq r \tag{6-31}$$

These formulas show that $v(t)$ satisfies Eq. (6-28). But instead of Eq. (6-27), it satisfies

$$P_+(D)v(t) = f(t) \qquad t > 0 \tag{6-32}$$

Thus, we have obtained a solution of Eqs. (6-27), (6-28) only for the case $P_+(D) = P(D)$, i.e., for the case when all of the roots of $P(z) = 0$ have positive imaginary parts. However, for this case, our solution is of the type desired. If $f \in S(0, \infty)$, then the same is true of $v(t)$. To see this, it suffices to show that

$$v_{jk}(t) = t^j \int_0^t f(s) D_t^k w_r(t-s)\, ds$$

is bounded for each j and k. Since $f \in S(0, \infty)$, there is a constant C, such that

$$|f(s)| \leq \frac{C}{(1+s)^{j+2}} \qquad s > 0$$

By inequality (6-24) there is a constant C, such that

$$|D_t^k w_r(t)| \leq \frac{C}{(1+t)^j} \qquad t > 0$$

Thus
$$|v_{jk}(t)| \leq Ct^j \int_0^t (1+t-s)^{-j}(1+s)^{-j+2}\, ds$$

Since $t - s \geq t/2$ for $0 \leq s \leq t/2$, this gives

$$|v_{jk}(t)| \leq Ct^j\left(1+\frac{t}{2}\right)^{-j} \int_0^{t/2} (1+s)^{-j-2}\, ds$$
$$+ Ct^j\left(1+\frac{t}{2}\right)^{-j} \int_{t/2}^{t} (1+t-s)^{-j}(1+s)^{-2}\, ds$$
$$\leq 2^j C t^j (2+t)^{-j} \int_0^\infty (1+s)^{-2}\, ds \tag{6-33}$$

Thus, $v_{jk}(t)$ is bounded for each j and k. This shows that $v(t)$ is in $S(0, \infty)$, and consequently we have found the desired solution for $r = m$.

What can we do when $r < m$? In this case, we make use of

Lemma 6-2 Let $P(D_t)$ be any constant-coefficient operator, and let $f(t)$ be any function in $S(E^1)$. Then there exists a function $u(t) \in S(E^1)$, such that

$$P(D_t)u(t) = f(t) \qquad t > 0 \tag{6-34}$$

PROOF It clearly suffices to prove the lemma for $P(D_t) = D_t - \alpha$, where α is any complex number. If α is real, we set

$$v(t) = -i \int_t^\infty e^{i\alpha(t-s)} f(s)\, ds \tag{6-35}$$

Clearly, v satisfies

$$(D_t - \alpha)v = f \tag{6-36}$$

Let $\rho(t)$ be a function in $C^\infty(E^1)$, which equals 1 for $t > 0$ and vanishes for $t < -1$. Then the function $u = \rho v$ is in $S(E^1)$ and satisfies Eq. (6-34). If α is not real, set

$$u(t) = \frac{1}{2\pi} \int_{-\infty}^{\infty} e^{it\tau} \frac{\tilde{f}(\tau)}{\tau - \alpha}\, d\tau \tag{6-37}$$

where
$$\tilde{f}(\tau) = \int_{-\infty}^{\infty} e^{-i\tau s} f(s)\, ds \tag{6-38}$$

Since f is in S, the same is true for $\tilde{f}$ and consequently for $\tilde{f}/(\tau - \alpha)$ (see Sec. 2-3). Thus $u \in S$. By taking Fourier transforms we see that u is a solution of Eq. (6-36). This completes the proof.

Now we return to Eqs. (6-27), (6-28). Let

$$P_-(z) = \frac{P(z)}{P_+(z)} \tag{6-39}$$

By the lemma just proved there is an $h \in S(E^1)$, such that

$$P_-(D_t)h(t) = f(t) \qquad t > 0 \tag{6-40}$$

We now let $u(t)$ be a solution of

$$P_+(D_t)u(t) = h(t) \qquad t > 0 \tag{6-41}$$

and Eq. (6-28). We know there exists a solution in $S(0, \infty)$. It is now obvious that this function is a solution of Eqs. (6-27), (6-28). Thus, we have proved

Theorem 6-3 Under the hypotheses of Theorem 6-1, for each $f \in S(E^1)$ and each choice of the constants g_k, the system (6-8), (6-9) has a solution in $S(0, \infty)$.

6-3 UNIQUENESS

In this section we shall show that the solution given by Theorem 6-3 is unique. We begin by proving

Theorem 6-4 Let $P(z)$ be a polynomial of degree m. Suppose $u(t)$ is a function in $C^m[a, b]$, such that

$$P(D_t)u(t) = 0 \qquad a < t < b \tag{6-42}$$

$$D_t^k u(a) = 0 \qquad 0 \le k < m \tag{6-43}$$

Then $u(t) \equiv 0$.

PROOF Suppose there is a t_1, such that $a < t_1 \le b$ and $u(t_1) \neq 0$.

Set

$$Y(t) = \sum_{k=0}^{m-1} |D_t^k u(t)|$$

and let t_0 be the highest point in $[a, t_1]$, such that $Y(t_0) = 0$. We may assume that $P(z)$ is of the form

$$P(z) = z^m + a_{m-1}z^{m-1} + \cdots + a_0$$

Thus, by Eqs. (6-42) and (6-43)

$$Y(t) = Y(t) - Y(t_0) \leq \sum_{k=1}^{m} \int_{t_0}^{t} |D_t^k u(s)|\, ds$$

$$\leq \sum_{k=0}^{m-1} \int_{t_0}^{t} (1 + |a_k|)\, |D_t^k u(s)|\, ds \tag{6-44}$$

Set

$$M(t) = \operatorname*{lub}_{t_0 < s < t} Y(s)$$

Then (6-44) gives

$$Y(t) \leq M(t)(t - t_0)\left(m + \sum_{k=0}^{m-1} |a_k|\right) \qquad t_0 < t < t_1 \tag{6-45}$$

This implies

$$M(t) \leq \tfrac{1}{2} M(t) \tag{6-46}$$

for t sufficiently close to t_0. But this is impossible since $M(t) > 0$ for $t_0 < t < t_1$. Thus, $Y(t) \equiv 0$ in $[a, b]$, and the proof is complete.

Corollary 6-5 The solutions of Eq. (6-42) form an m-dimensional space.

PROOF In Sec. 4-5 it was shown that, for each j satisfying $0 \leq j < m$, we can solve Eq. (6-42) and

$$D_t^k u(a) = \delta_{jk} \qquad 0 \leq k < m \tag{6-47}$$

where $\delta_{jk} = 0$ for $j \neq k$ and $\delta_{kk} = 1$. Denote the solution by $u_j(t)$. The u_j are linearly independent. For if

$$\sum \alpha_j u_j(t) \equiv 0 \qquad a \leq t \leq b$$

then

$$\sum \alpha_j D_t^k u_j(a) = 0 \qquad 0 \leq k < m$$

By Eq. (6-47) all of the α_j vanish. Now, let u be any solution of Eq. (6-42).

Set

$$v(t) = \sum_{0}^{m-1} D_t^k u(a) u_k(t)$$

By Eq. (6-47)

$$D_t^k v(a) = D_t^k u(a) \qquad 0 \leq k < m$$

Thus $u - v$ is a solution of Eqs. (6-42), (6-43). By Theorem 6-4, $u = v$, a linear combination of the u_j. Thus the corollary is proved.

Theorem 6-6 If $P(z)$ has exactly r roots (counting multiplicities) with positive imaginary parts, then the set of these solutions of Eq. (6-42) which are in $C^m(0, \infty) \cap L^2(0, \infty)$ form an r-dimensional space. They are precisely the solutions of

$$P_+(D_t)u(t) = 0 \tag{6-48}$$

PROOF Let $\tau_1, \ldots, \tau_m$ denote the roots of $P(z) = 0$. The functions $t^j e^{i\tau_k t}$ are solutions of Eq. (6-42), provided j is less than the multiplicity of τ_k. Since these functions are linearly independent and there are m of them, they form a basis for the solutions of Eq. (6-42). Only those for which $\operatorname{Im} \tau_k > 0$ are in $L^2(0, \infty)$, and no nonzero linear combination of the others can be in $L^2(0, \infty)$. Since precisely r of these functions are in $L^2(0, \infty)$ and they are all solutions of Eq. (6-48), the proof is complete.

Corollary 6-7 There is at most one solution of Eqs. (6-8), (6-9) in $L^2(0, \infty)$.

PROOF Every solution of Eq. (6-8) which is in $L^2(0, \infty)$ is a solution of Eq. (6-48) (Theorem 6-6). The result now follows from Theorem 6-4.

6-4 GENERAL BOUNDARY CONDITIONS

We now consider a slightly more general problem than (6-8), (6-9). Let $Q_1(z), \ldots, Q_r(z)$ be r polynomials of degree $<m$. We look for a function $u(t)$ satisfying

$$P(D_t)u(t) = f(t) \qquad t > 0 \tag{6-49}$$

$$Q_j(D_t)u(0) = g_j \qquad 1 \le j \le r \tag{6-50}$$

As before, we assume that $P(z)$ is of degree m and that the coefficient of z^m is 1. Moreover, $P(z)$ has exactly r roots (counting multiplicities) $\tau_1, \ldots, \tau_r$ satisfying Eq. (6-12).

In order to feel our way, let us first assume that all of the $Q_j(z)$ are of degree $<r$. Thus

$$Q_j(z) = \sum_{k=0}^{r-1} b_{jk} z^k \qquad 1 \le j \le r \tag{6-51}$$

Now, if the matrix (b_{jk}) is nonsingular, we can solve Eq. (6-51) for the z^k. Hence, if (b^{ij}) is the inverse matrix, we have

$$z^i = \sum_{j=1}^{r} b^{ij} Q_j(z) \qquad 0 \le i < r \tag{6-52}$$

Thus, Eq. (6-50) is equivalent to

$$D_t^k u(0) = \sum_{j=1}^{r} b^{kj} g_j \qquad 0 \le k < r \tag{6-53}$$

Thus, any solution of Eqs. (6-49), (6-50) is a solution of Eqs. (6-49), (6-53), and vice versa. Moreover, we know that the system (6-49), (6-53) has a unique solution (Theorem 6-3 and Corollary 6-7). On the other hand, if the matrix (b_{jk}) is singular, then there are constants α_j not all zero, such that

$$\sum_{j=1}^{r} \alpha_j b_{jk} = 0 \qquad 0 \le k < r$$

This means that

$$\sum_{1}^{r} \alpha_j Q_j(z) \equiv 0 \tag{6-54}$$

Thus, in order for there to be a solution of Eqs. (6-49), (6-50), we must have

$$\sum_{1}^{r} \alpha_j g_j = 0$$

which shows that we cannot have a solution for all choices of the g_k. Furthermore, if (b_{jk}) is singular there are constants g_k not all zero, such that

$$\sum_{k=0}^{r-1} b_{jk} g_k = 0 \qquad 1 \le j \le r \tag{6-55}$$

If we define $u(t)$ by Eqs. (6-25) and (6-26), it follows that it is a solution of

$$P(D_t)u(t) = 0 \qquad t > 0 \tag{6-56}$$

$$Q_j(D_t)u(0) = 0 \qquad 1 \le j \le r \tag{6-57}$$

To summarize, we find that if the matrix (b_{jk}) is nonsingular, the system (6-49), (6-50) has a unique solution for each choice of f and the g_k. If the matrix (b_{jk}) is singular, we cannot solve for all choices, and even when we can solve the solution is not unique.

Let us now see what can be done when the orders of the $Q_j(D_t)$ are permitted to be $\ge r$. Fortunately, this can be reduced to the previous case. To see this note that, by partial fractions, we have

$$Q_j(z) = S_j(z)P_+(z) + R_j(z) \tag{6-58}$$

where the degree of $R_j(z)$ is $< r$. If $m_j < m$ is the degree of $Q_j(z)$, then the degree of $S_j(z)$ is $m_j - r$. Let $S(z)$ be defined by Eq. (6-39). Then by Lemma 6-2 for each $f \in S(E^1)$ there is an $h \in S(E^1)$, such that Eq. (6-40) holds. Thus, (6-49), (6-50) is equivalent to

$$P_+(D_t)u(t) = h(t) \qquad t > 0 \tag{6-59}$$

$$R_j(D_t)u(0) = g_j - S_j(D_t)h(0) \qquad 1 \le j \le r \tag{6-60}$$

This is now in the form previously treated. If

$$R_j(z) = \sum_{k=0}^{r-1} c_{jk} z^k \qquad 1 \le j \le r \tag{6-61}$$

then Eqs. (6-59), (6-60) have a unique solution for all choices of h and the g_j, if and only if the matrix (c_{jk}) is nonsingular.

From Eq. (6-54) we see that the matrix (b_{jk}) is nonsingular if and only if the $Q_j(z)$ are linearly independent. Similarly, the matrix (c_{jk}) is nonsingular if and only if the $R_j(z)$ are linearly independent. We say that the $Q_j(z)$ are linearly independent *modulo* $P_+(z)$, if the $R_j(z)$ are linearly independent. Thus, we have proved

Theorem 6-8 If the $Q_j(z)$ are linearly independent modulo $P_+(z)$, then for each $f \in S(E^1)$ and each choice of the constants g_j, there is a function $u(t) \in S(0, \infty)$ satisfying Eqs. (4-49), (4-50).

Note that the solution is also unique, since any solution of Eq. (6-56) which is in $L^2(0, \infty)$ is also a solution of Eq. (6-48) (Theorem 6-6). Thus, a solution of Eqs. (6-56), (6-57) satisfies

$$R_j(D_t)u(0) = 0 \qquad 1 \leq j \leq r \tag{6-62}$$

But we have shown that the only solution of Eqs. (6-48) and (6-62) is $u = 0$.

6-5 ESTIMATES FOR A SIMPLE CASE

Now that we have solved Eqs. (6-49), (6-50), we ask if the solution u can be estimated in terms of f and the g_j. Since we have been working in $L^2(0, \infty)$, it appears reasonable to ask when is there an estimate in the form

$$\| u \| \leq C\left(\| f \| + \sum |g_j| \right) \tag{6-63}$$

for all solutions of (6-49), (6-50) in $L^2(0, \infty)$, where the constant depends only on $P(z)$ and the $Q_j(z)$ (the norm is that of $L^2(0, \infty)$)? Since we have no particular ideas on the subject, let us consider a very simple case.

Suppose $P(z) = z - \lambda$. Then we can solve Eq. (6-49) explicitly for u:

$$u(t) = e^{i\lambda t}\, u(0) + i \int_0^t e^{i\lambda(t-s)}\, f(s)\, ds \tag{6-64}$$

Assume first that λ is real. Then the function $f(s) = e^{i\lambda s}/(1 + s)$ is in $C^\infty[0, \infty] \cap L^2(0, \infty)$. For this particular f the solution u is

$$u(t) = e^{i\lambda t}\, [u(0) + i \ln (1 + t)] \tag{6-65}$$

Clearly, this function is not in $L^2(0, \infty)$ for any value of $u(0)$. Thus, no inequality of the form (6-63) could possibly hold. Next, let us suppose $\operatorname{Im} \lambda > 0$. In this case, the solution (6-64) is in $L^2(0, \infty)$ whenever f is. Clearly, the first term on the right-hand side is in $L^2(0, \infty)$. Thus, it suffices to show that

$$v(t) = \int_0^t e^{i\lambda(t-s)}\, f(s)\, ds \tag{6-66}$$

is in $L^2(0, \infty)$ whenever f is. Now, by Schwarz's inequality (1-62)

$$|v(t)|^2 \le \int_0^t e^{-\operatorname{Im}\lambda(t-s)}\,ds \int_0^t e^{-\operatorname{Im}\lambda(t-s)}\,|f(s)|^2\,ds$$

$$\le \frac{1}{\operatorname{Im}\lambda}\int_0^t e^{-\operatorname{Im}\lambda(t-s)}\,|f(s)|^2\,ds$$

Thus

$$\|v\|^2 \le \frac{1}{\operatorname{Im}\lambda}\int_0^\infty \int_0^t e^{-\operatorname{Im}\lambda(t-s)}\,|f(s)|^2\,ds\,dt$$

$$= \frac{1}{\operatorname{Im}\lambda}\int_0^\infty |f(s)|^2 \int_s^\infty e^{-\operatorname{Im}\lambda(t-s)}\,dt\,ds$$

$$\le \frac{1}{(\operatorname{Im}\lambda)^2}\int_0^\infty |f(s)|^2\,ds$$

Hence, we have the estimate

$$\|u\| \le \frac{1}{\operatorname{Im}\lambda}\|f\| + \frac{1}{\sqrt{2\operatorname{Im}\lambda}}|u(0)| \tag{6-67}$$

holding for all solutions of

$$(D_t - \lambda)u(t) = f(t) \qquad t > 0 \tag{6-68}$$

when $\operatorname{Im}\lambda > 0$. Finally, suppose $\operatorname{Im}\lambda < 0$. The only way Eq. (6-64) can be in $L^2(0, \infty)$ is when

$$u(0) = -i\int_0^\infty e^{-\lambda s} f(s)\,ds$$

Thus

$$u(t) = -i\int_t^\infty e^{i\lambda(t-s)} f(s)\,ds \tag{6-69}$$

In this case, we have

$$|u(t)|^2 \le \int_t^\infty e^{-\operatorname{Im}\lambda(t-s)}\,ds \int_t^\infty e^{-\operatorname{Im}\lambda(t-s)}|f(s)|^2\,ds$$

$$\le \frac{1}{|\operatorname{Im}\lambda|}\int_t^\infty e^{-\operatorname{Im}\lambda(t-s)}\,|f(s)|^2\,ds$$

by Schwarz's inequality.

Hence

$$\|u\|^2 \le \frac{1}{|\operatorname{Im}\lambda|}\int_0^\infty \int_t^\infty e^{-\operatorname{Im}\lambda(t-s)}\,|f(s)|^2\,ds\,dt$$

$$= \frac{1}{|\operatorname{Im}\lambda|}\int_0^\infty |f(s)|^2 \int_0^s e^{-\operatorname{Im}\lambda(t-s)}\,dt\,ds$$

$$\le \frac{1}{(\operatorname{Im}\lambda)^2}\int_0^\infty |f(s)|^2\,ds$$

Thus, we have the estimate

$$\| u \| \leq \frac{1}{|\operatorname{Im} \lambda|} \| f \| \tag{6-70}$$

for all solutions of (6-67) when $\operatorname{Im} \lambda < 0$.

To summarize, we note that an estimate of the form (6-63) does not hold for solutions of Eq. (6-68) when λ is real. If $\operatorname{Im} \lambda > 0$ it holds with one boundary condition (i.e., $u(0)$ given), and when $\operatorname{Im} \lambda < 0$, it holds without any boundary condition. We shall see in Sec. 6-6 that this situation is typical of the general case.

We now note a consequence of our work so far. In it we shall make use of the following family of norms

$$\| v \|_k^2 = \int_0^\infty \sum_{j=0}^{k} | D_t^j v(t) |^2 \, dt \qquad k = 0, 1, \ldots \tag{6-71}$$

We let $H^k = H^k(0, \infty)$ denote the completion of $S(0, \infty)$ with respect to the norm (6-71). We have

Theorem 6-9 Let $P(z)$ be a polynomial of degree m without real roots, and let k be a nonnegative integer. Then there is a constant C depending only on $P(z)$ and k, such that, for each $f \in C^k[0, \infty] \cap H^k(0, \infty)$, there is a function $u \in C^{m+k}[0, \infty] \cap H^{m+k}(0, \infty)$ satisfying Eq. (6-49) and

$$\| u \|_{m+k} \leq C \| f \|_k \tag{6-72}$$

In proving Theorem 6-9 we shall make use of the simple

Lemma 6-10 If $f \in H^{k-1}$ and $u \in L^2(0, \infty)$ is a solution of Eq. (6-68), then $u \in H^k$ and

$$\| D_t^k u \|^2 \leq 3^k |\lambda|^{2k} \| u \|^2 + 2 \sum_{j=0}^{k-1} 3^j |\lambda|^{2j} \| D_t^{k-j-1} f \|^2 \tag{6-73}$$

Proof We proceed by induction. Suppose inequality (6-73) holds for k. If $f \in H^k, u \in L^2(0, \infty)$, and Eq. (6-68) holds, then $u \in H^k$ and

$$D_t^{k+1} u = \lambda D_t^k u + D_t^k f$$

Consequently, $u \in H^{k+1}$ and

$$(D_t^{k+1} u, D_t^k u) = \lambda \| D_t^k u \|^2 + (D_t^k f, D_t^k u)$$

$$\begin{aligned} \| D_t^{k+1} u \|^2 &= \lambda (D_t^k u, D_t^{k+1} u) + (D_t^k f, D_t^{k+1} u) \\ &= \lambda [\bar{\lambda} \| D_t^k u \|^2 + (D_t^k u, D_t^k f)] + (D_t^k f, D_t^{k+1} u) \\ &= |\lambda|^2 \| D_t^k u \|^2 + \lambda (D_t^k u, D_t^k f) + (D_t^k f, D_t^{k+1} u) \\ &\leq \tfrac{3}{2} |\lambda|^2 \| D_t^k u \|^2 + \| D_t^k f \|^2 + \tfrac{1}{2} \| D_t^{k+1} u \|^2 \end{aligned}$$

where we used the notation

$$(u, v) = \int_0^\infty u(t)\overline{v(t)}\, dt \tag{6-74}$$

Thus $$\| D_t^{k+1}u \|^2 \leq 3|\lambda|^2 \| D_t^k u \|^2 + 2 \| D_t^k f \|^2 \qquad k = 0, 1, \ldots \tag{6-75}$$

Since (6-73) holds by the induction hypothesis, we have

$$\| D_t^{k+1}u \|^2 \leq 3|\lambda|^2[3^k|\lambda|^{2k} \| u \|^2 + \sum_{j=0}^{k-1} 3^j|\lambda|^{2j} \| D_t^{k-j-1}f \|^2] + 2 \| D_t^k f \|^2$$

$$= 3^{k+1}|\lambda|^{2k+2} \| u \|^2 + \sum_{j=0}^{k} 3^j|\lambda|^{2j} \| D_t^{k-j}f \|^2$$

This shows that (6-73) holds for $k + 1$. Furthermore, (6-75) shows that inequality (6-73) holds for $k = 1$. Thus, the lemma is proved by induction.

We are ready now for the

PROOF of Theorem 6-9 It clearly suffices to prove the theorem for the case $m = 1$. For we can always write

$$P(z) = \prod_{k=1}^{m} (z - \tau_k) \tag{6-76}$$

and, consequently

$$P(D_t) = \prod_{k=1}^{m} P_k(D_t)$$

where $$P_k(D_t) = D_t - \tau_k \tag{6-77}$$

Thus, if the theorem holds for $m = 1$, we know that there is a $u_1 \in C^{k+1}[0, \infty] \cap H^{k+1}(0, \infty)$, such that $P_1(D_t)u = f$ and

$$\| u_1 \|_{k+1} \leq C \| f \|_k$$

with C depending only on $P_1(z)$ and k. Likewise, there is a function $u_2 \in C^{k+2}[0, \infty] \cap H^{k+2}(0, \infty)$, such that $P_2(D_t)u_2 = u_1$ and

$$\| u_2 \|_{k+2} \leq C \| u_1 \|_{k+1}$$

If we repeat the procedure m times, we obtain the desired result.

To prove it for $m = 1$, suppose $f \in C^k[0, \infty] \cap H^k(0, \infty)$, and that $P(z) = z - \lambda$.

Set $$u(t) = i \int_0^t e^{i\lambda(t-s)} f(s)\, ds \qquad \text{if} \quad \operatorname{Im} \lambda > 0 \tag{6-78}$$

and $$u(t) = -i \int_t^\infty e^{i\lambda(t-s)} f(s)\, ds \qquad \text{if} \quad \operatorname{Im} \lambda < 0 \tag{6-79}$$

Clearly, $u \in C^{k+1}[0, \infty]$ and satisfies Eq. (6-68). By (6-67) and (6-70) we see that $u \in L^2(0, \infty)$. We now apply Lemma 6-10 to conclude that $u \in H^{k+1}(0, \infty)$ and satisfies

$$\| u \|_{k+1}^2 \le C(\| u \|^2 + \| f \|_k^2)$$

where the constant C depends only on λ and k. If we now apply inequalities (6-67) and (6-70), we obtain the desired result. This completes the proof.

Corollary 6-11 If $P(z)$ is a polynomial of degree m having all roots with negative imaginary parts, and k is a nonnegative integer, then there is a constant C depending only on $P(z)$ and k, such that

$$\| u \|_{m+k} \le C \| P(D_t)u \|_k \tag{6-80}$$

holds for all $u \in C^{m+k}[0, \infty] \cap H^{m+k}(0, \infty)$.

PROOF For each $f \in C^k[0, \infty] \cap H^k(0, \infty)$, there exists a $u \in C^{m+k}[0, \infty] \cap H^{m+k}(0, \infty)$, such that $P(D_t)u = f$ and (6-72) holds with the constant C depending on $P(z)$ and k (Theorem 6-9). If no roots of $P(z)$ have positive imaginary parts, the function u is the only solution of $P(D_t)u = f$ in $L^2(0, \infty)$ (Theorem 6-6). This gives (6-80).

6-6 ESTIMATES FOR THE GENERAL CASE

We now prove an inequality even stronger than (6-63), for the general case. Our main result is

Theorem 6-12 Let $P(z)$ be a polynomial of degree m having no real roots. Let r be the number of roots (counting multiplicities) with positive imaginary parts. Let $Q_1(z), \ldots, Q_r(z)$ be r polynomials which are linearly independent modulo $P_+(z)$, where $P_+(z)$ is defined by Eq. (6-16). Then, for each nonnegative integer k, there is a constant C, such that

$$\| u \|_{m+k} \le C\left(\| P(D_t)u \|_k + \sum_{j=1}^{r} | Q_j(D_t)u(0) | \right) \tag{6-81}$$

holds for all $u \in C^{m+k}[0, \infty] \cap H^{m+k}(0, \infty)$. In particular

$$\| u \|_{m+k} \le C\left(\| P(D_t)u \|_k + \sum_{j=0}^{r-1} | D_t^j u(0) | \right) \tag{6-82}$$

holds for all such u.

In proving this theorem we shall use two additional lemmas.

Lemma 6-13 There is a constant C, such that

$$\| u \|_{r+k} \le C \| P_+(D_t)u \|_k \tag{6-83}$$

holds for all $u \in C^{r+k}[0, \infty] \cap H^{r+k}(0, \infty)$ satisfying

$$D_t^j u(0) = 0 \qquad 0 \le j < r \tag{6-84}$$

PROOF Let $\tau_1, \ldots, \tau_r$ be the roots of $P_+(z) = 0$. Let u satisfy Eq. (6-84), and set $f = P_+(D)u$, and

$$u_1 = (D_t - \tau_2) \cdots (D_t - \tau_r)u$$

Then u_1 satisfies

$$(D_t - \tau_1)u_1(t) = f(t) \qquad t > 0 \tag{6-85}$$

$$u_1(0) = 0 \tag{6-86}$$

The only solution of this is

$$u_1(t) = i \int_0^t e^{i\tau_1(t-s)} f(s)\, ds \tag{6-87}$$

By (6-67) and (6-73) there is a constant C, such that

$$\| u_1 \|_{k+1} \le C \| f \|_k$$

Next, set

$$u_2 = (D_t - \tau_3) \cdots (D_t - \tau_r)u$$

Then u_2 satisfies

$$(D_t - \tau_2)u_2(t) = u_1(t) \qquad t > 0$$

$$u_2(0) = 0$$

Thus, there is a constant C, such that

$$\| u_2 \|_{k+2} \le C \| u_1 \|_{k+1}$$

Continuing in this way we eventually obtain the desired result.

Lemma 6-14 Functions in $H^1(0, \infty)$ are bounded and satisfy

$$| u(t) |^2 \le 2 \| u \|_1^2 \tag{6-88}$$

PROOF By Lemma 5-8 of Sec. 5-3, we have

$$| u(t) |^2 = | u(\lambda) |^2 - 2 \operatorname{Re} i \int_t^\lambda \bar{u}(s) D_t u(s)\, ds$$

$$\le | u(\lambda) |^2 + \int_t^{t+1} | u(s) |^2\, ds + \int_t^{t+1} | D_t u(s) |^2\, ds$$

for $t \le \lambda \le t + 1$. Integrating both sides with respect to λ from t to $t + 1$, we obtain

$$| u(t) |^2 \le 2 \int_t^{t+1} | u(s) |^2\, ds + \int_t^{t+1} | D_t u(s) |^2\, ds \tag{6-89}$$

which implies (6-88) for $u \in S(0, \infty)$. Since such functions are dense in $H^1(0, \infty)$, the result follows.

Now we can give the

PROOF of Theorem 6-12 Let u be a function in $C^{m+k}[0, \infty] \cap H^{m+k}(0, \infty)$, and set $h = P_+(D_t)u$. Let the polynomials $S_j(z), R_j(z)$ be defined by Eq. (6-58), and set

$$g_j = Q_j(D_t)u(0) - S_j(D_t)h(0) \qquad 1 \le j \le r \tag{6-90}$$

Let v_k be a solution of

$$P_+(D_t)v_k(t) = 0 \qquad t > 0 \tag{6-91}$$

$$R_j(D_t)v_k(0) = \delta_{jk} \qquad 1 \le j \le r \tag{6-92}$$

A solution exists, by Theorem 6-8. Set

$$v = u - \sum g_k v_k \tag{6-93}$$

Then v satisfies

$$P_+(D_t)v(t) = h(t) \qquad t > 0 \tag{6-94}$$

$$D_t^j v(0) = 0 \qquad 0 \le j < r \tag{6-95}$$

Statement (6-94) follows from Eqs. (6-91) and (6-93). To verify Eq. (6-95) note that, by Eq. (6-58)

$$\begin{aligned} R_j(D_t)v(0) &= Q_j(D_t)v(0) - S_j(D_t)h(0) \\ &= g_j - \sum g_k Q_j(D_t)v_k(0) = 0 \qquad 1 \le j \le r \end{aligned}$$

Since the $R_j(z)$ are linearly independent, this is equivalent to Eq. (6-95), by Eq. (6-53).

Since v satisfies Eqs. (6-94), (6-95), there is a constant C, independent of v, such that

$$\| v \|_{m+k} \le C \| h \|_{m-r+k} \tag{6-96}$$

Thus, we have

$$\begin{aligned} \| u \|_{m+k} &\le \| v \|_{m+k} + \sum |g_j| \; \| v_j \|_{m+k} \\ &\le C\left(\| h \|_{m-r+k} + \sum |g_j| \right) \end{aligned} \tag{6-96}$$

where the constant C does not depend on u. Since h satisfies

$$S(D_t)h(t) = P(D_t)u(t)$$

where $S(z)$ is given by Eq. (6-39), we have

$$\| h \|_{m-r+k} \le C \| P(D_t)u \|_k \tag{6-98}$$

by Corollary 6-11. Moreover, by Eq. (6-90)

$$|g_j| \leq |Q_j(D_j)u(0)| + |S_j(D_t)h(0)| \tag{6-99}$$

Since $S_j(z)$ is of degree $< m - r$, we have, by Lemma 6-14 and inequality (6-98)

$$\begin{aligned} |S_j(D_t)h(0)| &\leq 2 \| S_j(D_t)h \|_1 \leq C \| h \|_{m-r} \\ &\leq C' \| P(D_t)u \| \end{aligned} \tag{6-100}$$

If we now combine (6-97) to (6-100), we obtain (6-81). Clearly, (6-82) is a special case. The proof is complete.

We would like to make an important observation concerning inequality (6-81). The constant C there depends continuously upon the coefficients of $P(z)$ and the $Q_j(z)$. By this we mean that if $\tilde{P}(z)$ and $\tilde{Q}_j(z)$ are other polynomials, whose coefficients are close to those of $P(z)$ and the $Q_j(z)$, respectively, then the inequality

$$\| u \|_{m+k} \leq \tilde{C}\left(\| \tilde{P}(D_t)u \|_k + \sum |\tilde{Q}_j(D_t)u(0)| \right) \tag{6-101}$$

holds for all $u \in C^{m+k}[0, \infty] \cap H^{m+k}(0, \infty)$, with the constant $\tilde{C}$ close to C. To see this, note that

$$\| P(D_t)u - \tilde{P}(D_t)u \|_k \leq \varepsilon \| u \|_{m+k}$$

and

$$\begin{aligned} \sum |(Q_j(D_t) - \tilde{Q}_j(D_t))u(0)| &\leq 2 \sum \| Q_j(D_t)u - \tilde{Q}_j(D_t)u \|_1 \\ &\leq 2\varepsilon \| u \|_m \end{aligned}$$

where ε is the maximum difference between corresponding coefficients. Thus, by (6-81), we have

$$\| u \|_{m+k} \leq C\left(\| \tilde{P}(D_t)u \|_k + \sum |\tilde{Q}_j(D_t)u(0)| \right) + 3\varepsilon C \| u \|_{m+k}$$

Now, if ε is sufficiently small, we see that (6-101) holds, with

$$\tilde{C} = \frac{C}{1 - 3\varepsilon C}$$

Clearly, $\tilde{C}$ can be made as close to C as desired by taking ε sufficiently small.

6-7 ESTIMATES IN A HALF-SPACE

In this section we want estimates in a half-space similar to those obtained in Sec. 6-6 for ordinary differential equations. Our first result is

Theorem 6-15 Let $P(\xi, \tau)$ be a homogeneous polynomial of degree m, with the coefficient of τ^m equal to one. Assume that (*a*) For each $\xi \neq 0$ in E^n, the polynomial $P(\xi, \tau)$ has at most r roots $\tau_1(\xi), \ldots, \tau_r(\xi)$ (counting multiplicities), with positive imaginary parts and no real roots.

Let $Q_1(\xi, \tau), \ldots, Q_r(\xi, \tau)$ be r homogeneous polynomials of degrees $m_1, \ldots, m_r$, each less than m. Assume that

(*b*) For each $\xi \neq 0$ in E^n the polynomials $Q_j(\xi, \tau)$ are linearly independent modulo $P_+(\xi, \tau)$, where

$$P_+(\xi, \tau) = (\tau - \tau_1(\xi)) \cdots (\tau - \tau_r(\xi)) \tag{6-102}$$

Then, for each integer $k \geq 0$ there is a constant C, such that

$$\sum_{j=0}^{m+k} |\xi|^{2(m-j)} \int_0^\infty |D_t^j v(\xi, t)|^2 \, dt \leq C\left(\sum_{j=0}^{k} |\xi|^{-2j} \int_0^\infty |D_t^j P(\xi, D_t) v(\xi, t)|^2 \, dt + \sum_{j=1}^{r} |\xi|^{2(m-m_j)-1} |Q_j(\xi, D_t) v(\xi, 0)|^2 \right) \tag{6-103}$$

holds for all functions $v(\xi, t)$, such that for each fixed $\xi, v(\xi, t)$ is in $C^{m+k}[0, \infty] \cap H^{m+k}(0, \infty)$. The constant C does not depend on ξ or v.

PROOF For each $\xi \in E^n$ satisfying $|\xi| = 1$, there is a constant $C = C_\xi$ depending on ξ, such that

$$\sum_{j=0}^{m+k} \int_0^\infty |D_\lambda^j h(\xi, \lambda)|^2 \, d\lambda \leq C\left(\sum_{j=0}^{k} \int_0^\infty |D_\lambda^j P(\xi, D_\lambda) h(\xi, \lambda)|^2 \, dt + \sum_{j=1}^{r} |Q_j(\xi, D_\lambda) h(\xi, 0)|^2 \right) \tag{6-104}$$

holds for all functions h, such that $h(\xi, t) \in C^{m+k}[0, \infty] \cap H^{m+k}(0, \infty)$ for each $\xi \neq 0$, with C not depending on h (Theorem 6-12). Furthermore, the constant C depends continuously on ξ (see the end of Sec. 6-6). Since the set $|\xi| = 1$ is compact in E^n, there is a constant C, independent of ξ, such that (6-104) holds for all h and ξ with $|\xi| = 1$. Now, if the function $v(\xi, \lambda)$ is in $C^{m+k}[0, \infty] \cap H^{m+k}(0, \infty)$ for each $\xi \neq 0$, the same is true of the function $h(\xi, \lambda) = v(\xi, \lambda / |\xi|)$. Substituting into (6-104), we have for $\xi \neq 0$

$$\sum_{j=0}^{m+k} \int_0^\infty \left| D_\lambda^j v\left(\xi, \frac{\lambda}{|\xi|}\right) \right|^2 dt \leq C\left(\sum_{j=0}^{k} \int_0^\infty \left| D_\lambda^j P\left(\frac{\xi}{|\xi|}, D_\lambda\right) v\left(\xi, \frac{\lambda}{|\xi|}\right) \right|^2 d\lambda + \sum_{j=1}^{r} \left| Q_j\left(\frac{\xi}{|\xi|}, D_\lambda\right) v(\xi, 0) \right|^2 \right)$$

If we make the substitution $t = \lambda / |\xi|$, this gives (6-103). The proof is complete.

We now remove the restriction that $P(\xi, \tau)$ and the $Q_j(\xi, \tau)$ are homogeneous.

Theorem 6-16 Let $P(\xi,\tau)$ and $Q_1(\xi,\tau),\ldots,Q_r(\xi,\tau)$ be polynomials whose principal parts satisfy the hypotheses of Theorem 6-15. Then, for each integer $k \geq 0$, there is a constant C depending only on $P(\xi,\tau)$, the $Q_j(\xi,\tau)$, and k, such that

$$\sum_{j=0}^{m+k} |\xi|^{2(m-j)} \int_0^\infty |D_t^j v(\xi,t)|^2\,dt \leq C\Bigg(\sum_{j=0}^{k} |\xi|^{-2j} \int_0^\infty |D_t^j P(\xi,D_t)v(\xi,t)|^2\,dt$$

$$+ \sum_{j=1}^{r} |\xi|^{2(m-m_j)-1} |Q_j(\xi,D_t)v(\xi,0)|^2 + |\xi|^{-2k}\int_0^\infty |v(\xi,t)|^2\,dt\Bigg) \qquad (6\text{-}105)$$

holds for all functions v, such that for each fixed ξ, $v(\xi,t)$ is in $C^{m+k}[0,\infty] \cap H^{m+k}(0,\infty)$.

In proving this theorem we shall make use of the following lemmas.

Lemma 6-17 The inequality

$$|v(\xi,\lambda)|^2 \leq 2\sum_{j=0}^{1} |\xi|^{1-2j}\int_0^\infty |D_t^j v(\xi,t)|^2\,dt \qquad (6\text{-}106)$$

holds for all $\lambda \geq 0$ and all functions v, such that for fixed $\xi, v(\xi,t)$ is in $C^1[0,\infty] \cap H^1(0,\infty)$.

Proof If v is such a function, then for each fixed $\xi \neq 0$ the function $h(\xi,t) = v(\xi,t/|\xi|)$ is also of this type. Thus, by Lemma 6-14

$$|h(\xi,|\xi|\lambda)|^2 \leq 2\sum_0^1 \int_0^\infty |D_\tau^j h(\xi,\tau)|^2\,d\tau$$

If we make the transformation $t = \tau/|\xi|$, we obtain inequality (6-106).

Lemma 6-18 For $k \geq 0$ the inequality

$$\sum_{j=0}^{k} |\xi|^{-2j}|D_t^j v(\xi,t)|^2 \leq 4\sum_{j=0}^{k+1} |\xi|^{1-2j}\int_0^\infty |D_\tau^j v(\xi,\tau)|^2\,d\tau \qquad (6\text{-}107)$$

holds for all $t \geq 0$ and all v, such that for fixed $\xi, v(\xi,t)$ is in $C^k[0,\infty] \cap H^k(0,\infty)$.

Proof By Lemma 6-17

$$\sum_{j=0}^{k} |\xi|^{-2j}|D_t^j v(\xi,t)|^2 \leq 2\sum_{j=0}^{k} |\xi|^{1-2j}\int_0^\infty |D_\tau^j v(\xi,\tau)|^2\,d\tau$$

$$+ 2\sum_{j=0}^{k} |\xi|^{-1-2j}\int_0^\infty |D_\tau^{j+1} v(\xi,\tau)|^2\,d\tau$$

$$\leq 4\sum_{j=0}^{k+1} |\xi|^{1-2j}\int_0^\infty |D_\tau^j v(\xi,\tau)|^2\,d\tau$$

This is the desired inequality.

Lemma 6-19 If $u(t) \in L^2(-\infty, \infty)$ has continuous derivatives up to order two in $(-\infty, \infty)$, and these derivatives are in $L^2(-\infty, \infty)$, then

$$\int_{-\infty}^{\infty} |D_t u(t)|^2 \, dt = \int_{-\infty}^{\infty} u\overline{D_t^2 u} \, dt \tag{6-108}$$

Consequently, for each $\varepsilon > 0$

$$\int_{-\infty}^{\infty} |D_t u(t)|^2 \, dt \leq \varepsilon \int_{-\infty}^{\infty} |D_t^2 u(t)|^2 \, dt + \frac{1}{4\varepsilon} \int_{-\infty}^{\infty} |u(t)|^2 \, dt \tag{6-109}$$

PROOF For $R > 0$

$$\int_{-R}^{R} D_t u \overline{D_t u} \, dt = \int_{-R}^{R} u\overline{D_t^2 u} \, dt - iu(R)\overline{D_t u(R)} + iu(-R)\overline{D_t u(-R)} \tag{6-110}$$

Since u and its derivatives up to order two are in $L^2(-\infty, \infty)$, we have

$$\int_{R}^{R+1} (|u(t)|^2 + |D_t u(t)|^2 + |D_t^2 u(t)|^2) \, dt \to 0$$

as $R \to \infty$. We see, by (6-89), that $u(R) \to 0$ and $D_t u(R) \to 0$ as $R \to \infty$. Similarly, $u(-R) \to 0$ and $D_t u(-R) \to 0$ as $R \to \infty$. If we let $R \to \infty$ in Eq. (6-110), we obtain Eq. (6-108). Inequality (6-109) is a simple consequence.

Lemma 6-20 There is a constant K, such that

$$\int_{0}^{\infty} |D_t u(t)|^2 \, dt \leq \varepsilon \int_{0}^{\infty} |D_t u(t)|^2 \, dt + \frac{K}{\varepsilon} \int_{0}^{\infty} |u(t)|^2 \, dt \tag{6-111}$$

holds for each $u \in H^2(0, \infty)$ and $\varepsilon > 0$.

PROOF Since by definition $S(0, \infty)$ is dense in $H^2(0, \infty)$, it suffices to prove the inequality for $u \in S(0, \infty)$. Set

$$\begin{aligned} v(t) &= u(t) & t > 0 \\ &= 6u(-t) - 8u(-2t) + 3u(-3t) & t < 0 \end{aligned} \tag{6-112}$$

It follows that $v(t)$ has continuous derivatives up to order two in $(-\infty, \infty)$ and these derivatives are in $L^2(-\infty, \infty)$. Moreover, there is a constant C, not depending on u, such that

$$\int_{-\infty}^{0} |D_t^k v(t)|^2 \, dt \leq C \int_{0}^{\infty} |D_t^k u(t)|^2 \, dt \qquad k = 0, 1, 2 \tag{6-113}$$

Thus, by Lemma 6-19

$$\int_0^\infty |D_t u(t)|^2\,dt \le \int_{-\infty}^\infty |D_t v(t)|^2\,dt$$

$$\le \frac{\varepsilon}{C+1}\int_{-\infty}^\infty |D_t^2 v(t)|^2\,dt + \frac{C+1}{4\varepsilon}\int_{-\infty}^\infty |v(t)|^2\,dt$$

$$\le \varepsilon\int_0^\infty |D_t^2 u(t)|^2\,dt + \frac{(C+1)^2}{4\varepsilon}\int_0^\infty |u(t)|^2\,dt$$

This gives (6-111), and the proof is complete.

Lemma 6-21 For each $\varepsilon > 0$ and each integer $k \ge 1$, there is a constant K, such that

$$\sum_{j=0}^{k-1} |\xi|^{-2j-2}\int_0^\infty |D_t^j v(\xi,t)|^2\,dt \le \varepsilon\sum_{j=0}^{k} |\xi|^{-2j}\int_0^\infty |D_t^j v(\xi,t)|^2\,dt + K\varepsilon^{-k}|\xi|^{-2k}\int_0^\infty |v(\xi,t)|^2\,dt \tag{6-114}$$

holds for all v, such that for each fixed $\xi, v(\xi,t)$ is in $H^k(0,\infty)$.

PROOF By Lemma 6-20

$$\int_0^\infty |D_t^j v(\xi,t)|^2\,dt \le \varepsilon\int_0^\infty |D_t^{j+1} v(\xi,t)|^2\,dt + \frac{K}{\varepsilon}\int_0^\infty |D_t^{j-1} v(\xi,t)|^2\,dt$$

holds for any $j \ge 1$, where the constant K does not depend on j, ε, ξ, or v.

Thus

$$\begin{aligned}\sum_{j=1}^{k-1} |\xi|^{-2j-2}\int_0^\infty |D_t^j v(\xi,t)|^2\,dt &\le \varepsilon\sum_{j=1}^{k-1} |\xi|^{-2j-2}\int_0^\infty |D_t^{j+1} v(\xi,t)|^2\,dt \\ &\quad + \frac{K}{\varepsilon}\sum_{j=1}^{k-1} |\xi|^{-2j-2}\int_0^\infty |D_t^{j-1} v(\xi,t)|^2\,dt \\ &= \varepsilon\sum_{j=2}^{k} |\xi|^{-2j}\int_0^\infty |D_t^j v(\xi,t)|^2\,dt \\ &\quad + \frac{K}{\varepsilon}\sum_{j=0}^{k-2} |\xi|^{-2j-4}\int_0^\infty |D_t^j v(\xi,t)|^2\,dt\end{aligned} \tag{6-115}$$

Since $$1 \le \varepsilon|\xi|^2 + \frac{1}{4|\xi|^2\varepsilon}$$

we also have

$$|\xi|^{-2}\int_0^\infty |v(\xi,t)|^2\,dt \le \varepsilon\int_0^\infty |v(\xi,t)|^2\,dt + \frac{1}{4\varepsilon}|\xi|^{-4}\int_0^\infty |v(\xi,t)|^2\,dt \tag{6-116}$$

Combining (6-115) and (6-116), we have

$$\sum_{j=0}^{k-1}|\xi|^{-2j-2}\int_0^\infty |D_t^j v(\xi,t)|^2\,dt \le \varepsilon\sum_{j=0}^{k}|\xi|^{-2j}\int_0^\infty |D_t^j v(\xi,t)|^2\,dt + \frac{K_1}{\varepsilon}\sum_{j=0}^{k-2}|\xi|^{-2j-4}\int_0^\infty |D_t^j v(\xi,t)|^2\,dt \tag{6-117}$$

We can now prove (6-114), by induction. It clearly holds for $k = 1$. Assume that it holds for $k = m$. Then, by inequality (6-117)

$$\begin{aligned}\sum_{j=0}^{m}|\xi|^{-2j-2}\int_0^\infty |D_t^j v(\xi,t)|^2\,dt &\le \varepsilon\sum_{j=0}^{m+1}|\xi|^{-2j}\int_0^\infty |D_t^j v(\xi,t)|^2\,dt \\ &\quad + \frac{K_1}{\varepsilon}\sum_{j=0}^{m-1}|\xi|^{-2j-4}\int_0^\infty |D_t^j v(\xi,t)|^2\,dt \\ &\le \varepsilon\sum_{j=0}^{m+1}|\xi|^{-2j}\int_0^\infty |D_t^j v(\xi,t)|^2\,dt \\ &\quad + \frac{K_1}{\varepsilon}\left(\frac{\varepsilon}{2K_1}\sum_{j=0}^{m}|\xi|^{2-2j}\int_0^\infty |D_t^j v(\xi,t)|^2\,dt\right. \\ &\quad \left. + K_2\left(\frac{\varepsilon}{2K_1}\right)^{-m}|\xi|^{-2m-2}\int_0^\infty |v(\xi,t)|^2\,dt\right)\end{aligned}$$

where we have used the induction hypothesis. This gives

$$\sum_{j=0}^{m}|\xi|^{-2j-2}\int_0^\infty |D_t^j v(\xi,t)|^2\,dt \le 2\varepsilon\sum_{j=0}^{m+1}|\xi|^{-2j}\int_0^\infty |D_t^j v(\xi,t)|^2\,dt + K_3\varepsilon^{-m-1}|\xi|^{-2m-2}\int_0^\infty |v(\xi,t)|^2\,dt$$

Hence (6-114) holds for $k = m + 1$. The induction proof is complete.

We are now ready for the

PROOF of Theorem 6-16 Let $p(\xi,\tau)$, $q_1(\xi,\tau),\ldots,q_r(\xi,\tau)$ be the principal parts of the polynomials $P(\xi,\tau), Q_1(\xi,\tau),\ldots,Q_r(\xi,\tau)$, respectively. By Theorem 6-15, there is a constant C, such that

$$\sum_{j=0}^{m+k} |\xi|^{2(m-j)} \int_0^\infty |D_t^j v(\xi,t)|^2 \, dt \le C\left(\sum_{j=0}^{k} |\xi|^{-2j} \int_0^\infty |D_t^j p(\xi, D_t) v(\xi,t)|^2 \, dt \right.$$
$$\left. + \sum_{j=1}^{r} |\xi|^{2(m-m_j)-1} |q_j(\xi, D_t) v(\xi, 0)|^2 \right) \tag{6-118}$$

Since $P(\xi,\tau) - p(\xi,\tau)$ is of degree $<m$, we have

$$\sum_{j=0}^{k} |\xi|^{-2j} \int_0^\infty |D_t^j [P(\xi, D_t) - p(\xi, D_t)] v(\xi,t)|^2 \, dt$$
$$\le C \sum_{j=0}^{m+k-1} |\xi|^{2(m-j-1)} \int_0^\infty |D_t^j v(\xi,t)|^2 \, dt \tag{6-119}$$

Since $Q_j(\xi,\tau) - q_j(\xi,\tau)$ is of degree $<m_j$, we have

$$|\xi|^{2(m-m_j)-1} |[Q_j(\xi, D_t) - q_j(\xi, D_t)] v(\xi,0)|^2$$
$$\le C \sum_{j=0}^{m_j-1} |\xi|^{2(m-j)-3} |D_t^j v(\xi,0)|^2$$
$$\le C' \sum_{j=0}^{m_j} |\xi|^{2(m-j)} \int_0^\infty |D_t^j v(\xi,t)|^2 \, dt \tag{6-120}$$

by Lemma 6-18. Moreover, by Lemma 6-21, for each $\varepsilon > 0$

$$\sum_{j=0}^{m+k-1} |\xi|^{2(m-j-1)} \int_0^\infty |D_t^j v(\xi,t)|^2 \, dt \le \varepsilon \sum_{j=0}^{m+k} |\xi|^{2(m-j)} \int_0^\infty |D_t^j v(\xi,t)|^2 \, dt$$
$$+ K\varepsilon^{-m-k} |\xi|^{-2k} \int_0^\infty |v(\xi,t)|^2 \, dt \tag{6-121}$$

If we combine inequalities (6-118) to (6-121) and take ε sufficiently small, we obtain the desired inequality (6-105). This completes the proof.

Some of the results of this section can be expressed conveniently in terms of norms taken over the half-space Ω. For this purpose, we introduce the following families of norms, similar to those employed in Secs. 5-2 and 5-4. For k a nonnegative integer, s real, and $a \ge 0$, we set

$$[|u\|_{k,s}^{(a)}]^2 = \int \sum_{j=0}^{k} |\xi|^{2(k+s-j)} |D_t^j F u(\xi, a)|^2 \, d\xi \tag{6-122}$$

and

$$|u\|_{k,s}^2 = \int_0^\infty [|u\|_{k,s}^{(t)}]^2 \, dt \tag{6-123}$$

In terms of these norms, we have

Theorem 6-22 Under the hypotheses of Theorem 6-15

$$|u\|_{m+k,s} \le C\left(|P(D)u\|_{k,s} + \sum_{j=1}^{r} |Q_j(D)u\|^{(0)}_{0,m+k+s-m_j-1/2}\right) \tag{6-124}$$

holds for all $u \in H^{m+k,s}(\Omega)$, where C is the constant in (6-103).

Proof From the definition of the spaces $H^{k,s}(\Omega)$ (see Sec. 5-2) we see that it suffices to prove (6-124) for $u \in S(\Omega)$. We merely multiply both sides of (6-103) by $|\xi|^{2s}$, and integrate over E^n with respect to ξ. The result now follows from Eq. (2-22) of Sec. 2-3.

Theorem 6-23 Under the hypotheses of Theorem 6-16

$$|u\|_{m+k,s} \le C\left(|P(D)u\|_{k,s} + \sum_{j=1}^{r} |Q_j(D)u\|^{(0)}_{0,m+k+s-m_j-1/2} + |u\|_{0,s}\right) \tag{6-125}$$

holds for all $u \in H^{m+k,s}(\Omega)$, where C is the constant in (6-105).

Proof We multiply (6-105) by $|\xi|^{2k}$ and integrate with respect to ξ.

Theorem 6-24 For $k \ge 0$ an integer, and s real

$$|u\|^{(t)}_{k,s+1/2} \le 2|u\|_{k+1,s} \qquad t \ge 0 \qquad u \in H^{k+1,s} \tag{6-126}$$

Proof This follows from Lemma 6-18

Theorem 6-25 For each integer $k \ge 1$ there is a constant K, such that

$$|u\|_{k-1,s} \le \varepsilon|u\|_{k,s} + K\varepsilon^{-k}|u\|_{0,s} \tag{6-127}$$

holds for all $\varepsilon > 0$, real s, and $u \in H^{k,s}$.

Proof Apply Lemma 6-21.

We can also apply the norms $|u|_{k,s}$ given by Eq. (5-13) of Sec. 5-2. In order for them to apply to functions defined on Ω, we must take $b = \infty$ there. Thus, in the present situation

$$|u|^2_{k,s} = \int d\xi \int_0^\infty \sum_{j=0}^{k} (1 + |\xi|)^{2(k+s-j)} |D_t^j Fu(\xi, t)|^2\, dt \tag{6-128}$$

Note the distinction between this and Eq. (6-123). Clearly, we have

$$|u\|_{k,s} \le |u|_{k,s} \qquad s \ge 0 \tag{6-129}$$

We also have

$$|u|_{k,s} \le C(|u\|_{0,s} + |u\|_{k,s}) \qquad s \ge 0 \tag{6-130}$$

where the constant C depends only on k and s (see inequality (2-45) of Sec. 2-4). By (6-125) and (6-130), we have

Theorem 6-26 When $s \geq 0$, under the hypotheses of Theorem 6-16

$$|u|_{m+k,s} \leq C\left(\| P(D)u \|_{k,s} + \sum_{j=1}^{r} \| Q_j(D)u \|^{(0)}_{0,m+k+s-m_j-1/2} + \| u \|_{0,s} \right) \qquad (6\text{-}131)$$

holds for all $u \in H^{m+k,s}(\Omega)$.

6-8 EXISTENCE IN A HALF-SPACE

We now apply the inequalities obtained in Sec. 6-7 to prove existence of solutions of

$$P(D)u(x,t) = f(x,t) \qquad t > 0 \qquad (6\text{-}132)$$

$$Q_j(D)u(x,0) = g_j(x) \qquad 1 \leq j \leq r \qquad (6\text{-}133)$$

in the half-space $t \geq 0$. We shall prove

Theorem 6-27 Let $P(D)$ and the $Q_j(D)$ satisfy the hypotheses of Theorem 6-15. Let Ω denote the half-space $t > 0$, and suppose $f \in C^\infty(\bar{\Omega})$ with $D_x^\mu D_t^k f \in L^2(\Omega)$ for each k and μ. Assume that each $g_j \in C^\infty(E^n)$ with $D^\mu g \in L^2(E^n)$ for each μ. Then there exists a unique function $u \in C^\infty(\bar{\Omega})$, satisfying Eqs. (6-132), (6-133), and such that $D_x^\mu D_t^k \in L^2(\Omega)$ for each k and μ.

In proving the theorem we shall make use of some lemmas.

Lemma 6-28 If $v(x) \in C^{|\mu|}(E^n) \cap L^2(E^n)$ and $\xi^\mu F v \in L^2(E^n)$, then

$$F(D^\mu v) = \xi^\mu F v \qquad (6\text{-}134)$$

PROOF For $\varphi \in C_0^\infty(E^n)$, we have

$$(D^\mu v, \varphi) = (v, D^\mu \varphi) = (Fv, \xi^\mu F\varphi)$$
$$= (F^{-1}[\xi^\mu F v], \varphi)$$

by Eq. (2-22) of Sec. 2-3. Hence, by Lemma 4-17 of Sec. 4-6

$$D^\mu v = F^{-1}[\xi^\mu F v]$$

Applying the Fourier transform to both sides, we obtain Eq. (6-134).

Again we make use of the norms $|u|_{k,s}$ given by Eq. (6-128) (see Sec. 5-2).

Lemma 6-29 If $v(x,t)$ and $D_t v(x,t)$ are both in $L^2(\Omega)$, and $|v|_{1,s} < \infty$ for some $s > (n+1)/2$, then v is a bounded function on Ω with

$$\operatorname*{lub}_{\Omega} |v| \leq C |v|_{1,s}$$

The constant C does not depend on v.

PROOF Following the proof of inequality (2-34) of Sec. 2-3, we have

$$|v(x,t)| \leq C \int (1 + |\xi|)^{1/2-s}(1 + |\xi|)^{s-1/2} |Fv| \, d\xi$$

Thus
$$|v(x,t)|^2 \leq C \int (1 + |\xi|)^{1-2s} \, d\xi \int (1 + |\xi|)^{2s-1} |Fv|^2 \, d\xi$$

Moreover, by Lemma 6-17

$$|Fv|^2 \leq 2 \sum_0^1 (1 + |\xi|)^{1-2j} \int_0^\infty |D_t^j Fv|^2 \, dt$$

Combining the last two inequalities, we obtain

$$|v(x,t)|^2 \leq C \sum_{j=0}^1 \int_0^\infty \int (1 + |\xi|)^{2(s-j)} |D_t^j Fv|^2 \, dt$$

This gives the lemma.

Lemma 6-30 Let k, m be integers and s a real number satisfying $k > m \geq 0$ and $s - m > (n + 1)/2$. Suppose $u \in L^2(\Omega)$ and there is a sequence $\{v_j\}$ of functions in $C^m(\Omega)$, such that $|v_j|_{k,s} \leq C$ and $v_j \to u$ in $L^2(\Omega)$. Then, $u \in C^m(\Omega)$ and $D_x^\mu D_t^l u \in L^2(\Omega)$ for $|\mu| + l \leq m$.

PROOF Let M be the completion of the set of functions $w \in C^m(\Omega)$, such that $|w|_{k,s} < \infty$. Clearly, M is a Hilbert space and $\{v_j\}$ is a bounded sequence in M. Thus, by the Banach–Saks Lemma (Theorem 1-9 of Sec. 1-5), $\{v_j\}$ has a subsequence whose arithmetic means $\{w_j\}$ converge in M. They also converge to u in $L^2(\Omega)$. By Lemma 6-29

$$\operatorname{lub}_{\Omega} |D_x^\mu D_t^i(w_j - w_l)| \leq C |w_j - w_l|_{k,s}$$

for $|\mu| + i \leq m$. Thus, all derivatives up to order m of the w_j converge uniformly on Ω. Since u is the limit of the w_j in $L^2(\Omega)$, it must be their uniform limit as well. Thus, $u \in C^m(\Omega)$ and its derivatives up to order m are the uniform limits of those of the w_j. Since these derivatives of the w_j converge also in $L^2(\Omega)$, the corresponding derivatives of u must be their limits and, consequently, must be in $L^2(\Omega)$. This completes the proof.

We can now give the

PROOF of Theorem 6-27 By Theorem 6-8, for each $\xi \neq 0$ in E^m, there is a function $h(\xi, t) \in S(0, \infty)$, such that

$$P(\xi, D_t)h(\xi, t) = Ff(\xi, t) \qquad t > 0 \tag{6-135}$$

$$Q_j(\xi, D_t)h(\xi, 0) = Fg_j(\xi) \qquad 1 \leq j \leq r \tag{6-136}$$

(note that $Ff \in S(0, \infty)$ for each ξ). By Theorem 6-15, for each $k \geq 0$ there is a constant C, such that

$$\sum_{j=0}^{m+k} |\xi|^{2(m+s-j)} \int_0^\infty |D_t^j h(\xi,t)|^2 \, dt \le C\left(\sum_{j=0}^{k} |\xi|^{2(s-j)} \int_0^\infty |D_t^j F f(\xi,t)|^2 \, dt + \sum_{j=1}^{r} |\xi|^{2(m+s-m_j)-1} |Fg_j(\xi)|^2 \right) \tag{6-137}$$

holds for all ξ and s, and C does not depend on h, f, or the g_j. By hypothesis, $|\xi|^j D_t^k F f(\xi,t)$ is in $L^2(\Omega)$ for each $j \ge 0$ and $k \ge 0$, and $|\xi|^k Fg_j(\xi)$ is in $L^2(E^n)$ for each $1 \le j \le r$ and $k \ge 0$. Consequently, (6-137) implies that $|\xi|^j D_t^k h(\xi,t)$ is in $L^2(\Omega)$ for each $j \ge 0$ and $k \ge 0$.

For each $t > 0$, set $u(x,t) = F^{-1}h(\xi,t)$. Then $u \in L^2(\Omega)$, and by what we have just shown $|u|_{k,s} < \infty$ for each k and s. Let J_ε be the mollifier on E^n, defined in Sec. 2-2. We claim that

1. $J_\varepsilon u \in C^\infty(\Omega)$ for each $\varepsilon > 0$
2. $J_\varepsilon u \to u$ in $L^2(\Omega)$ as $\varepsilon \to 0$
3. $|J_\varepsilon u|_{k,s} \le |u|_{k,s}$ for each k and s.

From these statements and Lemma 6-30, it follows that $u \in C^\infty(\Omega)$ and that $D_x^\mu D_t^k u \in L^2(\Omega)$ for each μ and k. If we now apply the inverse Fourier transform to Eqs. (6-135) and (6-136), we obtain Eqs. (6-132), (6-133) in view of Lemma 6-28. Thus, $u(x,t)$ is the desired solution. Since $h(\xi,t) = Fu$ satisfies the conditions under which (6-103) holds, uniqueness follows from Theorem 6-15.

Thus, it remains only to verify 1 to 3. First, note that

$$J_\varepsilon u(x,t) = \int\int j_\varepsilon(x-y)\, e^{-i\xi y}\, h(\xi,t)\, d\xi\, dy$$

Since all derivatives of the integrand with respect to x,t are absolutely integrable, we can differentiate under the integral sign as much as we like. Thus, 1 is true. In proving 2, we note that

$$F(J_\varepsilon u) = Fj_\varepsilon Fu \tag{6-138}$$

To see this, let $\varphi(\xi)$ be any function in $C_0^\infty(E^n)$. For fixed t, we have

$$\begin{aligned}(F[J_\varepsilon u], \varphi) &= (u, J_\varepsilon F^{-1}\varphi) \\ &= (u, F^{-1}[Fj_\varepsilon \cdot \varphi]) = (Fj_\varepsilon Fu, \varphi)\end{aligned}$$

by Eqs. (2-24) and (2-62) of Chapter 2 (the scalar product is that of $L^2(E^n)$). Since this is true for each $\varphi \in C_0^\infty(E^n)$, we obtain Eq. (6-138). Now, by (2-65) of Sec. 2-5

$$\begin{aligned}|Fj_\varepsilon - 1| &= \left| \int [e^{-i\varepsilon\xi x} - 1] j(x)\, dx \right| \\ &\le \varepsilon |\xi| \int |x| j(x)\, dx\end{aligned}$$

Thus, by Eq. (6-138)

$$|J_\varepsilon u - u|_{0,0}^2 = \iint_\Omega |Fu|^2 |Fj_\varepsilon - 1|^2 \, d\xi \, dt$$

$$\leq \varepsilon^2 \left(\int |x| j(x) \, dz \right)^2 \iint_\Omega |\xi|^2 |Fu|^2 \, d\xi \, dt$$

Since $|\xi| h(\xi, t)$ is in $L^2(\Omega)$, this implies 2 by Parseval's identity. Finally, we note that, by Eq. (2-62) of Sec. 2-5 and Eq. (6-138)

$$|F(J_\varepsilon u)| \leq |Fu|$$

This immediately gives 3. The proof is complete.

6-9 SOME OBSERVATIONS

Since we have established existence and uniqueness of solutions of Eqs. (6-132), (6-133) under the hypotheses of Theorem 6-15, we should determine what kind of operators satisfy those hypotheses. Towards this end we note first that a homogeneous operator $P(D)$, satisfying the hypotheses of Theorem 6-15, must be elliptic. For suppose (ξ, τ) is real and a solution of

$$P(\xi, \tau) = 0 \tag{6-139}$$

If $\xi = 0$, then $P(\xi, \tau) = \tau^m$. Thus, τ must vanish as well. If $\xi \neq 0$, then Eq. (6-139) cannot hold for τ real by (*a*) of Theorem 6-15. Thus, the only real solution of Eq. (6-139) is $\xi = 0$ and $\tau = 0$.

We also note

Theorem 6-31 If $n > 1$, then every homogeneous elliptic operator $P(D)$ on E^{n+1} is of even order m, and satisfies condition (*a*) of Theorem 6-15 with $r = m/2$. Moreover, it cannot satisfy this condition for any smaller value.

Proof Let $\xi \neq 0$ be any vector in E^n. If only one component of ξ does not vanish, say $\xi_1 \neq 0, \xi_2 = \cdots = \xi_n = 0$, set

$$\xi^{(s)} = (s\xi_1, \quad \sqrt{(1 - s^2)}\, \xi_1, 0, \ldots, 0)$$

If more than one component of ξ does not vanish, say $\xi_1 \neq 0, \xi_2 \neq 0$, set

$$\xi^{(s)} = ([s + \sqrt{(1 - s^2)}]\, \xi_1, s\xi_2, \ldots, s\xi_n)$$

In either case, $\xi^{(s)}$ is a continuous function of s and does not vanish in the interval $-1 \leq s \leq 1$. Moreover

$$\xi^{(-1)} = -\xi \qquad \xi^{(1)} = \xi \tag{6-140}$$

Since

$$P(\xi^{(s)}, \tau) = 0 \tag{6-141}$$

cannot have real roots in this interval, and the roots depend continuously on $\xi^{(s)}$ (Lemma 4-14 of Sec. 4-4), the number of roots of Eq. (6-141) (counting multiplicities) with positive imaginary parts is constant in $-1 \le s \le 1$. Since

$$P(-\xi, -\tau) = (-1)^m P(\xi, \tau)$$

the roots of

$$P(-\xi, \tau) = 0 \tag{6-142}$$

are just the negatives of the roots of Eq. (6-139). Thus, the number of roots of (6-139) with negative imaginary parts, equals the number of roots of Eq. (6-142) with positive imaginary parts. But, by Eq. (6-140), Eqs. (6-139) and (6-142) have the same number of roots with positive imaginary parts. Hence, the number of roots of (6-139) with positive imaginary parts equals the number of roots with negative imaginary parts. Since it has m roots all together and no real roots, the number of roots with positive imaginary parts must equal $m/2$. Thus, m must be even and $r = m/2$. This completes the proof.

Theorem 6-31 is due to Lopatinski (1953). An elliptic operator $P(D)$ which is of even order and such that its principal part $p(\xi, \tau)$ has exactly half its roots above and half below the real axis, for each $\xi \neq 0$, is called *properly elliptic*. Theorem 6-31 says that in higher than two dimensions every elliptic operator is properly elliptic. Note also that an elliptic operator with real coefficients must be properly elliptic.

PROBLEMS

6-1 Derive Eq. (6-11).

6-2 Prove inequality (6-24).

6-3 Show that $v(t)$ given by Eq. (6-35) is in $S(0, \infty)$ and satisfies Eq. (6-36).

6-4 Show that (6-45) implies (6-46) for t sufficiently close to t_0.

6-5 Suppose the constants g_k satisfy Eq. (6-55). Show that the function $u(t)$ defined by means of Eqs. (6-25) and (6-26) is a solution of Eqs. (6-56), (6-57).

6-6 Show that Eqs. (6-40), (6-59), (6-60) imply Eqs. (6-49), (6-50). Also show that Eqs. (6-40), (6-49), (6-50) imply Eqs. (6-59), (6-60).

6-7 Derive Eq. (6-64).

6-8 Show that the function given by Eq. (6-65) is not in $L^2(0, \infty)$.

6-9 Derive inequality (6-109) from Eq. (6-108).

6-10 Compute the constant C in inequality (6-113) and, hence, the constant K in (6-111).

6-11 Show that the definition of proper ellipticity does not depend on which of the ξ_k was chosen to be τ (i.e., any choice other than $\xi_{n+1} = \tau$ would have given the same result).

6-12 Prove inequality (6-130).

CHAPTER

SEVEN

BOUNDARY VALUE PROBLEMS IN A HALF-SPACE (NON-ELLIPTIC)

7-1 INTRODUCTION

In Theorem 6-27 of Sec. 6-8 we used inequality (6-137) to conclude that the solution $h(\xi, t)$ of Eqs. (6-135), (6-136) has the property that $|\xi|^j D_t^k h(\xi, t) \in L^2(\Omega)$ for each j and k. If $P(D)$ and the $Q_j(D)$ are not homogeneous, we have only (6-105) to use in place of (6-103) (Sec. 6-7). This gives (6-137) with the added term

$$|\xi|^{2(m+s-k)} \int_0^\infty |h(\xi, t)|^2 \, dt$$

If we knew that $h(\xi, t)$ is in $L^2(\Omega)$, then it would follow as before that $|\xi|^j D_t^k h(\xi, t) \in L^2(\Omega)$ for each j and k. However, the development so far does not give us a way of determining this. We now give a different approach which is more general.

As before, it is convenient to consider first the problem for a half-line. We do this in Sec. 7-2, 7-3, and 7-5. The problem in a half-space is considered in Secs. 7-6 and 7-8. Examples are given in Sec. 7-7.

7-2 THE ESTIMATE IN A HALF-LINE

We shall prove an inequality which is more general than inequality (6-81) of Sec. 6-6. Let $P(z)$ be a polynomial in one variable, of degree m, having no real roots. Let r be the number of roots (counting multiplicities) with positive imaginary parts. Thus, if $\tau_1, \ldots, \tau_m$ are the roots, we may order them, so that

$$\operatorname{Im} \tau_k > 0 \qquad 1 \le k \le r \tag{7-1}$$

As before, we set

$$P_+(z) = (z - \tau_1) \cdots (z - \tau_r) \tag{7-2}$$

and

$$P_-(z) = \frac{P(z)}{P_+(z)} \tag{7-3}$$

For any polynomial $Q(z)$ of degree $<m$, we can resolve $Q(z)/P(z)$ into partial fractions:

$$\frac{Q(z)}{P(z)} = \frac{Q_+(z)}{P_+(z)} + \frac{Q_-(z)}{P_-(z)} \tag{7-4}$$

where $Q_+(z)$ is of degree $<r$ and $Q_-(z)$ is of degree $<m - r$. Suppose we are given r such polynomials $Q_1(z), \ldots, Q_r(z)$. Set

$$\alpha_{ij} = \int_{-\infty}^{\infty} \frac{Q_{i+}(\tau)\overline{Q_{j+}(\tau)}}{|P_+(\tau)|^2}\, d\tau \qquad 1 \le i, j \le r \tag{7-5}$$

and

$$\beta_{ij} = \int_{-\infty}^{\infty} \frac{Q_{i-}(\tau)\overline{Q_{j-}(\tau)}}{|P_-(\tau)|^2}\, d\tau \qquad 1 \le i, j \le r \tag{7-6}$$

Consider the $r \times r$ Hermitian matrices $A = (\alpha_{ij})$ and $B = (\beta_{ij})$. We shall prove

Theorem 7-1 Assume that A^{-1} exists and that there is a constant K_1, such that

$$BA^{-1}B \le K_1 B \tag{7-7}$$

If $R(z)$ is any polynomial satisfying

$$|R(\tau)| \le C_1 |P(\tau)| \qquad \tau \in E^1 \tag{7-8}$$

then there is a constant C depending only on C_1 and K_1, such that

$$\int_0^{\infty} |R(D_t)u|^2\, dt \le C\left[\int_0^{\infty} |P(D_t)u|^2\, dt + \sum_{i,j=1}^{r} \alpha^{ij} U_i \overline{U}_j\right] \qquad u \in S(0, \infty) \tag{7-9}$$

where $(\alpha^{ij}) = A^{-1}$, and

$$U_i = Q_i(D_t)u(0) \qquad 1 \le i \le r \tag{7-10}$$

Inequality (7-7) means that for each complex vector $U = (U_1, \ldots, U_r)$,

$$U^* BA^{-1} BU \le K_1 U^* BU$$

A brief review of matrix theory will be given in Sec. 7-4. The proof of Theorem 7-1 will be given in Sec. 7-3.

In proving Theorem 7-1 we shall make use of the following two lemmas. They will be proved in Sec. 7-5.

Lemma 7-2 Suppose $u \in S(0, \infty)$ and

$$\begin{aligned} w(t) &= u(t) \qquad t > 0 \\ &= 0 \qquad t < 0 \end{aligned} \tag{7-11}$$

If

$$\tilde{h}(\tau) = \int e^{-it\tau} h(t)\, dt \tag{7-12}$$

denotes the Fourier transform on E^1, then

$$\widetilde{D_t w} = \tau \tilde{w} + iu(0) \tag{7-13}$$

Similarly, if $v \in S(-\infty, 0)$ and

$$\begin{aligned} g(t) &= 0 \qquad t > 0 \\ &= v(t) \qquad t < 0 \end{aligned} \tag{7-14}$$

then

$$\widetilde{D_t g} = \tau \tilde{g} - iv(0) \tag{7-15}$$

Lemma 7-3 If w and g are given by Eqs. (7-11) and (7-14), respectively, then

$$\begin{aligned} u(0) &= \frac{1}{\pi} \lim_{R \to \infty} \int_{-R}^{R} \tilde{w}(\tau)\, d\tau \\ v(0) &= \frac{1}{\pi} \lim_{R \to \infty} \int_{-R}^{R} \tilde{g}(\tau)\, d\tau \end{aligned} \tag{7-16}$$

We shall sometimes write

$$\int_{-\infty}^{\infty} \quad \text{in place of} \quad \lim_{R \to \infty} \int_{-R}^{R}$$

when no confusion will result. Such an integral is taken in the Cauchy principal value sense.

The lemmas have the following consequences.

Corollary 7-4 Under the hypotheses of Lemma 7-2

$$\widetilde{D_t^k w} = \tau^k \tilde{w} + i \sum_{j=0}^{k-1} \tau^{k-j-1} D_t^j u(0) \qquad k = 1, 2, \ldots \tag{7-17}$$

and

$$\widetilde{D_t^k g} = \tau^k \tilde{g} - i \sum_{j=0}^{k-1} \tau^{k-j-1} D_t^j v(0) \qquad k = 1, 2, \ldots \tag{7-18}$$

PROOF Use Lemma 7-2 and induction on k.

Corollary 7-5 Under the same hypotheses, if $P(z)$ is a polynomial of degree m, then

$$\widetilde{P(D_t)w} = P(\tau)\tilde{w} + T_1(\tau) \tag{7-19}$$

$$\widetilde{P(D_t)g} = P(\tau)\tilde{g} + T_2(\tau) \tag{7-20}$$

where $T_1(z)$ and $T_2(z)$ are polynomials of degree $<m$.

Proof Use Corollary 7-4.

Corollary 7-6 Assume $u \in S(0, \infty)$, $v \in S(-\infty, 0)$, and

$$D_t^k u(0) = D_t^k v(0) \qquad 0 \le k < m \tag{7-21}$$

Set

$$\begin{aligned} h(t) &= u(t) \qquad t > 0 \\ &= v(t) \qquad t < 0 \end{aligned} \tag{7-22}$$

If $P(z)$ is a polynomial of degree $\le m$, then

$$\widetilde{P(D_t)h} = P(\tau)\tilde{h} \tag{7-23}$$

Proof Note that by Corollary 7-4 and Eq. (7-21)

$$\widetilde{D_t^k h} = \widetilde{D_t^k w} + \widetilde{D_t^k g} = \tau^k(\tilde{w} + \tilde{g}) = \tau^k \tilde{h}$$

for $k \le m$. This gives Eq. (7-23).

Corollary 7-7 If $j < k$, and w is given by Eq. (7-11), then

$$\tau^j \widetilde{D_t^k w} - \tau^k \widetilde{D_t^j w} = i \sum_{n=j}^{k-1} \tau^{j+k-n-1} D_t^n u(0) \tag{7-24}$$

Proof Use Eq. (7-17).

Corollary 7-8 Let

$$P(z) = \sum_{k=0}^{m} a_k z^k \qquad Q(z) = \sum_{j=0}^{m} b_j z^j$$

be polynomials of degrees $\le m$. If w is given by Eq. (7-11), then

$$Q(\tau)\widetilde{P(D_t)w} - P(\tau)\widetilde{Q(D_t)w}$$

is a polynomial of degree $<m$, with the coefficient of τ^{m-1} equal to $ia_m Q(D_t)u(0) - ib_m P(D_t)u(0)$.

Proof Use Corollary 7-7.

Corollary 7-9 Let $P(z)$ be a polynomial of degree r having all its roots contained in the half-plane $\operatorname{Im} z < 0$, and let $Q(z)$ be a polynomial of degree $<r$. If w is given by Eq. (7-11), then

$$\int_{-\infty}^{\infty} \frac{Q(\tau)}{P(\tau)} \widetilde{P(D_t)w}\, d\tau = 2\pi Q(D_t)u(0) \tag{7-25}$$

Similarly, if the roots of $P(z)$ are contained in $\operatorname{Im} z > 0$, then

$$\int_{-\infty}^{\infty} \frac{Q(\tau)}{P(\tau)} \widetilde{P(D_t)g}\, d\tau = 2\pi Q(D_t)v(0) \tag{7-26}$$

for g given by Eq. (7-14).

Proof Since the roots of $P(z)$ all have negative imaginary parts, we have, by contour integration

$$\begin{aligned} \int_{-R}^{R} \frac{\tau^k\, d\tau}{P(\tau)} &= -\int_{0}^{\pi} \frac{R^k\, e^{ik\theta}\, iR\, e^{i\theta}\, d\theta}{P(R\, e^{i\theta})} \\ &\to -\pi \frac{i}{a_r} \qquad \text{as } R \to \infty \qquad \text{when } k = r-1 \\ &\to 0 \qquad \text{as } R \to \infty \qquad \text{when } 0 \le k < r-1 \end{aligned} \tag{7-27}$$

where a_r is the coefficient of z^r in $P(z)$. Furthermore, by Corollary 7-8, the left-hand side of Eq. (7-25) equals

$$\int_{-\infty}^{\infty} \frac{T(\tau)}{P(\tau)}\, d\tau + \int_{-\infty}^{\infty} \widetilde{Q(D_t)w}\, d\tau$$

where $T(z)$ is a polynomial of degree $<r$, and the coefficient of z^{r-1} is $ia_r Q(D_t)u(0)$. The identity (7-25) follows from (7-27) and Lemma 7-3. The proof of Eq. (7-26) is similar.

7-3 PROOF OF THEOREM 7-1

We now give the proof of Theorem 7-1 based on Lemmas 7-2 and 7-3 and their corollaries. The lemmas will be proved in Sec. 7-5.

Let $u(t)$ be any function in $S(0, \infty)$. We find a function $v(t) \in S(-\infty, 0)$, such that

$$\overline{P}_+(D_t)P(D_t)v(t) = 0 \qquad t < 0 \tag{7-28}$$

$$D_t^k v(0) = D_t^k u(0) \qquad 0 \le k < m \tag{7-29}$$

Note that $\overline{P}_+(z)P(z)$ has m roots with negative imaginary parts. Thus, a solution of Eqs. (7-28), (7-29) exists, by Theorem 6-1 of Sec. 6-2 (or, more precisely, by the corresponding theorem for the half-line $t < 0$). Let w, g, and h be defined by Eqs. (7-11), (7-14), and (7-22), respectively. Thus

$$h = w + g \tag{7-30}$$

Now, by Eq. (7-20)

$$\widetilde{\bar{P}_{+}(D_t)P(D_t)g} = \bar{P}_{+}(\tau)\widetilde{P(D_t)g} - T(\tau)$$

where $T(z)$ is a polynomial of degree $<r$. Thus, by Eq. (7-28)

$$G = \widetilde{P(D_t)g} = \frac{T(\tau)}{\bar{P}_{+}(\tau)} \tag{7-31}$$

Next, we note that the polynomials $\{\bar{Q}_{j+}\}$ are linearly independent. For suppose there were constants $\gamma_1, \ldots, \gamma_r$, not all zero, such that

$$\sum_{1}^{r} \gamma_j \bar{Q}_{j+} = 0$$

This would imply that

$$\sum_{j=1}^{r} \alpha_{ij}\gamma_j = 0 \qquad 1 \le i \le r$$

contradicting the fact that the matrix A is nonsingular. Since the polynomials $\{\bar{Q}_{j+}\}$ are linearly independent and of degrees $<r$, and there are r of them, they form a basis for the vector space of all polynomials of degree $<r$ (see Sec. 3-3). Thus, there are constants λ_k, such that

$$T(z) = \sum_{k=1}^{r} \lambda_k \bar{Q}_{k+}(z)$$

Thus

$$\int_{-\infty}^{\infty} \frac{Q_{j+}}{P_{+}} G\, d\tau = \sum_{k=1}^{r} \alpha_{jk}\lambda_k \qquad 1 \le j \le r \tag{7-32}$$

On the other hand, we note that

$$\int_{-\infty}^{\infty} \frac{Q_{j+}}{P_{+}} G\, d\tau = \int_{-\infty}^{\infty} \frac{Q_{j+}}{P_{+}} \widetilde{P_{+}(D_t)P_{-}(D_t)g}\, d\tau$$

$$= 2\pi Q_{j+}(D_t)P_{-}(D_t)v(0) \equiv V_j \qquad 1 \le j \le r \tag{7-33}$$

By Eq. (7-26). Thus

$$\lambda_i = \sum_{j=1}^{r} \alpha^{ij} V_j \qquad 1 \le i \le r \tag{7-34}$$

As a consequence, we have

$$\int_{-\infty}^{\infty} |G|^2\, d\tau = \sum \bar{\lambda}_j \int_{-\infty}^{\infty} \frac{Q_{j+}}{P_{+}} W\, d\tau = \sum_{i,j=1}^{r} \alpha^{ij} V_i \bar{V}_j = V^* A^{-1} V \tag{7-35}$$

where V is the column vector with components V_j.

Since $P(z)$ has no real roots and Q_{j-} is of degree less than $m - r$, the functions $\bar{Q}_{j-}/\bar{P}_{-}$ are in $L^2(-\infty, \infty)$. The subspace of all functions of the form

$$\sum_{j=1}^{r} \gamma_j \frac{\bar{Q}_{j-}}{\bar{P}_-}$$

is finite dimensional (Lemma 3-8 of Sec. 3-3). It is closed, therefore (see Sec. 7-4). Thus, any element of $L^2(-\infty, \infty)$ can be expressed as the sum of two elements, one contained in the subspace and one orthogonal to it (Theorem 1-3 of Sec. 1-5). Set $W = \widetilde{P(D_t)w}$. Then

$$W = \sum_{j=1}^{r} \gamma_j \frac{\bar{Q}_{j-}}{\bar{P}_-} + \Phi \tag{7-36}$$

where

$$\int_{-\infty}^{\infty} \frac{Q_{j-}}{P_-} \Phi \, d\tau = 0 \qquad 1 \le j \le r \tag{7-37}$$

Thus

$$\int_{-\infty}^{\infty} |W|^2 \, d\tau = \int_{-\infty}^{\infty} |\Phi|^2 \, d\tau + \sum_{i,j=1}^{r} \beta_{ij}\gamma_i\bar{\gamma}_j \ge \gamma^* B \gamma \tag{7-38}$$

where γ is the column vector with components γ_j. Note also that, by Eqs. (7-36) and (7-37)

$$\int_{-\infty}^{\infty} \frac{Q_{j-}}{P_-} W \, d\tau = \sum_{k=1}^{r} \beta_{jk}\gamma_k \qquad 1 \le j \le r \tag{7-39}$$

On the other hand, by Eq. (7-25)

$$\int_{-\infty}^{\infty} \frac{Q_{j-}}{P_-} W \, d\tau = \int_{-\infty}^{\infty} \frac{Q_{j-}}{P_-} \widetilde{P_-(D_t)P_+(D_t)w} \, d\tau$$

$$= 2\pi Q_{j-}(D_t)P_+(D_t)u(0) \equiv Y_j \qquad 1 \le j \le r \tag{7-40}$$

Comparison of these identities shows that

$$B\gamma = Y \tag{7-41}$$

where Y is the column vector with components Y_j. We observe that

$$V_j + Y_j = 2\pi Q_j(D_t)u(0) \equiv U_j \qquad 1 \le j \le r \tag{7-42}$$

by Eqs. (7-4) and (7-29). Let U denote the column vector with components U_j.

We now tie the loose threads together. By Eq. (7-35)

$$\int_{-\infty}^{\infty} |G|^2 \, d\tau = V^* A^{-1} V = (U - Y)^* A^{-1}(U - Y) \le 2U^* A^{-1} U + 2Y^* A^{-1} Y \tag{7-43}$$

Moreover, by (7-41), (7-7) and (7-38)

$$Y^* A^{-1} Y = \gamma^* B A^{-1} B\gamma \le K_1 \gamma^* B \gamma \le K_1 \int_{-\infty}^{\infty} |W|^2 \, d\tau \tag{7-44}$$

By Eq. (7-30)

$$\int_{-\infty}^{\infty} |\widetilde{P(D_t)h}|^2 \, d\tau \le 2 \int_{-\infty}^{\infty} |W|^2 \, d\tau + 2 \int_{-\infty}^{\infty} |G|^2 \, d\tau \tag{7-45}$$

By Eq. (7-29) and Corollary 7-6

$$\widetilde{P(D_t)h} = P(\tau)\tilde{h} \qquad \widetilde{R(D_t)h} = R(\tau)\tilde{h}$$

Thus, by (7-8) and Parseval's identity (2-24)

$$\begin{aligned}\int_0^\infty |R(D_t)u|^2\,dt &\le \int_{-\infty}^\infty |R(D_t)h|^2\,dt \\ &= 2\pi \int_{-\infty}^\infty |R(\tau)\tilde{h}|^2\,d\tau \\ &\le C_1^2 \int_{-\infty}^\infty |P(\tau)\tilde{h}|^2\,d\tau \\ &= C_1^2 \int_{-\infty}^\infty |\widetilde{P(D_t)h}|^2\,d\tau \end{aligned} \tag{7-46}$$

Finally, we note, by Parseval's identity

$$\int_{-\infty}^\infty |W|^2\,d\tau = 2\pi \int_{-\infty}^\infty |P(D_t)u|^2\,dt \tag{7-47}$$

If we now combine (7-43) to (7-47), we obtain (7-9). Note that the resulting constant depends only on C_1 and K_1. The proof is complete.

7-4 HERMITIAN OPERATORS AND MATRICES

Let H be a (complex) Hilbert space, and let A be a bounded operator on H (i.e., A maps H into itself) defined everywhere (see Sec. 2-1). We call A *Hermitian* if

$$(u, Av) = (Au, v) \qquad u, v \in H \tag{7-48}$$

In this case (Au, u) is real for every $u \in H$, since the complex conjugate of (Au, u) is (u, Au), and this equals (Au, u), by Eq. (7-48). It is easily checked that an operator A is Hermitian if (Au, u) is always real.

For A, B Hermitian we write

$$A \le B \tag{7-49}$$

when we mean

$$(Au, u) \le (Bu, u) \qquad u \in H \tag{7-50}$$

Since (7-50) makes sense only when A and B are Hermitian, it is to be assumed that this is the case whenever an expression of the form (7-49) is used. We note

Lemma 7-10 If $A \ge 0$, then

$$|(Au, v)|^2 \le (Au, u)(Av, v) \tag{7-51}$$

and
$$(A[u + v], u + v) \le 2(Au, u) + 2(Av, v) \tag{7-52}$$

hold for all $u, v \in H$.

PROOF Set

$$a(u,v) = (Au,v) \qquad a(u) = (Au,u)$$

Then
$$a(u,v) = \overline{a(v,u)} \tag{7-53}$$

and
$$a(u \pm v) = a(u) \pm a(u,v) \pm a(v,u) + a(v) \tag{7-54}$$

Thus
$$a(u+v) + a(u-v) = 2a(u) + 2a(v) \tag{7-55}$$

This proves (7-52). To prove (7-51), let u, v be any elements of H such that $a(u,v)$ is real. By Eqs. (7-53) and (7-54)

$$4a(u,v) = a(u+v) - a(u-v)$$

Consequently $4|a(u,v)| \leq a(u+v) + a(u-v) = 2a(u) + 2a(v)$

by Eq. (7-55). Thus, for any real α

$$2|a(u,v)| \leq a(\alpha u) + a(v/\alpha) = \alpha^2 a(u) + \alpha^{-2} a(v)$$

If $a(u) = 0$, let $\alpha \to \infty$. This shows that $a(u,v) = 0$ and that (7-51) holds. Similarly, if $a(v) = 0$, let $\alpha \to 0$. Again this shows that $a(u,v) = 0$ and that (7-51) holds. If neither vanishes, set $\alpha^4 = a(v)/a(u)$. This gives (7-50).

Now let u, v be arbitrary. If $a(u,v)$ is not real, then

$$a(u,v) = e^{i\theta}|a(u,v)|$$

for some real θ. Thus, $a(e^{-i\theta}u, v)$ is real, and consequently, by the part just proved

$$|a(e^{-i\theta}u, v)|^2 \leq a(e^{-i\theta}u)a(v)$$

This gives (7-51), and the proof is complete.

Now suppose H has dimension $N < \infty$ (see Sec. 3-3). In fact, we let H be the collection of column vectors of the form

$$x = \begin{bmatrix} x_1 \\ x_2 \\ \vdots \\ x_N \end{bmatrix} \tag{7-56}$$

with the usual definitions of addition, multiplication by a scalar, and scalar product. We set

$$e_1 = \begin{bmatrix} 1 \\ 0 \\ \vdots \\ 0 \end{bmatrix}, \quad e_2 = \begin{bmatrix} 0 \\ 1 \\ \vdots \\ 0 \end{bmatrix}, \ldots, \quad e_N = \begin{bmatrix} 0 \\ 0 \\ \vdots \\ 1 \end{bmatrix}$$

It is easily seen that

$$(e_j, e_k) = \delta_{jk} \qquad 1 \leq j, k \leq N$$

Thus if x is given by Eq. (7-56), then

$$x = \sum_1^N x_j e_j$$

If

$$y = \sum_1^N y_j e_j$$

then

$$(x, y) = \sum_1^N x_j \bar{y}_j$$

If we define the row vector

$$y^* = [\bar{y}_1, \bar{y}_2, \ldots, \bar{y}_N]$$

then

$$y^* x = (x, y)$$

by the usual rule of matrix multiplication. Similarly, if A is a linear operator on H, we have

$$Ax = \sum_1^N x_k A e_k$$

Since the e_k span H, we know that there are coefficients α_{jk}, such that

$$Ae_k = \sum_{j=1}^N \alpha_{jk} e_j \qquad 1 \leq k \leq N$$

Thus

$$Ax = \sum_{j=1}^N \left(\sum_{k=1}^N \alpha_{jk} x_k \right) e_j \tag{5-57}$$

We can identify A with the matrix

$$A = (\alpha_{jk}) = \begin{bmatrix} \alpha_{11} \cdots \alpha_{1N} \\ \vdots \\ \alpha_{N1} \cdots \alpha_{NN} \end{bmatrix}$$

Matrix multiplication shows that Ax is the column vector

$$Ax = \begin{bmatrix} \alpha_{11} x_1 + \cdots + \alpha_{1N} x_N \\ \vdots \qquad\qquad \vdots \\ \alpha_{N1} x_1 + \cdots + \alpha_{NN} x_N \end{bmatrix}$$

This is precisely Eq. (7-57). We call the matrix A *Hermitian* if

$$y^* A x = (Ay)^* x \qquad x, y \in H \tag{7-58}$$

Since

$$(Ay)^* x = \sum_{j=1}^N \left(\sum_{k=1}^N \bar{\alpha}_{kj} x_k \right) \bar{y}_j$$

this is the same as saying

$$\alpha_{jk} = \bar{\alpha}_{kj} \qquad 1 \leq j, k \leq N$$

For A, B Hermitian we write again $A \leq B$ when we mean

$$x^*Ax \leq x^*Bx \qquad x \in H$$

Of course, all that is true for general Hilbert spaces is true here. In particular, we have

Lemma 7-11 If $A \geq 0$, then

$$|y^*Ax|^2 \leq (x^*Ax)(y^*Ay)$$

and

$$(x+y)^*A(x+y) \leq 2x^*Ax + 2y^*Ay$$

If the determinant det (α_{ij}) does not vanish, then we know from the theory of linear equations that the system

$$\sum_{k=1}^{N} \alpha_{jk} x_k = y_j$$

has a unique solution for each $y \in H$. This is the same as saying that

$$Ax = y \tag{7-59}$$

has a unique solution for each y. In this case, we say that A has an inverse $A^{-1} = (\alpha^{jk})$ defined by

$$A^{-1}y = x$$

when x and y satisfy Eq. (7-59). The elements α^{jk} of the matrix A^{-1} are easily computed (theoretically). If A is Hermitian, the same is true of A^{-1}.

We shall also need the following:

Lemma 7-12 A finite dimensional subspace of a Hilbert space is closed.

PROOF Let S be a subspace of dimension $N < \infty$, and let $e_1, \ldots, e_N$ be a basis for S. I claim that there is a constant C, such that

$$\sum_1^N |\alpha_k| \leq C \left\| \sum_1^N \alpha_k e_k \right\| \tag{7-60}$$

holds for all complex vectors $(\alpha_1, \ldots, \alpha_N)$. Before proving inequality (7-60), let me show how it can be used to imply that S is closed. Let

$$x_n = \sum_{k=1}^{N} \alpha_k^{(n)} e_k \qquad n = 1, 2, \ldots \tag{7-61}$$

be a sequence of vectors in S, such that

$$x_n \to x \quad \text{in } H \tag{7-62}$$

Since $\{x_n\}$ is a Cauchy sequence, we have by (7-60)

$$\sum_{k=1}^{N} |\alpha_k^{(n)} - \alpha_k^{(m)}| \leq C \|x_n - x_m\| \to 0$$

Thus, for each k, $\{\alpha_k^{(n)}\}$ is a Cauchy sequence of complex numbers. Thus, there are numbers $\alpha_1, \ldots, \alpha_N$, such that

$$\alpha_k^{(n)} \to \alpha_k \qquad 1 \le k \le N \tag{7-63}$$

Set
$$z = \sum_1^N \alpha_k e_k \tag{7-64}$$

Then
$$\| x_n - z \| = \left\| \sum_{k=1}^N (\alpha_k^{(n)} - \alpha_k) e_k \right\|$$

$$\le \max \| e_k \| \sum_{k=1}^N | \alpha_k^{(n)} - \alpha_k | \to 0 \quad \text{as} \quad n \to \infty \tag{7-65}$$

Thus $x_n \to z$. Since there can be only one limit, we must have $x = z \in S$.

To prove (7-60), assume that it does not hold. Then there is a sequence $\{x_n\}$ of the form (7-61), such that

$$\sum_{k=1}^N | \alpha_k^{(n)} | = 1 \tag{7-66}$$

and
$$x_n \to 0 \quad \text{in } H \tag{7-67}$$

Since the complex numbers $\alpha_k^{(n)}$ are bounded, there are a subsequence of $\{x_n\}$ (for convenience we assume it is the whole sequence) and complex numbers $\alpha_1, \ldots, \alpha_N$, such that (7-63) holds. By Eq. (7-66), we have

$$\sum_{k=1}^N | \alpha_k | = 1 \tag{7-68}$$

Define z by Eq. (7-64). Then by (7-65), $x_n \to z$. By (7-67), $z = 0$. Since the e_k are linearly independent, all of the α_k must vanish. But this contradicts Eq. (7-68). The proof is complete.

7-5 PROOFS OF THE LEMMAS

We now give proofs of Lemmas 7-2 and 7-3. We first start with Eq. (7-13). By Eq. (7-12)

$$\widetilde{D_t w} = \int_0^\infty e^{-it\tau} D_t u(t)\, dt$$

$$= \int_0^\infty D_t[e^{-it\tau} u(t)]\, dt - \int_0^\infty u(t) D_t[e^{-it\tau}]\, dt$$

$$= -i[-u(0)] + \tau \tilde{w}$$

(recall that $D_t = -i\, d/dt$). This gives Eq. (7-13). To prove Eq. (7-15), note that

$$\widetilde{D_t g} = \int_{-\infty}^{0} e^{-it\tau} D_t v(t)\, dt$$
$$= \int_{-\infty}^{0} D_t[e^{-it\tau} v(t)]\, dt - \int_{-\infty}^{0} v(t) D_t\, e^{-it\tau}\, dt$$
$$= -iv(0) + \tau\tilde{g}$$

Thus Lemma 7-2 is proved. In order to prove Lemma 7-3 we make use of

Lemma 7-13 If $f \in S(0, \infty)$, and

$$G(z) = \int_{0}^{\infty} e^{-itz} f(t)\, dt$$

then $G(z)$ is an entire function which is bounded in the half-plane $\operatorname{Im} z \le 0$. Similarly, if $h \in S(-\infty, 0)$ and $H(z)$ is given by

$$H(z) = \int_{-\infty}^{0} e^{-itz} h(t)\, dt$$

then $H(z)$ is an entire function which is bounded in the half-plane $\operatorname{Im} z \ge 0$.

PROOF That $G(z)$ and $H(z)$ are entire is easily verified by differentiating under the integral signs and noting that the integrals are absolutely convergent. Thus

$$G'(z) = -i \int_{0}^{\infty} e^{-itz}\, tf(t)\, dt$$
$$H'(z) = -i \int_{-\infty}^{0} e^{-itz}\, th(t)\, dt$$

Since
$$e^{-it(x+iy)} = e^{-itx+ty}$$
we see that $G(z)$ is bounded for $y \le 0$ and that $H(z)$ is bounded for $y \ge 0$.

We now give the proof of Lemma 7-3. Repeated applications of Eq. (7-13) give

$$\widetilde{(D_t - i)^2 w} = (\tau - i)[(\tau - i)\tilde{w} + iu(0)] + i(D_t - i)u(0)$$
$$= (\tau - i)^2\tilde{w} + i\tau u(0) + iD_t u(0) + 2u(0) \tag{7-69}$$

Now, by contour integration

$$\int_{-R}^{R} \frac{\tau^k\, d\tau}{(\tau - i)^2} = \int_{\pi}^{2\pi} \frac{R^k\, e^{ik\theta}\, iR\, e^{i\theta}\, d\theta}{(R\, e^{i\theta} - i)^2}$$

Thus
$$\int_{-\infty}^{\infty} \frac{d\tau}{(\tau - i)^2} = 0 \tag{7-70}$$

and
$$\int_{-\infty}^{\infty} \frac{\tau\, d\tau}{(\tau - i)^2} = \pi i \tag{7-71}$$

By Lemma 7-13 the function $G(z) = (D_t - i)^2 w$ is entire and bounded in the lower half-plane. Thus, we have

$$\int_{-\infty}^{\infty} \tilde{w}\, d\tau = \int_{-\infty}^{\infty} \frac{G(\tau)\, d\tau}{(\tau - i)^2} - iu(0) \int_{-\infty}^{\infty} \frac{\tau\, d\tau}{(\tau - i)^2} - [iD_t u(0) + 2u(0)] \int_{-\infty}^{\infty} \frac{d\tau}{(\tau - i)^2} \tag{7-72}$$

Another contour integration shows that

$$\int_{-\infty}^{\infty} \frac{G(\tau)\, d\tau}{(\tau - i)^2} = 0 \tag{7-73}$$

If we now substitute Eqs. (7-70), (7-71), and (7-73) into Eq. (7-72), we obtain the first identity in (7-16). To obtain the other, we use similar reasoning. By Eq. (7-15)

$$\widetilde{(D + i)^2 g} = (\tau + i)^2 \tilde{g} - i\tau v(0) - i(D_t + 2i)v(0)$$

Contour integrations give

$$\int_{-\infty}^{\infty} \frac{d\tau}{(\tau + i)^2} = 0$$

and

$$\int_{-\infty}^{\infty} \frac{\tau\, d\tau}{(\tau + i)^2} = -\pi i$$

Moreover, by Lemma 7-13, the function $H(z) = (D_t + i)^2 g$ is entire and bounded in the upper half-plane. Hence

$$\int_{-\infty}^{\infty} \frac{H(\tau)\, d\tau}{(\tau + i)^2} = 0$$

Combining the last four identities, we obtain the desired result.

7-6 EXISTENCE AND ESTIMATES IN A HALF-SPACE

We now return to the problem

$$P(D)u(x,t) = f(x,t) \qquad t > 0 \tag{7-74}$$

$$Q_j(D)u(x,0) = 0 \qquad 1 \le j \le r \tag{7-75}$$

(The case when the right-hand sides of the equations in (7-75) are prescribed arbitrary functions, will be considered later.) Our assumptions are as follows:

1. For almost every $\xi \in E^n$ the polynomial $P(\xi, z)$ is of degree m in z, and has precisely r roots with positive imaginary parts and no real roots. The $Q_j(\xi, z)$ are of degrees less than m in z.

   If $\tau_1(\xi), \ldots, \tau_r(\xi)$ are the roots of $P(\xi, z)$ with positive imaginary parts, set

$$P_+(\xi, \tau) = (\tau - \tau_1(\xi)) \cdots (\tau - \tau_r(\xi)) \tag{7-76}$$

This will not necessarily be a polynomial in the components of ξ. Set

$$P_-(\xi,\tau) = \frac{P(\xi,\tau)}{P_+(\xi,\tau)} \tag{7-77}$$

and for each ξ and j resolve the quotient $Q_j(\xi,z)/P(\xi,z)$ into partial fractions

$$\frac{Q_j(\xi,z)}{P(\xi,z)} = \frac{Q_{j+}(\xi,z)}{P_+(\xi,z)} + \frac{Q_{j-}(\xi,z)}{P_-(\xi,z)} \tag{7-78}$$

Set
$$\alpha_{ij}(\xi) = \int_{-\infty}^{\infty} \frac{Q_{i+}(\xi,\tau)\overline{Q_{j+}(\xi,\tau)}}{|P_+(\xi,\tau)|^2}\,d\tau \qquad 1 \le i,j \le r \tag{7-79}$$

and
$$\beta_{ij}(\xi) = \int_{-\infty}^{\infty} \frac{Q_{i-}(\xi,\tau)\overline{Q_{j-}(\xi,\tau)}}{|P_-(\xi,\tau)|^2}\,d\tau \qquad 1 \le i,j \le r \tag{7-80}$$

Set $A(\xi) = (\alpha_{ij})$ and $B(\xi) = (\beta_{ij})$. We assume

2. A^{-1} exists for almost all $\xi \in E^n$, and there is a constant K_1, such that

$$BA^{-1}B \le K_1 B \qquad \text{a.e.} \tag{7-81}$$

We have

Theorem 7-14 Let $R(\xi, z)$ be a polynomial satisfying

$$|R(\xi,\tau)| \le C_1 |P(\xi,\tau)| \qquad \xi \in E^n \qquad \tau \in E^1 \tag{7-82}$$

Under hypotheses 1 and 2 given above, there is a constant C depending only on C_1 and K_1, such that

$$\int_0^\infty |R(\xi,D_t)v(\xi,t)|^2\,dt \le C\left[\int_0^\infty |P(\xi,D_t)v(\xi,t)|^2\,dt + \sum_{i,j=1}^{r} \alpha^{ij}V_i\overline{V_j}\right] \qquad \text{a.e.} \tag{7-83}$$

holds for all functions $v(\xi,t)$, such that for almost all $\xi \in E^n$, $v(\xi,t) \in S(0,\infty)$. Here $(\alpha^{ij}) = A^{-1}$, and

$$V_j = Q_j(\xi,D_t)v(\xi,0) \qquad 1 \le j \le r \tag{7-84}$$

PROOF For each ξ this is a direct consequence of Theorem 7-1.

As before, we let Ω denote the half-space $t > 0$. If we take $b = \infty$ in Eq. (5-13) of Sec. 5-2, then the norms defined there apply to functions defined on Ω. In terms of these norms, we can state

Corollary 7-15 Under the hypotheses of Theorem 7-14 there is a constant C depending only on C_1 and K_1, such that for each real s

$$|R(D)u|_{0,s} \le C|P(D)u|_{0,s} \tag{7-85}$$

holds for all $u \in S(\Omega)$ satisfying Eq. (7-75).

PROOF We merely set $v(\xi, t) = Fu$ in (7-85), multiply by $(1 + |\xi|)^{2s}$, and integrate with respect to ξ.

For our existence theorem we make two more assumptions.

3. The coefficient $a(\xi)$ of z^m in $P(\xi, z)$ is bounded away from zero. Thus, there is a constant $c_0 > 0$, such that

$$|a(\xi)| \geq c_0 \qquad \xi \in E^n \tag{7-86}$$

4. There is a constant K_2, such that

$$\sum_\mu |P^{(\mu)}(\xi, \tau)| \leq K_2 |P(\xi, \tau)| \qquad \xi \in E^n \qquad \tau \in E^1 \tag{7-87}$$

We have

Theorem 7-16 Under hypotheses 1 to 4, for each $f \in C^\infty(\bar{\Omega})$ such that $D_x^\mu D_t^k f \in L^2(\Omega)$ for each μ and k, there is a unique solution $u \in C^\infty(\bar{\Omega})$ of Eqs. (7-74), (7-75), such that $D_x^\mu D_t^k u \in L^2(\Omega)$ for each μ and k.

PROOF As usual, let F denote the Fourier transform on E^n. For each ξ such that A^{-1} exists, the polynomials $Q_1(\xi, z), \ldots, Q_r(\xi, z)$ are linearly independent modulo $P_+(\xi, z)$. Thus, for each such ξ there is a function $h(\xi, t) \in S(0, \infty)$ satisfying

$$P(\xi, D_t)h(\xi, t) = Ff(\xi, t) \qquad t > 0 \tag{7-88}$$

$$Q_j(\xi, D_t)h(\xi, 0) = 0 \qquad 1 \leq j \leq r \tag{7-89}$$

(Theorem 6-8 of Sec. 6-14). By (7-87) and Theorem 7-14, $|\xi|^j P^{(\mu)}(\xi, D_t)h$ is in $L^2(\Omega)$ for each $j \geq 0$ and μ. Since every polynomial has a derivative which is a nonvanishing constant (see Problem **1-5** of Chapter 1), we see in particular that $|\xi|^j h \in L^2(\Omega)$ for each $j \geq 0$. Next, we note that

$$\frac{\partial^{m-1} P(\xi, \tau)}{\partial \tau^{m-1}} = m!\, a(\xi)\tau + p_1(\xi) \tag{7-90}$$

where $p_1(\xi)$ is a polynomial in the components of ξ only. Since $|\xi|^j p_1(\xi)h \in L^2(\Omega)$ for each j, the same is true of $|\xi|^j a(\xi) D_t h$. Since $a(\xi)$ is bounded away from zero, we see that $|\xi|^j D_t h \in L^2(\Omega)$ for each j. Next, note that

$$\frac{\partial^{m-2} P(\xi, \tau)}{\partial \tau^{m-2}} = \tfrac{1}{2} m!\, a(\xi)\tau^2 + p_1(\xi)\tau + p_2(\xi) \tag{7-90a}$$

where $p_2(\xi)$ is also a polynomial in the components of ξ only. Since $|\xi|^j(p_1 D_t h + p_2 h) \in L^2(\Omega)$ for each j, the same must be true for $|\xi|^j a(\xi) D_t^2 h$. Again, since $a(\xi)$ is bounded away from zero, $|\xi|^j D_t^2 h \in L^2(\Omega)$ for each j. Continuing in this way, we see that $|\xi|^j D_t^k h \in L^2(\Omega)$ for $j \geq 0$ and $0 \leq k \leq m$.

We can go even further. Since $h(\xi, t) \in S(0, \infty)$, we can differentiate both sides of Eq. (7-88) with respect to t. Thus

$$P(\xi, D_t)D_t h = D_t F f(\xi, t)$$

Since the left-hand side is of the form

$$a(\xi)D_t^{m+1}h + \sum_{k=1}^{m} q_k(\xi)D_t^k h$$

and $|\xi|^j q_k(\xi)D_t^k h \in L^2(\Omega)$ for each j, when $0 \le k \le m$, we see that $|\xi|^j a(\xi)D_t^{m+1}h \in L^2(\Omega)$ for each j. By differentiating Eq. (7-74) repeatedly with respect to t and applying the same reasoning, we conclude that $|\xi|^j D_t^k h \in L^2(\Omega)$ for each j and k.

For each $t > 0$, set $u(x, t) = F^{-1}h(\xi, t)$. Then $u \in L^2(\Omega)$, and for each $x \in E^n$ we know that $u(x, t)$ is in $S(0, \infty)$. Moreover, $|u|_{k,j} < \infty$ for each j and k. It now follows, as in the proof of Theorem 6-27 of Sec. 6-8, that $u \in C^\infty(\overline{\Omega})$ and that $D_x^\mu D_t^k u \in L^2(\Omega)$ for each μ and k. Uniqueness follows from Theorem 7-14. This completes the proof.

7-7 EXAMPLES

We consider some examples to illustrate our theorems.

1 $P(\xi, \tau) = a(\xi)(\tau - i)$, where $a(\xi)$ is a polynomial in the components of ξ. In this case $P_+(\xi, \tau) = \tau - i$, and $P_-(\xi, \tau) = a(\xi)$. Take $Q(\xi, \tau) = 1$. Then it can be checked easily that

$$Q_+(\xi, \tau) = \frac{1}{a(\xi)} \qquad Q_-(\xi, \tau) = 0$$

Thus

$$\alpha(\xi) = \frac{1}{|a(\xi)|^2}\int_{-\infty}^{\infty} \frac{d\tau}{(\tau - i)(\tau + i)} = \frac{\pi}{|a(\xi)|^2}$$

by contour integration.

Thus

$$A^{-1} = \frac{|a(\xi)|^2}{\pi} \qquad B = 0$$

and (7-81) is satisfied everywhere. Thus, by Theorem 7-14 if $R(\xi, \tau)$ is any polynomial satisfying

$$|R(\xi, \tau)| \le C_1 |a(\xi)(\tau - i)|$$

then

$$\int_0^\infty |R(\xi, D_t)v|^2\,dt \le C|a(\xi)|^2\left[\int_0^\infty |(D_t - i)v|^2\,dt + |v(\xi, 0)|^2\right]$$

holds for all $v(\xi, t)$ which are in $S(0, \infty)$ for each ξ. (This could also have been deduced from Theorem 6-12 of Sec. 6-6.) If $a(\xi)$ satisfies

$$\sum_\mu |a^{(\mu)}(\xi)| \le K_3|a(\xi)| \qquad \xi \in E^n \tag{7-91}$$

Then $P(\xi, \tau)$ satisfies hypotheses 3 and 4 of Sec. 7-6 as well. Thus, we may apply Theorem 7-16 to conclude that for each $f \in C^\infty(\overline{\Omega})$, such that $D_x^\mu D_t^k f \in L^2(\Omega)$ for each μ and k, there is a unique solution of

$$a(D_x)(D_t - i)u(x,t) = f(x,t) \qquad t > 0$$
$$u(x,0) = 0$$

with the same properties.

2 $P(\xi, \tau) = a(\xi)[D_t - b(\xi)]$, where $a(\xi)$ and $b(\xi)$ are polynomials. Assume that $\operatorname{Im} b(\xi) \geq 0$ for $\xi \in E^n$. In this case, $P_+(\xi, \tau) = \tau - b(\xi) = a(\xi)$. Again take $Q(\xi, \tau) = 1$. Then

$$Q_+(\xi, \tau) = \frac{1}{a(\xi)} \qquad Q_-(\xi, \tau) = 0$$

as before.

Thus $$\alpha(\xi) = \frac{1}{|a(\xi)|^2} \int_{-\infty}^{\infty} \frac{d\tau}{(\tau - b)(\tau - \bar{b})} = \frac{\pi}{|a(\xi)|^2 \operatorname{Im} b(\xi)}$$

Consequently $$A^{-1} = \frac{|a(\xi)|^2 \operatorname{Im} b(\xi)}{\pi} \qquad B = 0$$

and (7-81) is satisfied everywhere. Thus, if $R(\xi, \tau)$ satisfies

$$|R(\xi, \tau)| \leq C_1 |a(\xi)[\tau - b(\xi)]|$$

then we have the inequality

$$\int_0^{\infty} |R(\xi, D_t)v|^2 \, dt \leq C |a(\xi)|^2 \left\{ \int_0^{\infty} |[D_t - b(\xi)]v|^2 \, dt + \operatorname{Im} b(\xi) |v(\xi, 0)|^2 \right\}$$

where C depends only on C_1. If we assume that $a(\xi)$ satisfies (7-91) and $b(\xi)$ satisfies

$$\sum_{\mu} |b^{(\mu)}(\xi)| \leq K_4 \operatorname{Im} b(\xi) \qquad \xi \in E^n \tag{7-92}$$

then $P(\xi, \tau)$ satisfies (7-86) and (7-87). Thus, for each $f \in C^\infty(\bar{\Omega})$, such that $D_x^\mu D_t^k f \in L^2(\Omega)$ for all μ and k, there is a unique solution $u(x, t)$ of

$$a(D_x)[D_t - b(D_x)]u(x,t) = f(x,t) \qquad t > 0$$
$$u(x,0) = 0$$

with the same properties.

3 $$P(\xi, \tau) = \tau^2 + \xi_1^2 + \cdots + \xi_n^2 \qquad Q(\xi, \tau) = \tau + q(\xi)$$

where $q(\xi)$ is a polynomial with real coefficients. In this case, $P_+(\xi, \tau) = \tau - i|\xi|$ and $P_-(\xi, \tau) = \tau + i|\xi|$. Also

$$Q_+(\xi, \tau) = \frac{(|\xi| - iq)}{2|\xi|} \qquad Q_-(\xi, \tau) = \frac{(|\xi| + iq)}{2|\xi|}$$

Consequently $$\alpha(\xi) = \beta(\xi) = \frac{\pi(q^2 + |\xi|^2)}{4|\xi|^3}$$

Thus, A^{-1} exists a.e. and (7-81) holds. By Theorem 7-14

$$\int_0^{\infty} |R(\xi, D_t)v|^2 \, dt \leq C \left[\int_0^{\infty} |(D_t^2 + |\xi|^2)v|^2 \, dt + \frac{|\xi|^3}{q^2 + |\xi|^2} |[D_t + q(\xi)]v(\xi, 0)|^2 \right]$$

holds for any homogeneous operator $R(D)$ of order two. Since $P(\xi, \tau)$ does not satisfy (7-87) we cannot say more. (Note that Theorem 6-27 of Sec. 6-8 applies if $q(\xi)$ is homogeneous of degree one.)

4 $P(\xi, \tau) = \tau^2 + |\xi|^2 + 1$; $Q(\xi, \tau)$ same as in Example **3**. Here $P_+(\xi, \tau) = \tau - ie$, $P_-(\xi, \tau) = \tau + ie$, where e is the positive square root of $1 + |\xi|^2$. We now have

$$Q_+(\xi, \tau) = \frac{(e - iq)}{2e} \qquad Q_-(\xi, \tau) = \frac{(e + iq)}{2e}$$

and consequently

$$\alpha(\xi) = \beta(\xi) = \frac{\pi(q^2 + e^2)}{4e^3}$$

Again A^{-1} exists (everywhere) and (7-81) holds. By Theorem 7-14

$$\int_0^\infty |R(\xi, D_t)v|^2\, dt \le C\left[\int_0^\infty |(D_t^2 + |\xi|^2 + 1)v|^2\, dt + \frac{e^3}{q^2 + e^2}|[D_t + q(\xi)]v(\xi, 0)|^2\right]$$

holds for each second order operator $R(D)$. Now, $P(\xi, \tau)$ satisfies (7-86) and (7-87), so we know that for each $f \in C^\infty(\bar{\Omega})$ satisfying $D_x^\mu D_t^k f \in L^2(\Omega)$ for each μ and k, there is a unique solution $u(x, t)$ of

$$(1 - \Delta)u(x, t) = f(x, t) \qquad t > 0$$

$$[D_t + q(D_x)]u(x, 0) = 0$$

with the same properties.

5 $P(\xi, \tau) = (\tau - \tau_1(\xi))(\tau - \tau_2(\xi))$, where

$$\operatorname{Im} \tau_1(\xi) > 0 \qquad \operatorname{Im} \tau_2(\xi) < 0$$

and $Q(\xi, \tau) = \tau + q(\xi)$ (the coefficients of $q(\xi)$ need not be real). We have

$$P_+(\xi, \tau) = \tau - \tau_1(\xi) \qquad P_-(\xi, \tau) = \tau - \tau_2(\xi)$$

$$Q_+(\xi, \tau) = \frac{(\tau_1 + q)}{(\tau_1 - \tau_2)} \qquad Q_-(\xi, \tau) = \frac{(\tau_2 + q)}{(\tau_2 - \tau_1)}$$

$$\alpha(\xi) = \frac{\pi}{\operatorname{Im} \tau_1}\frac{|\tau_1 + q|^2}{|\tau_1 - \tau_2|^2}$$

$$\beta(\xi) = \frac{-\pi}{\operatorname{Im} \tau_2}\frac{|\tau_2 + q|^2}{|\tau_2 - \tau_1|^2}$$

Thus, A^{-1} exists a.e. provided $\tau_1 + q \neq 0$ a.e. In this case, (7-81) is equivalent to

$$\frac{|\tau_2 + q|^2}{|\operatorname{Im} \tau_2|} \le K_1 \frac{|\tau_1 + q|^2}{\operatorname{Im} \tau_1} \tag{7-93}$$

Thus, we have

$$\int_0^\infty |R(\xi, D_t)v|^2\, dt \le C\left[\int_0^\infty |P(\xi, D_t)v|^2\, dt + \frac{|\tau_1 - \tau_2|^2 \operatorname{Im} \tau_1}{|\tau_1 + q|^2}|[D_t + q(\xi)]v(\xi, 0)|^2\right]$$

for each $R(D)$ satisfying (7-82) whenever (7-93) holds. If $q(\xi)$ has real coefficients and the τ_k are pure imaginary, then (7-93) reduces to

$$|\tau_2| + \frac{q^2}{|\tau_2|} \le K_1\left(|\tau_1| + \frac{q^2}{|\tau_1|}\right) \tag{7-94}$$

This will hold if there is a constant K, such that

$$|\tau_1| \leq K|\tau_2| \tag{7-95}$$

and

$$|\tau_1 \tau_2| \leq K(q^2 + |\tau_1|^2) \tag{7-96}$$

Another set of conditions which implies (7-94) is

$$q^2 \leq K|\tau_1 \tau_2| \tag{7-97}$$

$$|\tau_1 \tau_2| \leq K(q^2 + |\tau_1|^2) \tag{7-98}$$

Without further knowledge of $P(\xi, \tau)$ we cannot verify (7-87).

6 $P(\xi, \tau) = (\tau - i|\xi|^4)(\tau + i|\xi|^2)$, $Q(\xi, \tau) = 1$, and $R(\xi, \tau) = (\tau + i|\xi|^4)(\tau - i|\xi|^2)$. Note that $P(D)$ is hypoelliptic (see Sec. 3-6). Also (7-82) holds (in fact, we have $|R(\xi, \tau)| = |P(\xi, \tau)|$ for $(\xi, \tau) \in E^{n+1}$). But

$$Q_+(\xi, \tau) = -Q_-(\xi, \tau) = \frac{1}{i|\xi|^4 + i|\xi|^2}$$

and consequently

$$\alpha(\xi) = \frac{\pi|Q_+|^2}{|\xi|^4} \qquad \beta(\xi) = \frac{\pi|Q_-|^2}{|\xi|^2}$$

This shows that (7-81) does not hold. As a matter of fact, inequality (7-83) does not hold either. To see this set

$$v(\xi, t) = e^{-t|\xi|^4} - e^{-t|\xi|^2}$$

Clearly, $v(\xi, t) \in S(0, \infty)$ for each $\xi \neq 0$, and

$$v(\xi, 0) = 0 \qquad \xi \neq 0$$

Thus, $Q(\xi, D_t)v(\xi, 0) = 0$. Simple calculations give

$$(D_t - i|\xi|^2)v = i|\xi|^2(|\xi|^2 - 1)\,e^{-t|\xi|^4}$$

and

$$(D_t - i|\xi|^4)v = i|\xi|^2(|\xi|^2 - 1)\,e^{-t|\xi|^2}$$

Hence

$$P(\xi, D_t)v = -2|\xi|^4(|\xi|^2 - 1)\,e^{-t|\xi|^2}$$

and

$$R(\xi, D_t)v = -2|\xi|^6(|\xi|^2 - 1)\,e^{-t|\xi|^4}$$

Consequently, we have

$$\int_0^\infty |P(\xi, D_t)v|^2\,dt = 2|\xi|^6(|\xi|^2 - 1)^2$$

and

$$\int_0^\infty |R(\xi, D_t)v|^2\,dt = 2|\xi|^8(|\xi|^2 - 1)^2$$

Clearly, the ratio of these expressions cannot be bounded for $\xi \in E^n$.

7-8 NONVANISHING BOUNDARY CONDITIONS

We now consider the problem

$$P(D)u(x, t) = f(x, t) \qquad t > 0 \tag{7-99}$$

$$Q_j(D)u(x, 0) = g_j(x) \qquad 1 \leq j \leq r \tag{7-100}$$

under the hypothesis of Theorem 7-16. We assume that $D_x^\mu D_t^k f \in L^2(\Omega)$ for each μ and k, and that $D_x^\mu g_j \in L^2(E^n)$ for each μ, $1 \leq j \leq r$. An obvious statement is

Theorem 7-17 Suppose there is a function $v(x, t)$, such that $D_x^\mu D_t^k v \in L^2(\Omega)$ for each μ and k, and

$$Q_j(D)v(x, 0) = g_j(x) \qquad 1 \leq j \leq r \tag{7-101}$$

Then under hypotheses 1 to 4 of Sec. 7-6, there is a solution $u(x, t)$ of Eqs. (7-99), (7-100) having the same properties.

Proof By Theorem 7-16 there is a solution $w(x, t)$ of

$$P(D)w(x, t) = f(x, t) - P(D)v(x, t) \qquad t > 0 \tag{7-102}$$

$$Q_j(D)w(x, 0) = 0 \qquad 1 \leq j \leq r \tag{7-103}$$

having the desired properties. Set $u = w + v$. Then it is easily checked that u gives a desired solution.

Naturally, Theorem 7-17 leads us to ask the question when is there a function $v(x, t)$ having the desired properties satisfying Eq. (7-101)? The answer is not as difficult as it may seem. First, we have

Theorem 7-18 If $g_1(x), \ldots, g_r(x)$ are any functions such that $D_x^\mu g_j \in L^2(E^n)$ for each μ and j, then there exists a function $v(x, t)$, such that

$$D_t^{j-1} v(x, 0) = g_j(x) \qquad 1 \leq j \leq r \tag{7-104}$$

and $D_x^\mu D_t^k v \in L^2(\Omega)$ for each μ and k.

Proof Let $\zeta(t)$ be a function in $C_0^\infty(-\infty, \infty)$ which equals one near $t = 0$. Set

$$v(x, t) = \zeta(t) \sum_{k=1}^{r} \frac{(it)^{k-1} g_k(x)}{(k-1)!} \tag{7-105}$$

Then, near $t = 0$, we have

$$D_t^j v(x, t) = \sum_{k=j}^{r} \frac{(it)^{k-j-1} g_k(x)}{(k-j-1)!}$$

This gives Eq. (7-104).

We now turn to the general operators $Q_j(D)$. Let m_j denote the order of the $Q_j(D)$. We shall say that the set $Q_1(D), \ldots, Q_r(D)$ is *normal* (to the hyperplane $t = 0$) if the orders m_j are all different, and if the coefficient of τ^{m_j} in $Q_j(\xi, \tau)$ does not vanish. Another way of saying it is that the orders of the $Q_j(\xi, \tau)$ are distinct, and that the hyperplane $t = 0$ is not characteristic for any of them (see Sec. 4-1). We have

Theorem 7-19 If the set $Q_1(D), \ldots, Q_r(D)$ is normal, then for each set $g_1(x), \ldots, g_r(x)$ of functions satisfying $D_x^\mu g_j \in L^2(E^n)$ for each μ and j, there is a function $v(x, t)$, such that $D_x^\mu D_t^k v \in L^2(\Omega)$ for each μ and k, and Eq. (7-101) holds.

PROOF Let q be such that all the $Q_j(D)$ are of order $<q$. Let $(n_1,\ldots,n_{q-r})$ be the numbers which when combined with $(m_1,\ldots,m_r)$ complete the set $(0,\ldots,q-1)$. If we put

$$Q_{r+j}(D) = D_t^{n_j} \qquad 1 \le j \le q-r$$

then it is easily checked that the set $Q_1(D),\ldots,Q_q(D)$ is normal. Let us rearrange them so that the order of $Q_j(D)$ is $j-1$. Thus

$$Q_j(\xi,\tau) = \sum_{k=1}^{j} R_{jk}(\xi)\tau^{k-1} \qquad 1 \le j \le q \tag{7-106}$$

where $R_{jk}(\xi)$ is a polynomial in ξ of degree $\le j-k$, and $R_{jj}(\xi) = \text{constant} \ne 0$ (this is from the definition of normalicy). I claim that all we need show is that there are polynomials $S_{kl}(\xi)$ in ξ, such that $S_{kl}(\xi)$ is of degree $\le k-l$, $S_{kk}(\xi) = \text{constant} \ne 0$, and

$$\sum_{i=j}^{k} R_{ki}(\xi)S_{ij}(\xi) = \delta_{jk} \qquad 1 \le j \le k \le q \tag{7-107}$$

For suppose Eqs. (7-106) and (7-107) hold, and suppose $g_1,\ldots,g_q$ satisfy the hypotheses of the theorem. Then, by Theorem 7-18 there is a function $v(x,t)$ with the desired properties, such that

$$D_t^{k-1}v(x,0) = \sum_{l=1}^{k} S_{kl}(D_x)g_l(x) \qquad 1 \le k \le q$$

Thus, by Eqs. (7-106) and (7-107)

$$\begin{aligned} Q_j(D)v(x,0) &= \sum_{k=1}^{j} R_{jk}(D_x) \sum_{l=1}^{k} S_{kl}(D_x)g_l(x) \\ &= \sum_{l=1}^{j} \left[\sum_{k=l}^{j} R_{jk}(D_x)S_{kl}(D_x) \right] g_l(x) \\ &= g_j(x) \qquad 1 \le j \le q \end{aligned}$$

Thus, the function $v(x,t)$ satisfies more than Eq. (7-101). It remains, therefore, only to prove Eq. (7-107). We do this by an easy induction. It clearly holds for $k=1$. We merely take $S_{11}(\xi) = 1/R_{11}(\xi)$. Now assume that (7-107) holds for $k < p \le q$. We must show that it holds for $k = p$.

Define
$$S_{pp}(\xi) = \frac{1}{R_{pp}(\xi)}$$

and
$$S_{pl}(\xi) = -S_{pp}(\xi) \sum_{k=l}^{p-1} R_{pk}(\xi)S_{kl}(\xi) \qquad l < p$$

Thus, for $j < p$

$$\sum_{i=j}^{p} R_{pi}(\xi)S_{ij}(\xi) = R_{pp}(\xi)S_{pj}(\xi) + \sum_{i=j}^{p-1} R_{pi}(\xi)S_{ij}(\xi)$$

$$= -\sum_{i=j}^{p-1} R_{pi}(\xi)S_{ij}(\xi) + \sum_{i=j}^{p-1} R_{pi}(\xi)S_{ij}(\xi) = 0$$

Thus, Eq. (7-107) holds for $k = p$ as well, and the proof is complete.

PROBLEMS

7-1 Prove an identity corresponding to (7-24) for g given by Eq. (7-14).

7-2 Same problem for Eq. (7-25).

7-3 Prove Corollary 7-8.

7-4 Prove Eq. (7-27).

7-5 Prove identity (7-38).

7-6 If A is a linear operator on a complex Hilbert space and (Au, u) is real for all u, show that A is Hermitian.

7-7 If $A = (\alpha_{jk})$ and $A^{-1} = (\alpha^{jk})$, find the α^{jk} in terms of the α_{jk}.

7-8 If $A = (\alpha_{jk})$ is Hermitian, show that A^{-1} is also Hermitian.

7-9 Prove Eqs. (7-90) and (7-90a).

7-10 Show that (7-91) implies (7-87) when $P = a(\tau - i)$.

7-11 Show that (7-91) and (7-92) imply (7-87) when $P = a(\tau - b)$.

7-12 Verify the statements made concerning Example **3** of Sec. 7-7.

7-13 Same problem for Example **4** of Sec. 7-7.

7-14 Show that (7-95) and (7-96) imply (7-94).

7-15 Show that (7-97) and (7-98) imply (7-94).

CHAPTER

EIGHT

THE DIRICHLET PROBLEM

8-1 INTRODUCTION

Let $P(D)$ be a homogeneous, properly elliptic operator of order $m = 2r$ in E^{n+1} (see Sec. 6-9), and consider the problem

$$P(D)u(x,t) = f(x,t) \qquad x \in E^n \qquad t > 0 \tag{8-1}$$

$$D_t^k u(x,0) = g_k(x) \qquad x \in E^n \qquad 0 \le k < r \tag{8-2}$$

This is a special case of problem (6-132), (6-133) of Sec. 6-8. In fact, we have

$$Q_j(\xi,\tau) = \tau^{j-1} \qquad 1 \le j \le r \tag{8-3}$$

Note that condition (*b*) of Theorem 6-15 of Sec. 6-7 is satisfied. This is obvious since all of the $Q_j(\xi,\tau)$ are of degree $< r$ (= the degree of $P_+(\xi,\tau)$) and they are linearly independent (see Sec. 3-3 and Sec. 6-4). Thus, by Theorem 6-27 of Sec. 6-8, there is a unique solution of Eqs. (8-1), (8-2) for each set f, g_k of functions having all their derivatives in L^2. The name of *Dirichlet* is associated with this problem.

We now wish to consider the problem for operators with variable coefficients, and not necessarily homogeneous. In dealing with variable coefficients it is prudent to begin with a small domain. Thus, instead of a half-space we shall confine our setting to the region σ_R given by $|x|^2 + t^2 < R^2$, $t > 0$. Our operator will be of the form

$$P(x,t,D) = \sum_{|\mu|+k \le m} a_{\mu,k}(x,t) D_x^\mu D_t^k \tag{8-4}$$

The coefficients are assumed to be in $C^\infty(\bar{\sigma}_R)$. The operator is to be elliptic in $\bar{\sigma}_R$, and

$$P_0(D) = \sum_{|\mu|+k\le m} a_{\mu,k}(0,0)D_x^\mu D_t^k \tag{8-5}$$

is assumed to be properly elliptic.

Theorem 7-18 of Sec. 7-8 shows that we can always satisfy Eq. (8-2) for any choice of the g_k. Reasoning as in Theorem 7-17 there, we see that we need only consider the case when the g_k vanish. Thus, our problem takes the form

$$P(x,t,D)u(x,t) = f(x,t) \qquad (x,t)\in\sigma_R \tag{8-6}$$

$$D_t^k u(x,0) = 0 \qquad |x| < R \qquad 0 \le k < r \tag{8-7}$$

We are willing to take $f \in C^\infty(\bar{\sigma}_R)$ and hope that we can find a solution in $C^m(\bar{\sigma}_R)$. We shall see that even this modest problem requires some work.

8-2 A WEAK SOLUTION

As we usually do when we have no leads concerning a problem, let us assume that Eqs. (8-6), (8-7) have a solution, and see what stipulations this implies concerning f and $P(x,t,D)$. Suppose $u(x,t)\in C^m(\bar{\sigma}_R)$ is a solution of (8-6), (8-7). Our first instinct is to integrate by parts. Let $\partial_0\sigma_R$ denote that part of the boundary of σ_R contained in the hyperplane $t=0$, and let $\partial_1\sigma_R$ denote the rest (i.e., the part contained on the surface $|x|^2+t^2=R^2$). Let $\varphi\in C^m(\bar{\sigma}_R)$ be any function in $C^m(\bar{\sigma}_R)$ which vanishes near $\partial_1\sigma_R$ (we do this to avoid any boundary integrals over $\partial_1\sigma_R$ where boundary conditions are not given for u). Integrating by parts, we have

$$\begin{aligned}(P(x,t,D)u,\varphi) &= \sum_{|\mu|+k\le m} (a_{\mu,k}D_x^\mu D_t^k u, \varphi)\\ &= \sum_{|\mu|+k\le m} (D_t^k u, D_x^\mu[\bar{a}_{\mu,k}\varphi])\\ &= \sum_{k=0}^{m} (D_t^k u, \varphi_k)\end{aligned}$$

where $$\varphi_k = \sum_{|\mu|\le m-k} D_x^\mu[\bar{a}_{\mu,k}\varphi]$$

Thus, by Eq. (4-20) of Sec. 4-2

$$\begin{aligned}(P(x,t,D)u,\varphi) &= (u, P'(x,t,D)\varphi) - i\sum_{k=1}^{m}\sum_{j=1}^{k}\int_{\partial_0\sigma_R} D_t^{j-1}u\overline{D_t^{k-j}\varphi_k}\,dx\\ &= (u, P'(x,t,D)\varphi) - i\sum_{j=1}^{m}\int_{\partial_0\sigma_R} D_t^{j-1}u\sum_{k=j}^{m}\overline{D_t^{k-j}\varphi_k}\,dx\end{aligned} \tag{8-8}$$

where
$$P'(x,t,D)\varphi = \sum_{|\mu|+k\le m} D_x^\mu D_t^k[\bar{a}_{\mu,k}\varphi] \tag{8-9}$$

is the formal adjoint of $P(x,t,D)$ (see Sec. 1-3). Consequently, we have

$$(P(x,t,D)u,\varphi) = (u, P'(x,t,D)\varphi) - i\sum_{j=1}^{m} \int_{\partial_0\sigma_R} D_t^{j-1}\overline{N_j\varphi}\,dx \tag{8-10}$$

where
$$N_j\varphi = \sum_{k=0}^{m-j} \sum_{|\mu|\le m-j-k} D_x^\mu D_t^k[\bar{a}_{\mu,j+k}\varphi] \qquad 1\le j\le m \tag{8-11}$$

If φ also satisfies

$$D_t^k\varphi(x,0) = 0 \qquad |x|<R \qquad 0\le k<r \tag{8-12}$$

we have, by Eqs. (8-7) and (8-12)

$$(P(x,t,D)u,\varphi) = (u,P'(x,t,D)\varphi) \tag{8-13}$$

Thus, if u satisfies Eq. (8-6), we have

$$|(f,\varphi)| \le C\,\|P'(x,t,D)\varphi\| \tag{8-14}$$

for all $\varphi\in C^m(\bar{\sigma}_R)$ which vanish near $\partial_1\sigma_R$ and satisfy Eq. (8-12). Consequently (8-14) is a necessary condition for (8-6), (8-7) to have a solution (compare Sec. 1-6).

Conversely, suppose (8-14) holds. Can we deduce the existence of a solution from this? We try. Let W be the set of those functions $\psi(x,t)$ which are continuous in $\bar{\sigma}_R$, and for which there is a $\varphi\in C^m(\bar{\sigma}_R)$ vanishing near $\partial_1\sigma_R$, satisfying Eq. (8-12), and such that

$$P'(x,t,D)\varphi = \psi \tag{8-15}$$

Set
$$G(\psi) = (\varphi, f)$$

As we noted in Sec. 1-6, $G(\psi)$ does not depend on φ as long as Eq. (8-15) holds (inequality (8-14) is used for this purpose). By (8-14), $G(\psi)$ is a bounded linear functional on W considered as a subspace of $L^2(\sigma_R)$. Since the closure $\overline{W}$ of W in $L^2(\sigma_R)$ is a Hilbert space, we know by Theorems 1-5 and 1-6 of Sec. 1-5 that there is a $u\in\overline{W}$ satisfying

$$G(\psi) = (\psi, u) \qquad \psi\in\overline{W}$$

Hence
$$(u, P'(x,t,D)\varphi) = (f,\varphi) \tag{8-16}$$

holds for all $\varphi\in C^m(\bar{\sigma}_R)$ which vanish near $\partial_1\sigma_R$ and satisfy Eq. (8-12). In keeping with the terminology of Sec. 1-6 and Sec. 4-2, we call u a *weak* solution of (8-6), (8-7). In order to justify this name, we must show that if $u\in C^m(\bar{\sigma}_R)$ and satisfies Eq. (8-16) for the prescribed functions φ, then u is indeed a solution of Eqs. (8-6), (8-7).

To see this, first note that u satisfies Eq. (8-16) for all $\varphi\in C_0^\infty(\sigma_R)$. Integrating by parts, we get

$$(P(x,t,D)u - f,\varphi) = 0 \qquad \varphi\in C_0^\infty(\sigma_R)$$

This implies that u is a solution of (8-6) (see Sec. 1-3). Next, we have, by Eq. (8-10)

$$\sum_{j=1}^{m} \int_{\partial_0 \sigma_R} D_t^{j-1} u \overline{N_j \varphi}\, dx = 0 \tag{8-17}$$

for all $\varphi \in C^m(\bar{\sigma}_R)$ vanishing near $\partial_1 \sigma_R$ and satisfying Eq. (8-12). An examination of Eq. (8-11) shows that

$$N_j \varphi(x, 0) = 0 \qquad r < j \leq m \tag{8-18}$$

holds for such φ. Thus, Eq. (8-17) reduces to

$$\sum_{j=1}^{r} \int_{\partial_0 \sigma_R} D_t^{j-1} u \overline{N_j \varphi}\, dx = 0 \tag{8-19}$$

holding for the φ prescribed above. We shall show in Sec. 8-3 that for any functions $g_1(x), \ldots, g_r(x) \in C_0^\infty(E^n)$ which vanish near $|x| = R$, there is a function φ of the prescribed type, such that

$$N_j \varphi(x, 0) = g_j(x) \qquad |x| < R \qquad 1 \leq j \leq r \tag{8-20}$$

(Lemma 8-4). From this, it follows that

$$\sum_{j=1}^{r} \int_{\partial_0 \sigma_R} D_t^{j-1} u \bar{g}_j\, dx = 0 \tag{8-21}$$

holds for all $g_j \in C_0^\infty(E^n)$ which vanish near $|x| = R$. This implies that u satisfies Eq. (8-7) (Lemma 4-2 of Sec. 4-2). Thus, the term "weak solution" will be appropriate.

We summarize the results of this section.

Theorem 8-1 A necessary and sufficient condition for problem (8-6), (8-7) to have a weak solution is for inequality (8-14) to hold. A weak solution which is in $C^m(\bar{\sigma}_R)$ is an actual solution.

Thus, we have no hope for solving problem (8-6), (8-7) unless (8-14) holds. We must determine, therefore, when this inequality is valid. This is done quite easily in Sec. 8-4. The more involved task of obtaining an actual solution is tackled in Sec. 8-7.

8-3 NORMAL BOUNDARY OPERATORS

Let $Q_1(x, t, D), \ldots, Q_r(x, t, D)$ be a set of partial differential operators with coefficients in $C^\infty(\bar{\sigma}_R)$. We call the set *normal* (with respect to the surface $t = 0$) if their orders m_j are all different, and if the coefficient of $D_t^{m_j}$ in $Q_j(x, 0, D)$ is a function which does not vanish in $|x| \leq R$ (see Sec. 7-8 for the constant-coefficient case). It is called a *Dirichlet system* of order r if all of the operators are of order less than r. Since there are r of them and their orders are distinct, all of the orders $0, 1, \ldots, r-1$ are found among them. By rearranging the operators if necessary, we may assume that $Q_j(x, t, D)$ is of order $j-1$. We shall make this assumption whenever we discuss Dirichlet systems.

Theorem 8-2 If $Q_1(x, t, D), \ldots, Q_r(x, t, D)$ is a Dirichlet system of order r, then

$$Q_j(x, 0, D) = \sum_{k=1}^{j} \Gamma_{jk}(x, D_x) D_t^{k-1} \qquad 1 \le j \le r \tag{8-22}$$

and

$$D_t^{j-1} = \sum_{k=1}^{j} \Lambda_{jk}(x, D_x) Q_k(x, 0, D) \qquad 1 \le j \le r \tag{8-23}$$

where Γ_{jk} and Λ_{jk} are partial differential operators with respect to the x variables, with infinitely differentiable coefficients, and of orders $\le j-k$. For each j, Γ_{jj} and Λ_{jj} are functions which do not vanish for $|x| \le R$. Also

$$\sum_{k=l}^{j} \Gamma_{jk} \Lambda_{kl} = \delta_{jl} \qquad 1 \le l \le j \le r \tag{8-24}$$

and

$$\sum_{k=l}^{j} \Lambda_{jk} \Gamma_{kl} = \delta_{jl} \qquad 1 \le l \le j \le r \tag{8-25}$$

Proof Equation (8-22) is a direct consequence of the definition. We prove Eqs. (8-23) to (8-25) by induction. They are certainly true for $j = 1$. Assume they are true for all $j < i \le r$. By Eq. (8-22) and the induction hypothesis

$$\begin{aligned} Q_i(x, 0, D) - \Gamma_{ii} D_t^{i-1} &= \sum_{k=1}^{i-1} \Gamma_{ik} D_t^{k} \\ &= \sum_{k=1}^{i-1} \sum_{l=1}^{k} \Gamma_{ik} \Lambda_{kl} Q_l(x, 0, D) \\ &= \sum_{l=1}^{i-1} \left(\sum_{k=l}^{i-1} \Gamma_{ik} \Lambda_{kl} \right) Q_l(x, 0, D) \end{aligned}$$

We can take $\Lambda_{ii} = 1/\Gamma_{ii}$ and

$$\Lambda_{il} = -\Lambda_{ii} \sum_{k=l}^{i-1} \Gamma_{ik} \Lambda_{kl} \qquad 1 \le l < i \tag{8-26}$$

This implies (8-23) for $j = i$. It also implies (8-24) and (8-25). In fact, we have

$$\sum_{k=l}^{i} \Gamma_{ik} \Lambda_{kl} = \Gamma_{ii} \Lambda_{il} + \sum_{k=l}^{i-1} \Gamma_{ik} \Lambda_{kl} = \delta_{il}$$

and

$$\begin{aligned} \sum_{k=l}^{i} \Lambda_{ik} \Gamma_{kl} &= \Lambda_{ii} \Gamma_{il} + \sum_{k=l}^{i-1} \Lambda_{ik} \Gamma_{kl} \\ &= \Lambda_{ii} \Gamma_{il} - \Lambda_{ii} \sum_{k=l}^{i-1} \sum_{j=k}^{i-1} \Gamma_{ij} \Lambda_{jk} \Gamma_{kl} \end{aligned}$$

$$= \Lambda_{ii}\Gamma_{il} - \Lambda_{ii} \sum_{j=l}^{i-1} \Gamma_{ij} \sum_{k=l}^{j} \Lambda_{jk}\Gamma_{kl}$$

$$= \delta_{il}$$

by the induction hypothesis. This completes the proof.

Corollary 8-3 If $Q'_1(x, t, D), \ldots, Q'_r(x, t, D)$ is any other Dirichlet system of order r, then

$$Q'_j(x, 0, D) = \sum_{k=1}^{j} \Theta_{jk}(x, D_x)Q_k(x, 0, D) \qquad 1 \le j \le r \tag{8-27}$$

where Θ_{jk} is a partial differential operator of order $\le j - k$ and Θ_{jj} is a function which does not vanish.

Proof By Theorem 8-2 there are operators Γ'_{jk} and Λ'_{jk} behaving, with respect to the Q'_j, as the operators Γ_{ij} and Λ_{jk} behave with respect to the Q_j. Thus

$$Q'_j(x, 0, D) = \sum_{k=1}^{j} \Gamma'_{jk} D_t^{k-1}$$

$$= \sum_{k=1}^{j} \Gamma'_{jk} \sum_{i=1}^{k} \Lambda_{ki} Q_i(x, 0, D)$$

$$= \sum_{i=1}^{j} \left(\sum_{k=i}^{j} \Gamma'_{jk}\Lambda_{ki} \right) Q_i(x, 0, D)$$

It is easily checked that this is of the required form.

Lemma 8-4 If $Q_1(x, t, D), \ldots, Q_r(x, t, D)$ is a Dirichlet system of order r, and $g_1(x), \ldots, g_r(x)$ are arbitrary functions which are infinitely differentiable in $|x| < R$ and vanish near $|x| = R$, then there is a function $v \in C^\infty(\bar{\sigma}_R)$ which vanishes near $\partial_1\sigma_R$, and satisfies

$$Q_j(x, 0, D)v(x, 0) = g_j(x) \qquad 1 \le j \le r \tag{8-28}$$

Proof By Theorem 8-2, it suffices to find a function $v \in C^\infty(\bar{\sigma}_R)$ vanishing near $\partial_1\sigma_R$, and satisfying

$$D_t^{j-1}v(x, 0) = \sum_{k=1}^{j} \Lambda_{jk} g_k(x) = h_j(x) \qquad 1 \le j \le r \tag{8-29}$$

Now, there is an $R_1 < R$ such that all the $g_k(x)$ vanish for $|x| \ge R_1$. Let $\delta > 0$ be so small that the cylinder $|x| \le R_1$, $0 \le t \le \delta$ is contained in σ_R. Let $\zeta(t)$ be a function in $C_0^\infty(-\delta, \delta)$ which equals 1 near $t = 0$. If we now set

$$v(x, t) = \zeta(t) \sum_{k=1}^{r} \frac{(it)^{k-1} h_k(x)}{(k-1)!} \tag{8-30}$$

we obtain a function with the desired properties.

The statement made at the end of Sec. 8-2, that Eq. (8-20) has a solution of the required type, follows from Lemma 8-4 and the fact that the set $N_1, \ldots, N_r$ given by Eq. (8-11) is a Dirichlet system.

The results of this section are due to Aronszajn–Milgram (1953).

8-4 THE ESTIMATE

We now consider the question of when inequality (8-14) holds for the φ described in Sec. 8-2. We examine this inequality in more detail.

Let $\mathscr{D}_R$ be the set of those $\varphi \in C^m(\bar{\sigma}_R)$ which vanishes near $\partial_1 \sigma_R$ and satisfy Eq. (8-12). Define them to vanish outside σ_R, and let H_R be the completion of $\mathscr{D}_R$ with respect to the norm $| \ |_{m,0}$ (see Eq. (6-128) of Sec. 6-7). If $R(x, t, D)$ is any operator of order $\leq m$ with coefficients bounded on σ_R, we can extend the definition of $R(x, t, D)$ to apply to functions in H_R. This can be done by means of the inequality

$$\| R(x,t,D)\varphi \| \leq C \, |\varphi|_{m,0} \qquad \varphi \in \mathscr{D}_R \tag{8-31}$$

which will be proved later in this section. In fact, if u is any function in H_R, there is a sequence $\{\varphi_k\}$ of functions in $\mathscr{D}_R$ converging to u in H. By (8-31), the sequence $\{R(x, t, D)\varphi_k\}$ converges in $L^2(\sigma_R)$ to a function w. We define $R(x, t, D)u$ to be w. In order that this definition make sense, we must show that it does not depend on the particular sequence $\{\varphi_k\}$. This is indeed the case. For if $\{\psi_k\}$ is another sequence of functions in $\mathscr{D}_R$ converging to u in H, then

$$\| R(x,t,D)[\varphi_k - \psi_k] \| \leq C \, |\varphi_k - \psi_k|_{m,0} \to 0$$

by (8-31). Thus, $R(x, t, D)\psi_k \to w$ in $L^2(\sigma_R)$ as well. Note that $H_R \subset L^2(\sigma_R)$.

Next, we note that (8-31) implies

$$\| R(x,t,D)v \| \leq C \, |v|_{m,0} \qquad v \in H_R \tag{8-32}$$

For if $\{\varphi_k\}$ is a sequence of functions in $\mathscr{D}_R$ converging to v in H_R, then

$$\| R(x,t,D)\varphi_k \| \leq C \, |\varphi_k|_{m,0}$$

by (8-31). Taking the limit as $k \to \infty$, we obtain (8-32). Similarly, the same reasoning shows that (8-14) is equivalent to

$$|(f, v)| \leq C \, \| P'(x,t,D)v \| \qquad v \in H_R \tag{8-33}$$

Let N'_R be the set of those $v \in H_R$ satisfying

$$P'(x,t,D)v = 0 \tag{8-34}$$

From (8-33) we see that, in order for (8-14) to hold, it is necessary that $(f, N'_R) = 0$ (written $f \perp N'_R$ for short). We shall examine this set.

Clearly, N'_R is a subspace of $L^2(\sigma_R)$. We shall prove in Sec. 8-6:

Theorem 8-5 For R sufficiently small, N'_R is finite dimensional.

It follows from this that N'_R is a closed subspace of $L^2(\sigma_R)$ (see Sec. 7-4). As a consequence, we have

Lemma 8-6 If $f \perp N'_R$ and

$$|(f, v)| \leq C \, \| P'(x, t, D)v \| \qquad v \in H_R \qquad v \perp N'_R \tag{8-35}$$

then (8-33) holds.

PROOF If $v \in H_R$, then $v = v_1 + v_2$, where $v_2 \in N'_R$ and $v_1 \perp N'_R$ in $L^2(\sigma_R)$ (Theorem 1-3 of Sec. 1-5). Since v and v_2 are both in H, the same is true of v_1. Thus, if $f \perp N'_R$ and (8-35) holds, we have

$$|(f, v)| = |(f, v_1)| \leq C \, \| P'(x, t, D)v_1 \| = C \, \| P'(x, t, D)v \|$$

This is precisely (8-33).

Corollary 8-7 If

$$\| v \| \leq C \, \| P'(x, t, D)v \| \qquad v \in H_R \qquad v \perp N'_R \tag{8-36}$$

then (8-33) holds for each $f \perp N'_R$.

From Corollary 8-7, we see that if we want (8-14) to hold for each $f \perp N'_R$, we should prove (8-36). For this purpose, we use

Theorem 8-8 Under the assumption stated in Sec. 8-1, there are constants $R > 0$ and C, such that

$$|\varphi|_{m,0} \leq C \, (\| P'(x, t, D)\varphi \| + \| \varphi \|) \qquad \varphi \in \mathscr{D}_R \tag{8-37}$$

and

Lemma 8-9 If $\{u_k\}$ is a sequence of functions in H_R, such that

$$|u_k|_{m,0} \leq K$$

then it has a subsequence converging in $L^2(\sigma_R)$.

As we noted before, (8-37) implies

$$|v|_{m,0} \leq C(\| P'(x, t, D)v \| + \| v \|) \qquad v \in H_R \tag{8-38}$$

We shall prove Theorem 8-8 at the end of this section and Lemma 8-9 in Sec. 8-6. We now use them to prove

Theorem 8-10 Under the assumption of Sec. 8-1, there are constants $R > 0$ and C, such that

$$|v|_{m,0} \leq C \, \| P'(x, t, D)v \| \qquad v \in H_R \qquad v \perp N'_R \tag{8-39}$$

PROOF Suppose (8-39) did not hold. Then there would be a sequence $\{v_k\}$ of functions in H_R which satisfy $(v_k, N'_R) = 0$, and

$$|v_k|_{m,0} = 1 \qquad \| P'(x, t, D)v_k \| \to 0 \tag{8-40}$$

By Lemma 8-9, there is a subsequence which converges in $L^2(\sigma_R)$. We are only interested in this subsequence. Thus, we may assume that all other terms have been discarded, and that the entire sequence converges in $L^2(\sigma_R)$. By inequality (8-38)

$$|v_j - v_k|_{m,0} \le C(\| P'(x, t, D)[v_j - v_k] \| + \| v_j - v_k \|) \to 0$$

Hence, v_k converges in H_R to a function $v \in H_R$. Since each of the functions v_k is orthogonal to N'_R, the same is true for v. By (8-32)

$$\| P'(x, t, D)[v_j - v_k] \| \le C |v - v_k|_{m,0} \to 0$$

Thus $P'(x, t, D)v = 0$. This means that $v \in N'_R$. The only way this can happen is if $v = 0$. But

$$|v|_{m,0} = \lim |v_k|_{m,0} = 1$$

This contradiction proves (8-39).

Corollary 8-11 Under the assumptions of Sec. 8-1, there are constants $R > 0$ and C, such that (8-33) holds for each $f \perp N'_R$.

PROOF Combine Corollary 8-7 and Theorem 8-10.

Corollary 8-12 There exists an $R_0 > 0$ such that Eqs. (8-6), (8-7) have a weak solution for each $f \perp N'_R$ whenever $R \le R_0$.

PROOF By Theorem 8-8 there is a constant $R > 0$ for which (8-37) holds. Note that $\mathscr{D}_R$ contains $\mathscr{D}_{R_1}$ for any $R_1 < R$. Thus, (8-37) holds for any smaller value of R. This in turn implies (8-39) for any smaller value of R, and this in turn implies (via (8-33)) that Eqs. (8-6), (8-7) have a weak solution for each $f \perp N'_R$ for each smaller value of R.

We now give the proofs promised earlier. Inequality (8-31) is a consequence of

$$\sum_{|\mu| + k \le m} \| D_x^\mu D_t^k \varphi \| \le C_0 |\varphi|_{m,0} \qquad \varphi \in \mathscr{D}_R \tag{8-41}$$

and the fact that the coefficients of $R(x, t, D)$ are bounded. To prove (8-41), note that

$$\begin{aligned} \| D_x^\mu D_t^k \varphi \|^2 &= (2\pi)^{-n} \int\int_0^\infty | \xi^\mu D_t^k F\varphi(\xi, t) |^2 \, d\xi \, dt \\ &\le C \int\int_0^\infty |\xi|^{2|\mu|} | D_t^k F\varphi(\xi, t) |^2 \, d\xi \, dt \\ &\le C |\varphi|_{m,0}^2 \end{aligned}$$

for $|\mu| + k \leq m$, by Parseval's identity (2-24) and Eq. (6-128) of Sec. 6-7. This gives (8-41). Note that the constant C_0 does not depend on R.

We also give the

PROOF of Theorem 8-8 We refer to Theorem 6-26 of Sec. 6-7. Since the coefficients of $P(x, t, D)$ are in $C^\infty(\bar{\sigma}_R)$, we have

$$P'(x,t,D)\varphi = \sum D_x^\mu D_t^k[\bar{a}_{\mu,k}(x,t)\varphi]$$

$$= \sum b_{\mu,k}(x,t) D_x^\mu D_t^k \varphi \qquad (8\text{-}42)$$

where the $b_{\mu,k}$ are in $C^\infty(\bar{\sigma}_R)$. Note that

$$b_{\mu,k}(x,t) = \bar{a}_{\mu,k}(x,t) \qquad |\mu| + k = m \qquad (8\text{-}43)$$

Thus, $P'(x, t, D)$ is elliptic in $\bar{\sigma}_R$, and

$$P_1(D) = \sum b_{\mu,k}(0,0) D_x^\mu D_t^k \qquad (8\text{-}44)$$

is properly elliptic. Set

$$c_{\mu,k}(x,t) = b_{\mu,k}(x,t) - b_{\mu,k}(0,0) \qquad |\mu| + k \leq m$$

and

$$R_1(x,t,D) = \sum c_{\mu,k}(x,t) D_x^\mu D_t^k$$

Then

$$P'(x,t,D) = P_1(D) + R_1(x,t,D) \qquad (8\text{-}45)$$

and the coefficients of $R_1(x, t, D)$ vanish at the origin. Since $P_1(D)$ is properly elliptic and the operators given by Eq. (8-3) satisfy hypothesis (*b*) of Theorem 6-15 of Sec. 6-7, we know that there is a constant C_1, such that

$$|\varphi|_{m,0} \leq C_1(\| P_1(D)\varphi \| + \| \varphi \|) \qquad \varphi \in \mathscr{D}_R \qquad (8\text{-}46)$$

Moreover, we know that there is a constant C_0 independent of R, such that (8-41) holds. Since the $c_{\mu,k}(x, t)$ vanish at the origin, we can take $R > 0$ so small, that

$$|c_{\mu,k}(x,t)| \leq \frac{1}{2C_0C_1} \qquad |\mu| + k \leq m$$

Thus

$$\| R_1(x,t,D)\varphi \| \leq \frac{|\varphi|_{m,0}}{2C_1} \qquad \varphi \in \mathscr{D}_R \qquad (8\text{-}47)$$

Combining (8-46) and (8-47), we have

$$|\varphi|_{m,0} \leq C_1(\| P'(x,t,D)\varphi \| + \| \varphi \|) + \tfrac{1}{2}|\varphi|_{m,0}$$

This gives the desired inequality, and the proof is complete.

Since $P'(x, t, D)$ is elliptic if and only if $P(x, t, D)$ is, $P'(0, 0, D)$ is properly elliptic if and only if $P(0, 0, D)$ is, and the formal adjoint of $P'(x, t, D)$ is $P(x, t, D)$, we have the following immediate consequence of Theorem 8-8:

Theorem 8-13 Under the assumptions of Sec. 8-1 there are constants $R > 0$ and C, such that

$$|\varphi|_{m,0} \leq C(\| P(x, t, D)\varphi \| + \| \varphi \|) \qquad \varphi \in \mathscr{D}_R \tag{8-48}$$

Of course the proof given for Theorem 8-8 applies as well for Theorem 8-13.

8-5 COMPACT OPERATORS

Let H_1, H_2 be Hilbert spaces. A linear operator T from H_1 to H_2 defined on the whole of H_1 is called *compact*, if for each sequence $\{x_k\}$ of elements of H_1 satisfying

$$\| x_k \| \leq C \tag{8-49}$$

there is a subsequence $\{y_j\}$ of $\{x_k\}$, such that $\{Ty_j\}$ converges in H_2. It is easy to show that if T_1 and T_2 are compact operators, so is $\alpha_1 T_1 + \alpha_2 T_2$ for each α_1 and α_2. We shall need the following results concerning compact operators.

Lemma 8-14 Compact operators are bounded.

PROOF If T is not a bounded operator, then there is a sequence $\{x_k\}$, such that

$$\| x_k \| = 1 \qquad \| Tx_k \| \to \infty \qquad k \to \infty \tag{8-50}$$

Thus, the sequence $\{Tx_k\}$ cannot have a convergent subsequence. Hence, T cannot be compact.

Theorem 8-15 An operator T from H_1 to H_2 defined everywhere, is compact if $\| Tx_n \| \to 0$ for each sequence $\{x_k\}$ converging weakly to zero.

PROOF Suppose T is compact and $\{x_k\}$ converges weakly to zero. Suppose there were a subsequence $\{z_n\}$ of $\{x_k\}$ and an $\varepsilon > 0$, such that

$$\| Tz_n \| \geq \varepsilon \qquad n = 1, 2, \ldots \tag{8-51}$$

By Corollary 1-11 of Sec. 1-5 there is a constant C such that inequality (8-49) holds. Consequently, by the definition of compactness, there is a subsequence $\{y_j\}$ of $\{z_n\}$, such that Ty_j converges in H_2 to an element w. By Lemma 8-14, $|(Ty, w)| \leq C \| y \|$ for all $y \in H_1$. Hence (Ty, w) is a bounded linear functional on H_1. This implies that there is an element $T^*w \in H_2$, such that

$$(Ty, w) = (y, T^*w) \qquad y \in H_1$$

(Theorem 1-5 of Sec. 1-5). In particular

$$(Ty_j, w) = (y_j, T^*w) \qquad j = 1, 2, \ldots$$

Since $Ty_j \to w$ and y_j converges weakly to 0 as $j \to \infty$

$$(w, w) = 0$$

Thus $w = 0$. But, by (8-51), we have $\| w \| \geq \varepsilon$. This contradiction shows that there can be no subsequence $\{z_n\}$ satisfying (8-51). This means that $\| Tx_k \| \to 0$.

Conversely, suppose T has the property mentioned in the theorem, and let $\{x_k\}$ be a sequence satisfying (8-49). By Theorem 1-8 of Sec. 1-5 there is a subsequence $\{y_j\}$ of $\{x_k\}$, which converges weakly to some element $y \in H_1$. Set $z_j = y_j - y$. Then z_j converges weakly to zero. Hence, there is a subsequence $\{u_n\}$ of $\{z_j\}$, such that $\| Tu_n \| \to 0$. Thus, if $v_n = u_n + y$, then $\{v_n\}$ is a subsequence of $\{x_k\}$ and $\{Tv_n\}$ converges in H_2 to Ty. Hence T is compact. This completes the proof.

A subset B of a Hilbert space is called *totally bounded* if, for each $\varepsilon > 0$, there is a finite subset G of B, such that every element in B is within a distance ε from some element of G. Another way of stating it is that for each $\varepsilon > 0$ there are elements $x_1, \ldots, x_n$ of B, such that B is contained in the union of the balls of radii ε and centers x_k.

Theorem 8-16 Suppose B is a totally bounded subset of a Hilbert space H, and suppose $\{x_k\}$ is a sequence of elements of H satisfying (8-49), and such that

$$(x_k, y) \to 0 \qquad \text{as} \qquad k \to \infty \tag{8-52}$$

for each $y \in B$. Then the convergence is uniform. Thus, for each $\varepsilon > 0$, there is an N not depending on y, such that

$$|(x_k, y)| < \varepsilon \qquad \text{for} \qquad k > N \qquad y \in B$$

Proof Let $\varepsilon > 0$ be given. Since B is totally bounded, there exist elements $y_1, \ldots, y_n$ such that every element of B is within $\varepsilon/2C$ of one of the y_j. Let N be so large, that

$$|(x_k, y_j)| < \frac{\varepsilon}{2} \qquad k > N \qquad 1 \leq j \leq n$$

Let y be any element of B. Then there is a y_j, such that

$$\| y - y_j \| < \frac{\varepsilon}{2C}$$

Thus
$$|(x_k, y)| \leq |(x_k, y - y_j)| + |(x_k, y_j)|$$
$$\leq \| x_k \| \, \| y - y_j \| + |(x_k, y_j)| < \frac{\varepsilon}{2} + \frac{\varepsilon}{2} = \varepsilon$$

for $k > N$. Since N does not depend on y, the proof is complete.

We shall also use

Theorem 8-17 If H is a Hilbert space such that every sequence satisfying (8-49) has a convergent subsequence, then H is finite dimensional.

PROOF Suppose H were infinite dimensional. Let e_1 be a *unit* vector (i.e., has norm one), and let V_1 be the subspace *spanned* by e_1 (i.e., all vectors are of the form αe_1, α a scalar). Since V_1 is finite dimensional (Lemma 3-7 of Sec. 3-3), it is closed (Lemma 7-12 of Sec. 7-4) and not the whole of H. Thus, by Corollary 1-14 of Sec. 1-5, there is a vector $e_2 \neq 0$ such that $(e_2, V_1) = 0$. We may take e_2 to be a unit vector. Let V_2 be the subspace spanned by e_1 and e_2 (i.e., the set of all vectors of the form $\alpha_1 e_1 + \alpha_2 e_2$). Again V_2 is finite dimensional and, therefore, closed but not the whole of H. Thus, there is a unit vector e_3, such that $(e_3, V_2) = 0$. Continuing in this manner, we obtain a sequence $\{e_k\}$ of vectors, such that

$$(e_j, e_k) = \delta_{jk} \qquad j, k = 1, 2, \ldots$$

(Such a sequence is called *orthonormal*). It never ends because the subspace V_n spanned by the first n of them is finite dimensional and, consequently, closed but not the whole of H. The sequence $\{e_k\}$ satisfies (8-49), but it has no convergent subsequence. In fact, for $j \neq k$, we have

$$\| e_j - e_k \|^2 = \| e_j \|^2 - 2 \operatorname{Re}(e_j, e_k) + \| e_k \|^2 = 2$$

This contradiction shows that H must be finite dimensional.

8-6 COMPACT EMBEDDING

We now show that certain operators are compact in the function spaces we have been using. Our first result is

Theorem 8-18 Let φ be a function in C_0^∞. Suppose $s > 0$ and that there is a sequence $\{v_k\}$ of elements of H^s, such that

$$|v_k|_s \leq C \qquad k = 1, 2, \ldots \tag{8-53}$$

Then for each t satisfying $0 \leq t < s$ there is a subsequence $\{u_j\}$ of $\{v_k\}$, such that $\{\varphi u_j\}$ converges in H^t.

PROOF The theorem states that the operator T, defined by

$$Tv = \varphi v \qquad v \in H^s \tag{8-54}$$

is compact from H^s to H^t (see Sec. 8-5). By Theorem 8-15, it suffices to show that

$$(v_k, h)_s \to 0 \qquad \text{as} \qquad k \to \infty \qquad h \in H^s \tag{8-55}$$

implies

$$|Tv_k|_t \to 0 \quad \text{as} \quad k \to \infty \tag{8-56}$$

We first observe that (8-55) implies

$$(Fv_k, w) \to 0 \quad \text{as} \quad k \to \infty \qquad w \in S \tag{8-57}$$

In fact, if $w \in S$ and we put $h = F^{-1}[w/(1 + |\xi|)^{2s}]$, then $h \in S$ and (8-55) implies (8-57). Next, we observe that (8-55) also implies that there is a constant C, such that (8-53) holds (Corollary 1-11 of Sec. 1-5). Thus, by Lemma 3-4 of Sec. 3-2, there is a constant C_1, such that

$$|\varphi v_k|_s \leq C_1 \tag{8-58}$$

For $\eta \in E^n$, set

$$w_\eta(\xi) = \overline{F\varphi(\eta - \xi)}$$

Then, for each η, the function w_η is in S, and

$$(Fv_k, w_\eta) \to 0 \quad \text{as} \quad k \to \infty$$

by (8-57). By Theorem 2-10 of Sec. 2-5, this is equivalent to

$$F(\varphi v_k) \to 0 \quad \text{as} \quad k \to \infty \ \text{ for each } \xi \tag{8-59}$$

Now, for any $R > 0$

$$\begin{aligned} |\varphi v_k|_t^2 &= \left(\int_{|\xi|>R} + \int_{|\xi|<R} \right) (1 + |\xi|)^{2t} |F[\varphi v_k]|^2 \, d\xi \\ &\leq (1 + R)^{2(t-s)} \int_{|\xi|>R} (1 + |\xi|)^{2s} |F[\varphi v_k]|^2 \, d\xi \\ &\quad + \int_{|\xi|<R} (1 + |\xi|)^{2t} |F[\varphi v_k]|^2 \, d\xi \\ &\leq C_1(1 + R)^{2(t-s)} + \int_{|\xi|<R} (1 + |\xi|)^{2t} |F[\varphi v_k]|^2 \, d\xi \end{aligned}$$

The first term tends to zero as $R \to \infty$ and, hence, can be made as small as desired. If $\varepsilon > 0$ is given, we can take R so large that the first term is $< \varepsilon/2$. Once R is fixed, we hope that the second term will tend to zero as $k \to \infty$. We know that the integrand tends to zero for each fixed ξ by (8-59). However, this does not always imply that the integrals converge. For this purpose it would be convenient to know that the convergence is uniform. Then it would follow that the integrals converge and the theorem would be proved.

Fortunately, help is at hand. The set $\{w_\eta\}$ for $|\eta| \leq R < \infty$ is totally bounded in L^2 (see Sec. 8-5). In fact, we have

$$\|w_\eta - w_\zeta\|^2 = \int |F\varphi(\eta - \xi) - F\varphi(\zeta - \xi)|^2 \, d\xi$$

$$= \int |F\varphi(\eta - \zeta + z) - F\varphi(z)|^2 \, dz = \int |F[\varphi \, e^{-i(\eta-\zeta)x} - \varphi]|^2 \, d\xi$$

$$= \int |\varphi(x)|^2 |e^{-i(\eta-\zeta)x} - 1|^2 \, dx \leq |\eta - \zeta|^2 \int |\varphi(x)|^2 |x|^2 \, dx$$

by Parseval's identity, (2-24) of Sec. 2-3, and (2-66) of Sec. 2-5. Clearly, for each $\delta > 0$ there is a finite number of points $\eta^{(1)}, \ldots, \eta^{(m)}$, such that the spheres

$$|\eta - \eta^{(j)}| < \frac{\delta}{\| |x| \varphi \|} \qquad 1 \leq j \leq m \tag{8-60}$$

cover the set $|\eta| \leq R$. This shows that the set $\{w_\eta\}$ is totally bounded for $|\eta| \leq R$. We can now apply Theorem 8-16, to conclude that the convergence in (8-59) is uniform in $|\xi| \leq R$. This completes the proof.

Before we continue we shall state and prove two lemmas.

Lemma 8-19 Suppose $m \geq 0$ and $u \in C^m(E^n)$ is such that

$$D^\mu u \in L^2(E^n) \quad \text{for} \quad |\mu| \leq m \tag{8-61}$$

and

$$|x|^k u(x) \text{ is bounded in } E^n \text{ for each } k. \tag{8-62}$$

Then

1. $J_\varepsilon u \in S$ for each $\varepsilon > 0$
2. $|u|_k < \infty$ for $0 \leq k \leq m$
3. $|J_\varepsilon u|_k \leq |u|_k \qquad 0 \leq k \leq m$
4. $u \in H^m$.

Proof Clearly $J_\varepsilon u \in C^\infty(E^n)$. Now

$$|D^\mu J_\varepsilon u| = \left| \int D^\mu j_\varepsilon(x - y) u(y) \, dy \right|$$

$$\leq \int |D^\mu j_\varepsilon(z)| \, dz \cdot \max_{|x-y| \leq \varepsilon} |u(y)|$$

$$\leq K_{\varepsilon,\mu,k} \max_{|x-y| \leq \varepsilon} \frac{1}{|y|^k}$$

where $K_{\varepsilon,\mu,k}$ depends on ε, μ, and k. If we take $2\varepsilon \leq |x|$, this gives

$$|D^\mu J_\varepsilon u| \leq \frac{2^k K_{\varepsilon,\mu,k}}{|x|^k}$$

This shows that $J_\varepsilon u \in S$. Next, note that (8-61) implies that $\xi^\mu Fu \in L^2(E^n)$ for $|\mu| \le m$ (see Eq. (2-22) of Sec. 2-3). Thus, $(1 + |\xi|)^k Fu \in L^2(E^n)$ for each k. This gives 2. We prove 3 by noting, as in Sec. 6-8, that

$$F(J_\varepsilon u) = Fj_\varepsilon Fu \tag{8-63}$$

By Eq. (2-62) of Sec. 2-5, this implies

$$|F(J_\varepsilon u)| \le |Fu|$$

which in turn implies 3. Finally, we note that since $|J_\varepsilon u|_m$ is uniformly bounded, there is a sequence $\{\varepsilon_k\}$ converging to zero, such that the arithmetical means of the functions $J_{\varepsilon_k} u$ converge in H^m (Theorem 1-9 of Sec. 1-5). Since $J_\varepsilon u$ converges to u in $L^2(E^n)$ (Theorem 2-3 of Sec. 2-2), the limit must equal u almost everywhere. Since each $J_\varepsilon u$ is in S, we see that $u \in H^m$, and the proof is complete.

As usual we let Ω denote the half-space $t > 0$ in E^{n+1}.

Lemma 8-20 For each integer $m \ge 0$, there is a bounded linear mapping L from $H^{m,0}(\Omega)$ to $H^m(E^{n+1})$, such that

$$Lu = u \text{ a.e. in } \Omega \tag{8-64}$$

Moreover, L is bounded from $H^{k,0}(\Omega)$ to $H^k(E^{n+1})$ for $0 \le k \le m$.

PROOF For $u \in S(\Omega)$ define

$$\begin{aligned} Lu(x,t) &= u(x,t) && t > 0 \quad x \in E^n \\ &= \sum_{k=1}^{m+1} \lambda_k u(x, -kt) && t < 0 \quad x \in E^n \end{aligned} \tag{8-65}$$

where the constants λ_k satisfy

$$\sum_{k=1}^{m+1} \lambda_k (-k)^j = 1 \qquad 0 \le j \le m \tag{8-66}$$

It is easily checked that such constants exist. Moreover, Eq. (8-66) guarantees that derivatives of Lu, up to order m, are continuous across $t = 0$. Thus, Lu is in $C^m(E^{n+1})$. Since $u \in S(\Omega)$, it follows from Lemma 8-19 that $u \in H^m(E^{n+1})$. Now

$$|Lu|_k^2 = \int\!\!\int (1 + [|\xi|^2 + \tau^2]^{1/2})^{2k} |F\widetilde{Lu}(\xi, \tau)|^2 \, d\xi \, d\tau$$

where F is the Fourier transform with respect to the variables $x \in E^n$, and $\tilde{h}$ is the Fourier transform with respect to t. There is a constant depending only on n and k, such that

$$(1 + |\xi| + |\tau|)^{2k} \le C \sum_{j=0}^{k} (1 + |\xi|)^{2(k-j)} \tau^{2j} \tag{8-67}$$

Hence, for $k \leq m$

$$|Lu|_k^2 \leq C \sum_{j=0}^{k} \iint (1 + |\xi|)^{2(k-j)} |D_t^j F Lu(\xi, \tau)|^2 \, d\xi \, d\tau$$

$$= C \sum_{j=0}^{k} \iint (1 + |\xi|)^{2(k-j)} |D_t^j F Lu(\xi, t)|^2 \, d\xi \, dt \tag{8-68}$$

by Corollary 7-6 of Sec. 7-2 and Parseval's identity (2-24). We now note that for $q \geq 1$

$$\int ds \int_{-\infty}^{0} (1 + |\xi|)^{2(k-j)} |D_t^j F u(\xi, -qt)|^2 \, dt$$

$$= q^{2j-1} \int ds \int_{0}^{\infty} (1 + |\xi|)^{2(k-j)} |D_s^j F u(\xi, s)|^2 \, ds$$

$$\leq q^{2j-1} |u|_{k,0}^2$$

(see Eq. (6-128) of Sec. 6-7). This combined with (8-68) gives

$$|Lu|_k \leq C |u|_{k,0} \qquad 0 \leq k \leq m \tag{8-69}$$

where the constant depends only on k, m, and n. We can now extend L to the whole of $H^{m,0}(\Omega)$, by using (8-69) and the fact that $S(\Omega)$ is dense in $H^{m,0}(\Omega)$ (see Sec. 8-1). This completes the proof.

Lemma 8-21 For each $k \geq 0$ there is a constant C, such that

$$|v|_{k,0} \leq C |v|_k \qquad v \in S(E^{n+1}) \tag{8-70}$$

Thus, the restriction to Ω of each $v \in H^k(E^{n+1})$ is in $H^{k,0}(\Omega)$, and (8-70) holds for such v.

Proof By Eq. (6-128) of Sec. 6-7

$$|v|_{k,0}^2 = \sum_{j=0}^{k} \int d\xi \int_{0}^{\infty} (1 + |\xi|)^{2(k-j)} |D_t^j F v(\xi, t)|^2 \, dt$$

$$\leq \sum_{j=0}^{k} \int d\xi \int_{-\infty}^{\infty} (1 + |\xi|)^{2(k-j)} |D_t^j F v(\xi, t)|^2 \, dt$$

$$= \sum_{j=0}^{k} \int d\xi \int_{-\infty}^{\infty} (1 + |\xi|)^{2k-j} \tau^{2j} |\widetilde{Fv}(\xi, \tau)|^2 \, d\tau$$

$$\leq C \int d\xi \int_{-\infty}^{\infty} (1 + [|\xi|^2 + \tau^2]^{1/2})^{2k} |\widetilde{Fv}(\xi, \tau)|^2 \, d\tau$$

$$= C |v|_k^2$$

Where we used Eq. (2-22) of Sec. 2-3, Parseval's identity (2-24), and the inequality

$$\sum_{j=0}^{k} (1 + |\xi|)^{2(k-h)}\tau^{2j} \le C(1 + [|\xi|^2 + \tau^2]^{1/2})^{2k} \tag{8-71}$$

If $v \in H^k(E^{n+1})$, there is a sequence of functions in $S(E^{n+1})$ converging to it in $H^k(E^{n+1})$. By (8-70), the restrictions of these functions converge in $H^{k,0}(\Omega)$. Since these restrictions are in $S(\Omega)$, and they converge to v in $L^2(\Omega)$, we see that the restriction of v to Ω is in $H^{k,0}(\Omega)$ and (8-70) holds for v. This completes the proof.

Theorem 8-22 If $m > 0$ and $\{v_j\}$ is a sequence of functions in $H^{m,0}(\Omega)$, such that

$$|v_j|_{m,0} \le C \tag{8-72}$$

then for each $\varphi \in C_0^\infty(E^{n+1})$ there is a subsequence of $\{\varphi v_j\}$ which converges in $H^{k,0}(\Omega)$ for each $k < m$. In particular, it converges in $L^2(\Omega)$.

PROOF Let L be the mapping described in Lemma 8-20. Then (8-72) implies

$$|Lv_k|_m \le C'$$

By Theorem 8-18 there is a subsequence $\{g_i\}$ of $\{v_k\}$, such that $\varphi L g_i$ converges in $H^k(E^{n+1})$. By Lemma 8-21 the φg_i (which are the restrictions of $\varphi L g_i$ to Ω) converge in $H^{k,0}(\Omega)$. This completes the proof.

We now return to the spaces of functions defined on σ_R (see Sec. 8-4).

Lemma 8-23 $H_R \subset H^{m,0}(\Omega)$. (Recall that the functions of H_R vanish outside σ_R.)

PROOF Since $\mathscr{D}_R$ is dense in H_R and the norm is the same, it suffices to show that $\mathscr{D}_R \subset H^{m,0}(\Omega)$. If $v \in \mathscr{D}_R$, then Lv is in $C^m(E^{n+1})$ (see the proof of Lemma 8-20) and vanishes for $|x| > R$. Thus, Lv is in $H^m(E^{n+1})$ by Lemma 8-19. Hence, its restriction to Ω (which is v itself) is in $H^{m,0}(\Omega)$ (Lemma 8-21).

We can now give the

PROOF of Theorem 8-5 Clearly N'_R is closed. For if $\{v_k\}$ is a sequence of functions in N'_R which converges to v in H_R, then $P'(x,t,D)v_k \to P'(x,t,D)v$ in $L^2(\sigma_R)$. Thus $v \in N'_R$. This shows that N'_R is itself a Hilbert space. Now, suppose $\{v_k\}$ is a sequence of functions in N'_R satisfying (8-72). Let ζ be a function in C_0^∞ which equals one on σ_R. Then $\zeta v_k = v_k$. Thus, by Theorem 8-22, there is a subsequence $\{g_j\}$ of $\{v_k\}$ which converges in $L^2(\sigma_R)$. Applying Theorem 8-8, we get

$$|g_j - g_k|_{m,0} \le C \| g_j - g_k \| \to 0 \quad \text{as} \quad j,k \to \infty$$

and consequently $\{g_j\}$ converges in H_R as well. This shows that N'_R is finite dimensional, by Theorem 8-17.

8-7 SOLVING THE PROBLEM

In Sec. 8-4 we showed that for R sufficiently small the Dirichlet problem (8-6), (8-7) has a weak solution for each $f \in L^2(\sigma_R)$, such that $f \perp N'_R$ (Corollary 8-12). In this section we shall show that we can obtain an actual solution, provided f is sufficiently smooth. For this purpose we use a regularity theorem, which we state as

Theorem 8-24 Let $P(x, t, D)$ satisfy the assumptions of Sec. 8-1. Suppose $R_1 \geq R > 0$, $f \in C^\infty(\bar{\sigma}_R)$, $u \in H_{R_1}$, and

$$(P(x,t,D)u, P(x,t,D)v) = (f, v) \qquad v \in \mathscr{D}_R \tag{8-73}$$

then for some $R' > 0$, we have $u \in C^\infty(\bar{\sigma}_{R'})$ after correction on a set of measure zero.

The proof of Theorem 8-24 is not trivial; it will be given in Sec. 8-9. We now use it to solve Eqs. (8-6), (8-7). An immediate consequence is

Corollary 8-25 If $u \in N'_R$ then, for some $R' > 0$, $u \in C^\infty(\bar{\sigma}_{R'})$

Another consequence is

Theorem 8-26 If $f \in C^\infty(\bar{\sigma}_R)$ and $f \perp N'_R$, then for some $R' > 0$ the problem (8-6), (8-7) has a solution $u \in C^\infty(\bar{\sigma}_{R'})$.

PROOF Let M be the set of all $v \in H_R$ such that $v \perp N'_R$. It is easily checked that M is a closed subspace of H_R and, hence, a Hilbert space itself. Take R so small that (8-39) holds, and set

$$((u, v)) = (P'(x,t,D)u, P'(x,t,D)v) \qquad u, v \in H_R \tag{8-74}$$

It can be seen from (8-31) and (8-40) that $((u, v))$ has all of the properties of a scalar product on M, and that M is a Hilbert space with respect to this scalar product. Next, consider the functional

$$Fv = (v, f)$$

on M. It is clearly linear. It is also bounded, since

$$|(v, f)| \leq \|v\| \ \|f\| \leq \|f\| \ \|P'(x,t,D)v\| \qquad v \in M$$

by (8-39). Thus, by the Fréchet–Riesz theorem (Theorem 1-5 of Sec. 1-5), there is a function $w \in M$, such that

$$Fv = ((v, w)) \qquad v \in M$$

Thus
$$(P'(x, t, D)w, P'(x, t, D)v) = (f, v) \qquad v \in M \tag{8-75}$$

We claim that this holds not only for all $v \in M$, but also for all $v \in H_R$. For, if v is any function in H_R, we have $v = v_1 + v_2$, where $v_1 \in M$ and $v_2 \in N'_R$ (here we use the Projection theorem [Theorem 1-3 of Sec. 1-5] and the fact that N'_R is finite dimensional [Theorem 8-5]). Since $P'(x, t, D)v_2 = 0$ and $(f, v_2) = 0$, we have

$$(P'(x, t, D)w, P'(x, t, D)v) = (P'(x, t, D)w, P'(x, t, D)v_1) = (f, v_1) = (f, v)$$

showing that Eq. (8-75) holds for all $v \in H_R$. We now note that $P'(x, t, D)$ satisfies all of the assumptions of Sec. 8-1. Therefore, we can apply Theorem 8-24 to conclude that $w \in C^\infty(\bar{\sigma}_{R'})$ for some $R' > 0$. Set $u = P'(x, t, D)w$. Then $u \in C^\infty(\bar{\sigma}_{R'})$, and it is a weak solution of Eqs. (8-6), (8-7). Hence, by Theorem 8-1 it is an actual solution. This completes the proof.

We conclude this section with the following observation.

Corollary 8-27 If u satisfies the hypotheses of Theorem 8-24, then

$$D_t^k u(x, 0) = 0 \qquad |x| < R \qquad 0 \le k < r$$

PROOF Since $u \in H_{R_1}$, there is a sequence $\{u_k\}$ of functions in $\mathscr{D}_{R_1}$, converging to u in $H^{m,0}(\Omega)$. In particular, u_k converges to u and $P(x, D)u_k$ converges to $P(x, D)u$ in $L^2(\Omega)$. Thus, if $v \in \mathscr{D}_R$, we have

$$(u, P'(x, D)v) = \lim\, (u_k, P'(x, D)v) = \lim\, (P(x, D)u_k, v) = (P(x, D)u, v)$$

Thus, u is a weak solution of (8-6), (8-7), with $f(x, t)$ replaced by $P(x, D)u$. Since $u \in C^\infty(\bar{\sigma}_{R'})$ for some $R' > 0$, it follows, from the reasoning of Sec. 8-2, that u is an actual solution.

8-8 SOME THEOREMS IN HALF-SPACE

Before we give the proof of Theorem 8-24 we pause to gather some information which will be needed.

Lemma 8-28 Let k be a positive integer, and suppose $v \in C^k(E^n)$ is such that $D^\mu v \in L^2$ for $|\mu| \le k$. Then $v \in H^k$.

PROOF Let $\psi(x)$ be a function in C_0^∞, such that

$$0 \le \psi(x) \le 1 \qquad x \in E^n$$
$$\psi(x) = 1 \qquad |x| < 1$$

Set $\psi_R(x) = \psi(x/R)$ and $v_{R,\varepsilon} = \psi_R J_\varepsilon v$. Then, clearly $v_{R,\varepsilon} \in S$ for each R and ε.

Now

$$\int |v_{R,\varepsilon}(x) - J_\varepsilon v(x)|^2\,dx = \int (1-\psi_R)^2 |J_\varepsilon v|^2\,dx \to 0 \quad \text{as} \quad R \to \infty$$

Thus, $v_{R,\varepsilon} \to v$ as $R \to \infty$ and $\varepsilon \to 0$. Now, if $P(D)$ is any partial differential operator of order $\le k$, we have, by Eq. (1-102) of Sec. 1-7

$$P(D)v_{R,\varepsilon} = \sum \frac{P^{(\mu)}(D)J_\varepsilon v D^\mu v_R}{\mu!}$$

$$= \sum \frac{(J_\varepsilon P^{(\mu)}(D)v)\psi_\mu(x/R)}{R^{|\mu|}\mu!}$$

where $\psi_\mu(x) = D^\mu \psi(x)$. Since the ψ_μ are bounded functions and $P^{(\mu)}(D)v \in L^2$ for each μ, there is a constant K independent of R and ε, such that

$$\| P(D)v_{R,\varepsilon} \| \le K$$

Since this is true of any operator of degree $\le k$, there is a constant C independent of R and ε, such that

$$|v_{R,\varepsilon}|_k \le C$$

Since $v_{R,\varepsilon}$ are in S and converge to v in L^2, we see that $v \in H^k$. The proof is complete.

Lemma 8-29 Let s, k be integers such that $0 < s < k$. Assume that $v(x, t)$ is a function on E^{n+1}, such that $D_j v \in H^{s-1}(E^{n+1}) \cap C^s(E^{n+1})$ for $1 \le j \le n$, and

$$|(v, D_t^k \varphi)| \le K |\varphi|_{k-s} \qquad \varphi \in C_0^\infty(E^{n+1}) \tag{8-76}$$

Then $v \in H^s(E^{n+1})$.

Proof For $w(x, t)$ defined on E^{n+1}, let $\tilde{w}$ denote the Fourier transform with respect to all the $n+1$ variables. By Lemma 8-28, it suffices to show that $D_x^\mu D_t^j v \in L^2$ for $|\mu| + j \le s$. By Lemma 6-28 of Sec. 6-8 this is equivalent to showing that $\xi^\mu \tau^j \tilde{v} \in L^2$ for such μ and j. By hypothesis, this is true if $j < s$. Thus, it suffices to show that $\tau^s \tilde{v} \in L^2$.

Set $\eta = (\xi, \tau)$. By (8-76) the linear functional $(D_t^k \varphi, v)$ is bounded on H^{k-s}. Thus, by the Fréchet–Riesz representation theorem (Theorem 1-5 of Sec. 1-5), there is a $w \in H^{k-s}$, such that

$$(v, D_t^k \varphi) = \int (1 + |\eta|)^{2k-2s} \tilde{w}\bar{\tilde{\varphi}}\,d\eta$$

Since C_0^∞ is dense in L^2, this gives

$$\tau^k \tilde{v} = (1 + |\eta|)^{2k-2s} \tilde{w} \tag{8-77}$$

Now
$$|\tau|^s |\tilde{v}| = \frac{|\tau|^{s+2k} |\tilde{v}|}{|\tau|^{2k} + |\xi|^{2k} + 1} + \frac{|\tau|^s (|\xi|^{2k} + 1) |\tilde{v}|}{|\tau|^{2k} + |\xi|^{2k} + 1}$$

Hence, by Eq. (8-77)

$$|\tau|^s|\tilde{v}| = \frac{|\tau|^{s+k}(1+|\eta|)^{2k-2s}|\tilde{w}|}{|\tau|^{2k}+|\xi|^{2k}+1} + \frac{|\tau|^s(|\xi|^{2k}+1)|\tilde{v}|}{|\tau|^{2k}+|\xi|^{2k}+1}$$

$$\leq \frac{(1+|\eta|)^{2k}}{|\tau|^{2k}+|\xi|^{2k}+1}(1+|\eta|)^{k-s}|\tilde{w}|$$

$$+ \text{const.}\,(|\xi|\,|\tau|^{s-1}+|\xi|^s+1)|\tilde{v}| \tag{8-78}$$

This last inequality follows from the following facts:

1. $|\xi|\,|\tau|^{2k+s-1}+|\xi|^{2k+1}|\tau|^{s-1}+|\tau|^{2k}+1$
   $\leq (|\xi|\,|\tau|^{s-1}+|\xi|^s+1)(|\tau|^{2k}+|\xi|^{2k}+1)$
2. $|\tau|^s \leq \text{const.}\,(|\tau|^{2k}+1)$
3. The function $\alpha^{2k-1}+\alpha^{-1}$ has a positive minimum on $(0,\infty)$, for it is positive on this interval and tends to ∞ as $\alpha\to 0$ and as $\alpha\to\infty$.
4. Hence

$$|\tau|^s|\xi|^{2k} \leq \text{const.}\,(|\xi|\,|\tau|^{2k+s-1}+|\xi|^{2k+1}|\tau|^{s-1}) \tag{8-79}$$

This follows from 3 if we take $\alpha = |\tau|/|\xi|$.

We now note that all of the terms on the right of (8-78) are in L^2, by hypothesis. Since

$$(1+|\eta|)^s \leq \text{const.}\,(|\tau|^s+|\xi|^s+1) \tag{8-80}$$

we see that $v \in H^s$. The proof is complete.

For any real $h \neq 0$, set

$$x_i^h = (x_1,\ldots,x_i+h,\ldots,x_n) \tag{8-81}$$

and

$$\delta_i^h u(x,t) = \frac{[u(x_i^h,t)-u(x,t)]}{h\sqrt{-1}} \tag{8-82}$$

Lemma 8-30 For $u \in H^{0,1}(\Omega)$

$$\|\delta_i^h u\|^\Omega \leq \|D_i u\|^\Omega \tag{8-83}$$

and

$$\|\delta_i^h u - D_i u\|^\Omega \to 0 \quad \text{as} \quad h\to 0 \tag{8-84}$$

Proof Clearly (8-84) holds if $u \in C^\infty(\bar{\Omega})$ vanishes outside a bounded set (for then the difference quotient converges uniformly to the derivative). Moreover, for such functions

$$\delta_i^h u(x,t) = \int_0^1 D_i u(x_i^{\lambda h},t)\,d\lambda \tag{8-85}$$

Schwarz's inequality (1-62) yields

$$|\delta_i^h u(x,t)|^2 \le \int_0^1 |D_i u(x_i^{\lambda h},t)|^2 \, d\lambda$$

If we now integrate over Ω, we obtain (8-83).

Now, suppose $u \in H^{0,1}(\Omega)$. Then there is a sequence of functions $\{u_k\}$, of the type mentioned above, which converge to u in $H^{0,1}(\Omega)$. Thus

$$\| \delta_i^h u \|^\Omega = \lim \| \delta_i^h u_k \|^\Omega \le \lim \| D_i u_k \|^\Omega = \| D_i u \|^\Omega$$

and $\| \delta_i^h u - D_i u \|^\Omega \le \| \delta_i^h (u - u_k) \|^\Omega + \| \delta_i^h u_k - D_i u_k \|^\Omega + \| D_i(u_k - u) \|^\Omega$

$$= A + B + C$$

Since (8-83) is known to hold for u, for any $\varepsilon > 0$ we can take k so large that A and C are each less than $\varepsilon/3$. Once k is fixed, we take h so small that B is less than $\varepsilon/3$. This completes the proof.

Next, we let $H_0^{k,s}(\Omega)$ denote the completion of $C_0^\infty(\Omega)$ with respect to the norm of $H^{k,s}(\Omega)$ (i.e., $H_0^{k,s}(\Omega)$ is the closure of $C_0^\infty(\Omega)$ in $H^{k,s}(\Omega)$). We have

Lemma 8-31 If $v \in C_0^\infty(E^{n+1})$ and

$$D_t^j v(x,0) = 0 \qquad 0 \le j < k \tag{8-86}$$

then $v \in H_0^{k,s}(\Omega)$ for any s.

PROOF Set

$$v_\varepsilon(x,t) = v(x, t - 3\varepsilon) \qquad t > 3\varepsilon$$
$$= 0 \qquad t < 3\varepsilon$$

Then, by Eq. (8-86), $v_\varepsilon \in C^{k-1}(\Omega)$, and it is easily checked that

$$\| v_\varepsilon - v \|^\Omega \to 0 \quad \text{as} \quad \varepsilon \to 0$$

Moreover, $J_\varepsilon v_\varepsilon \in C_0^\infty(\Omega)$ for each $\varepsilon > 0$, and

$$\| J_\varepsilon v_\varepsilon - v \|^\Omega \le \| J_\varepsilon(v_\varepsilon - v) \|^\Omega + \| J_\varepsilon v - v \|^\Omega$$
$$\le \| v_\varepsilon - v \|^\Omega + \| J_\varepsilon v - v \|^\Omega \to 0 \quad \text{as} \quad k \to \infty$$

By the Banach–Saks theorem (Theorem 1-9 of Sec. 1-5), it suffices to show that

$$|J_\varepsilon v_\varepsilon|_{k,s} \le M_s$$

for any s. By Eq. (2-19) of Sec. 2-2

$$\iint_\Omega |D^\mu J_\varepsilon v_\varepsilon|^2 \, dx\, dt = \iint_{t>2\varepsilon} |J_\varepsilon D^\mu v_\varepsilon|^2 \, dx\, dt$$
$$\le \iint_{t>2\varepsilon} |D^\mu v_\varepsilon|^2 \, dx\, dt = \iint_\Omega |D^\mu v|^2 \, dx\, dt$$

Hence
$$|J_\varepsilon v_\varepsilon|_{k,s} \le |v|_{k,s}$$
and the proof is complete.

Lemma 8-32 Let s be an integer, and let k be a positive integer greater than s. If $v \in H^{s-1,1}(\Omega)$, and
$$|(v, D_t^k \varphi)| \le K|\varphi|_{k-s,0} \qquad \varphi \in C_0^\infty(\Omega) \tag{8-87}$$
then $v \in H^{s,0}(\Omega)$.

PROOF By completion, (8-87) holds for $\varphi \in H_0^k(\Omega)$. Let ψ be any function in $C_0^\infty(E^{n+1})$. If L is given by Eq. (8-65) with $m = k$, we have
$$(Lv, D_t^k\psi) = \iint_\Omega v(x,t)\overline{D_t^k\psi(x,t)}\,dx\,dt$$
$$+ \sum_{j=1}^{k+1} \lambda_j \iint_{t<0} v(x,-jt)D_t^k\psi(x,t)\,dx\,dt$$
$$= \iint_\Omega v(x,t)D_t^k\varphi(x,t)\,dx\,dt$$
where
$$\varphi(x,t) = \psi(x,t) - \sum_{j=1}^{k+1} \lambda_j(-j)^{k-1}\psi\left(x, -\frac{t}{j}\right)$$
Now, by Eq. (8-66), we see that φ satisfies the hypotheses of Lemma 8-31. Hence $\varphi \in H_0^{k,0}(\Omega)$. By (8-87), we have
$$|(Lv, D_t^k\psi)| = |(v, D_t^k\varphi)| \le K|\varphi|_{k-s,0}$$
$$\le \text{const. } |\psi|_{k-s}$$
Applying Lemma 8-29 to Lv, we see that $Lv \in H^s(E^{n+1})$. Since $Lv = v$ on Ω, this gives $v \in H^s(\Omega)$, and the proof is complete.

Lemma 8-33 For each $k \ge 0$ and $\varepsilon > 0$ there is a constant K, such that
$$|v|_{k,s} \le \varepsilon \sum_{|\mu|=k+1} |D^\mu v|_{0,s} + K|v|_{0,s}$$
holds for any s and $v \in H^{k+1,s}(\Omega)$.

PROOF The lemma follows from repeated applications of Theorem 6-25 of Sec. 6-7.

8-9 REGULARITY AT THE BOUNDARY

We now give the proof of Theorem 8-24. First we note that Eq. (8-73) holds for all $v \in C_0^\infty(\sigma_R)$. Thus

$$(u, P'(x,t,D)P(x,t,D)v) = (f, v) \qquad v \in C_0^\infty(\sigma_R)$$

It is easily checked that the operator $P'(x,t,D)P(x,t,D)$ is elliptic. Thus, by Theorem 3-1 of Sec. 3-1, $u \in C^\infty(\sigma_R)$ after correction on a set of measure zero. Thus, all we must show is that u is infinitely differentiable up to the boundary $t = 0$.

To this end, let ζ be a real-valued function in $C^\infty(\bar{\sigma}_R)$, which vanishes near $\partial_1 \sigma_R$, and is identically one in $\sigma_{R-\varepsilon}$ for some positive ε. We set

$$[u, v] = (P(x,t,D)u, P(x,t,D)v) \tag{8-88}$$

Lemma 8-34 There is a constant K, such that

$$|[\delta_i^h(D_x^\mu \zeta u), v] - [u, D_x^\mu \zeta \delta_i^{-h} v]| \le K |v|_{m,0} \sum_{|\nu| \le |\mu|} |D_x^\nu u|_{m,0} \tag{8-89}$$

holds for h sufficiently small.

PROOF The proof of this lemma is elementary but a bit tedious. It is based on the following simple facts

1. If w vanishes near $\partial_1 \sigma_R$ and h is sufficiently small, then

$$\iint_{\sigma_R} \delta_i^h w(x,t)\, dx\, dt = 0 \tag{8-90}$$

2. $\delta_i^h D^\mu = D^\mu \delta_i^h$
3. $\delta_i^h [w(x)v(x)] = w(x)\delta_i^h v(x) + v(x_i^h)\delta_i^h w(x)$
4. If wv vanishes near $\partial_1 \sigma_R$ and h is sufficiently small

$$(\delta_i^h w, v) = (w, \delta_i^{-h} v) \tag{8-91}$$

If $|A - B|$ is bounded by the right-hand side of (8-89), we shall write $A \sim B$. To prove the lemma, we must show that

$$(aD^\rho \delta_i^h D_x^\mu \zeta u, D^\sigma v) \sim (aD^\rho u, D^\sigma D_x^\mu \zeta \delta_i^{-h} v)$$

for h small, whenever $a(x,t) \in C^\infty(\bar{\sigma}_R)$ and $|\rho|, |\sigma| \le m$. By 1 to 4, we have

$$\begin{aligned}
(aD^\rho \delta_i^h D_x^\mu \zeta u, D^\sigma v) &= (D^\rho D_x^\mu \zeta u, \delta_i^{-h} \bar{a} D^\sigma v) \\
&= (D_x^\mu D^\rho \zeta u, \bar{a}\delta_i^{-h} D^\sigma v + D^\sigma v(x_i^{-h})\delta_i^{-h}\bar{a}) \\
&\sim (\delta_i^h a D_x^\mu D^\rho \zeta u, D^\sigma v) \sim (\delta_i^h \zeta a D_x^\mu D^\rho u, D^\sigma v) \\
&\sim (\delta_i^h \zeta D_x^\mu a D^\rho u, D^\sigma v) = (D_x^\mu a D^\rho u, \zeta \delta_i^{-h} D^\sigma v) \\
&\sim (D_x^\mu a D^\rho u, D^\sigma \zeta \delta_i^{-h} v) = (aD^\rho u, D^\sigma D_x^\mu \zeta \delta_i^{-h} v)
\end{aligned}$$

We have used Lemma 8-30 here as well. This completes the proof.

We are now ready for the

PROOF of Theorem 8-24 By completion, Eq. (8-73) holds for all $v \in H_R$ (see the beginning of Sec. 8-4). Let ζ be defined as at the beginning of this section (Sec. 8-9). Then we have, by Lemma 8-34

$$|[\delta_i^h(\zeta u), v]| \leq |(f, \zeta\delta_i^{-h}v)| + K|u|_{m,0}|v|_{m,0} \tag{8-92}$$

(we took $\mu = 0$). Now, it is easily verified that $\delta_i^h(\zeta u)$ is in H_R, for h sufficiently small. In fact, if $\{u_k\}$ is a sequence of functions in $C^\infty(\bar{\sigma}_R)$ satisfying Eq. (8-12) and $|u_k - u|_{m,0} \to 0$, then $\delta_i^h(\zeta u_k)$ is in $\mathscr{D}_R$, and

$$|\delta_i^h(\zeta u_k) - \delta_i^h(\zeta u)|_{m,0} \to 0$$

Thus, we may take $v = \delta_i^h(\zeta u)$ in (8-92). This gives

$$\| P(x,t,D)\delta_i^h(\zeta u) \|^2 \leq \text{const.} \, |\delta_i^h(\zeta u)|_{m,0}$$

We may assume that R is so small that Theorem 8-8 holds. Thus

$$|\delta_i^h(\zeta u)|_{m,0}^2 \leq \text{const.} \, (|\delta_i^h(\zeta u)|_{m,0} + \|\delta_i^h(\zeta u)\|^2)$$

where the constant is independent of h. Since $\|\delta_i^h(\zeta u)\| \leq \|D_i(\zeta u)\|$ (Lemma 8-30), we have

$$|\delta_i^h(\zeta u)|_{m,0} \leq \text{const.}$$

the constant being independent of h. Since $\delta_i^h(\zeta u)$ converges to $D_i(\zeta)$ in L^2 (again by Lemma 8-30) it follows from the Banach–Saks theorem (Theorem 5-9 of Sec. 1-5) that $D_i(\zeta u)$ is in $H^m(\Omega)$. This is true for any $i \leq n$.

Next, we apply Lemma 8-34 for $|\mu| = 1$. Let $\varepsilon > 0$ be given, and let R_1 satisfy $R - \varepsilon < R_1 < R$. Let $\zeta \in C^\infty(\bar{\sigma}_R)$ be such that it vanishes outside of σ_{R_1}, and equals one in $\sigma_{R-\varepsilon}$. Then, we have, for $i, j \leq n$

$$|[\delta_i^h D_j(\zeta u)v]| \leq |(\delta_i^h \zeta D_j f, v)| + K|v|_{m,0} \sum_{j=1}^{n} |D_j u|_{m,0}$$

The right-hand side is finite, because $D_j u$ is in $H^m(\Omega)$ for each $j \leq n$. As before, we note that $\delta_i^h D_j(\zeta u)$ is in H for h sufficiently small. Hence, we can take $v = \delta_i^h D_j(\zeta u)$. This gives

$$\| P(x,t,D)\delta_i^h D_j(\zeta u) \|^2 \leq C|\delta_i^h D_j(\zeta u)|_{m,0}$$

from which we conclude, via Theorem 6-26 of Sec. 6-7, that

$$|\delta_i^h D_j(\zeta u)|_{m,0} \leq \text{const.}$$

where the constant is independent of h. Since $\delta_i^h D_j(\zeta u)$ converges to $D_i D_j(\zeta u)$ in L^2, we see that the latter function is in $H^m(\Omega)$ for any $i, j \leq n$ and any ζ described above. Continuing in this way, we see that $D_x^\mu(\zeta u)$ is in $H^m(\Omega)$ for any μ and any ζ.

Now the proof is almost finished. We know that $u \in C^\infty(\sigma_R)$ and satisfies

$$P'(x,t,D)P(x,t,D)u = f$$

an elliptic equation of order $2m$. If $a(x,t)$ is the coefficient of D_t^{2m} in this equation, we know that $a(x,t) \neq 0$ or $\bar{\sigma}_R$. Let $\varepsilon > 0$ be given, and let $\zeta \in C^\infty(\bar{\sigma}_R)$ vanish near $\partial_1 \sigma_R$ and equal one on $\sigma_{R-\varepsilon}$. Setting $w = a\zeta u$, we have

$$D_t^{2m} w = \zeta f + \sum_{k=0}^{2m-1} D_t^k B_k u \tag{8-43}$$

where the B_k are partial differential operators involving only derivatives with respect to the x_k, and having infinitely differentiable coefficients in $\bar{\sigma}_R$ which vanish near $\partial_1 \sigma_R$. Thus, for any μ

$$D_t^{2m} D_x^\mu w = D_x^\mu \zeta f + \sum_{k=0}^{2m-1} D_t^k D_x^\mu B_k u \tag{8-94}$$

Now, we know that $D_x^\mu w \in H^m(\Omega)$ for all μ. Moreover, by Eq. (8-94)

$$|(D_x^\mu w, D_t^{2m} \varphi)| \leq \text{const.} |\varphi|_{m-1,0} \qquad \varphi \in C_0^\infty(\sigma_R) \tag{8-95}$$

This follows from the fact that $D_x^\mu B_k u \in H^m(\Omega)$ for each k and μ. Thus, $D_t^j D_x^\mu B_k u$ is in $L^2(\sigma_R)$ for $j \leq m$. Moreover we have for $k > m$

$$\begin{aligned} |(D_t^k D_x^\mu B_k u, \varphi)| &= |(D_t^m D_x B_k u, D_t^{k-m} \varphi)| \\ &\leq \| D_t^m D_x B_k u \| \; \| D_t^{k-m} \varphi \| \leq \text{const.} |\varphi|_{m-1,0} \end{aligned}$$

Since $D_x^\mu w \in H^m(\Omega)$ for every μ, this means that $D_x^\mu w \in H^{m,s}(\Omega)$ for every s. This fact and (8-95) allow us to apply Lemma 8-32, to conclude that $D_x^\mu w \in H^{m+1}(\Omega)$. Thus, $D_x^\mu(\zeta u)$ is in $H^{m+1}(\Omega)$ for every μ and ζ. But now Eq. (8-94) allows us to conclude

$$|(D_x^\mu w, D_t^{2m} \varphi)| \leq \text{const.} |\varphi|_{m-2,0} \qquad \varphi \in C_0^\infty(\sigma_R)$$

Another application of Lemma 8-32 shows us that $D_x^\mu w \in H^{m+2}(\Omega)$. Thus, $D_x^\mu(\zeta u) \in H^{m+2}(\Omega)$ for each μ. Continuing in this manner we eventually obtain $D_x^\mu(\zeta u) \in H^{2m}(\Omega)$ for each μ and ζ. We do not need Lemma 8-32 to continue. To go further we merely differentiate Eq. (8-94) with respect to t. Thus

$$D_t^{2m+1} D_x^\mu w = D_t D_x^\mu \zeta f + \sum_{k=0}^{2m-1} D_t^{k+1} D_x^\mu B_k u$$

Every term on the right-hand side is in $L^2(\sigma_R)$. Hence, $D_x^\mu(\zeta u)$ is in $H^{2m+1}(\Omega)$ for every μ and ζ. Continuing this process, we have $|\zeta u|_{k,s} < \infty$ for each k, s, and ζ. In Sec. 6-8 it was shown that this implies $\zeta u \in C^\infty(\bar{\Omega})$. Since for any $\varepsilon > 0$ we can find a ζ of the required type which equals one on $\sigma_{R-\varepsilon}$, we have $u \in C^\infty(\bar{\sigma}_{R-\varepsilon})$. The proof is complete.

PROBLEMS

8-1 Show that the set of operators given by Eq. (8-11) is a Dirichlet system.

8-2 Show that the functions defined by Eq. (8-30) satisfy Eq. (8-28).

8-3 Show that $P_1(D)$ given by Eq. (8-44) is properly elliptic.

8-4 Prove that $\alpha_1 T_1 + \alpha_2 T_2$ is compact when T_1 and T_2 are compact.

8-5 Show that T compact and B bounded imply that BT and TB are compact.

8-6 Prove that, for each $\delta > 0$ and $R > 0$, the set $|\xi| \leq R$ in E^n can be covered by a finite number of spheres of radius δ.

8-7 Prove (8-63).

8-8 Show that there always exist constants λ_k such that Eq. (8-66) holds.

8-9 Show that $Lu \in C^n(E^{n+1})$ when Eq. (8-66) holds.

8-10 Prove inequalities (8-67) and (8-71).

8-11 In the proof of Theorem 8-26, show that M is a Hilbert space with respect to the scalar product of $H^{m,0}(\Omega)$, or the one given by Eq. (8-74).

8-12 Prove statement 3 in the proof of Lemma 8-29.

8-13 Prove inequality (8-80).

8-14 If $\varphi \in C_0^\infty$, show that $\delta_i^h \varphi \rightarrow D_i \varphi$ uniformly.

8-15 Prove statements 1 to 4 of Sec. 8-9.

8-16 Prove Eq. (8-93).

8-17 Prove Eq. (8-85).

CHAPTER

NINE

GENERAL DOMAINS

9-1 THE BASIC THEOREM

In Chapters 4 to 8 we were able to solve various boundary problems in a slab, half-space, or semisphere. In particular, we were able to solve the Dirichlet problem for an elliptic equation with variable coefficients in a region σ_R, provided that R was sufficiently small and that we did not prescribe boundary conditions on $\partial_1 \sigma_R$, the "curved" part of the boundary.

In this chapter we shall show that the Dirichlet problem can be solved for an elliptic operator with variable coefficients in an arbitrary bounded domain, provided the boundary is sufficiently smooth. We shall not specify how smooth. In fact, we shall assume that it is infinitely differentiable throughout. But it will be clear, if anyone wants to take the trouble of checking, that the same results could be obtained by assuming much less.

First we must define our terms. Let G be a bounded domain in E^n. We shall call its boundary ∂G *smooth* if, for each $x \in \partial G$, there is a neighborhood $\mathcal{N}_x$ of x and a mapping Φ_x of $\mathcal{N}_x$ on to the set $|x| \leq 1$, such that

(*a*) Φ_x is one-to-one.
(*b*) $\Phi_x(x) = 0$.
(*c*) Φ_x and Φ_x^{-1} are both infinitely differentiable.
(*d*) $\Phi_x(\mathcal{N}_x \cap \partial G)$ is the set of those $x \in E^{n-1}$, such that $|x| < 1$.

Here, as usual, E^{n-1} denotes the hyperplane $x_n = 0$ in E^n.

Next, we consider functions defined on G. We define the set $C^\infty(\overline{G})$ as the set of functions which are infinitely differentiable in G, and all of whose derivatives are continuous up to the boundary ∂G. It is easy to show that, when ∂G is smooth, every function in $C^\infty(\overline{G})$ is the restriction to G of functions in C_0^∞, but this fact will not be needed.

For a partial differential operator $P(x, D)$, we have already defined ellipticity at a point and in a domain (see Sec. 3-1). Similarly, we shall say that $P(x, D)$ is elliptic in $\overline{G}$ if it is elliptic at each point of $\overline{G}$. For a homogeneous constant-coefficient operator we defined proper ellipticity in Sec. 6-9. We generalize this to operators with variable coefficients and to general domains. We shall call an operator $P(x, D)$ of order m *properly elliptic* in $\overline{G}$, if

(i) m is even.
(ii) $P(x, D)$ is elliptic in $\overline{G}$.
(iii) For each $x \in \overline{G}$ and each pair of linearly independent vectors $\xi, \eta \in E^n$, the polynomial

$$p(z) = p(x, \xi + z\eta) \tag{9-1}$$

has exactly $m/2$ roots with positive imaginary parts (recall the definition of $p(x, \xi)$ given by Eq. (3-5) of Sec. 3-1).

By Theorem 6-31 of Sec. 6-9, we have

Lemma 9-1 If $n > 2$, every operator $P(x, D)$ which is elliptic in $\overline{G}$ is also properly elliptic.

We also have

Lemma 9-2 If $P(x, D)$ is elliptic in $\overline{G}$, and there is an $x \in G$ and vectors $\xi, \eta \in E^n$, such that $p(x)$ given by Eq. (9-1) has exactly $m/2$ roots with positive imaginary parts, then $P(x, D)$ is properly elliptic in $\overline{G}$.

We leave the proof of this lemma as an exercise. The main result of this chapter will be

Theorem 9-3 Let G be a smooth, bounded domain, and let $P(x, D)$ be an operator of order $m = 2r$, which is properly elliptic in $\overline{G}$, and having coefficients in $C^\infty(\overline{G})$. Let N be the set of those $v \in C^\infty(\overline{G})$ satisfying $P(x, D)v = 0$ and

$$D^\mu v = 0 \quad \text{on} \quad \partial G \qquad |\mu| < r \tag{9-2}$$

and let N' be the set of those $v \in C^\infty(\overline{G})$ satisfying Eq. (9-2) and $P'(x, D)v = 0$ (as usual, $P'(x, D)$ denotes the formal adjoint of $P(x, D)$, see Sec. 1-3). Then, we have

1. N and N' are finite dimensional subspaces of $L^2(G)$

2. For $f \in C^\infty(\overline{G})$, there exists a solution $u \in C^\infty(\overline{G})$ of

$$P(x, D)u = f \text{ in } G \tag{9-3}$$

and

$$D^\mu u = 0 \quad \text{on} \quad \partial G \qquad |\mu| < r \tag{9-4}$$

if and only if $f \perp N'$.

3. If $u \in L^2(G)$, $f \in C^\infty(\overline{G})$, and

$$(u, P'(x, D)v) = (f, v) \tag{9-5}$$

for all $v \in C^\infty(\overline{G})$ satisfying (9-2), then $u \in C^\infty(\overline{G})$ after correction on a set of measure zero. Moreover, u satisfies Eqs. (9-3), (9-4)

4. If $v \in L^2(G)$, $g \in C^\infty(\overline{G})$, and

$$(v, P(x, D)u) = (g, u) \tag{9-6}$$

holds for all $u \in C^\infty(\overline{G})$ satisfying Eq. (9-4), then $v \in C^\infty(\overline{G})$ after correction on a set of measure zero. Moreover, v satisfies (9-2) and $P'(x, D)v = g$.

The rest of the chapter will be devoted to the proof of this theorem. In Sec. 9-2 we show that an inequality and a regularity theorem can be used to prove Theorem 9-3. Then, of course, we have to prove the inequality and regularity theorem.

9-2 AN INEQUALITY AND A REGULARITY THEOREM

In this section we shall show how the proof of Theorem 9-3 can be reduced to an inequality and a regularity theorem similar to Theorem 8-24 of Sec. 8-7. In stating the inequality, we shall use a family of norms, similar to those we have been using. For k a nonnegative integer and G a bounded domain, set

$$\| u \|_k^G = \left(\int_G \sum_{|\mu| \le m} | D^\mu u(x) |^2 \, dx \right)^{1/2} \tag{9-7}$$

and let $H^k(G)$ denote the completion of $C^\infty(\overline{G})$ with respect to this norm. When no confusion will result, we shall drop the superscript G.

We shall also modify the notation of Sec. 8-4 slightly. We shall let $\mathscr{D}$ denote the set of those $u \in C^\infty(\overline{G})$ satisfying Eq. (9-4), and let H denote the closure of $\mathscr{D}$ in $H^m(G)$. If $P(x, D)$ is a partial differential operator of order $\le m$ with bounded coefficients, then

$$\| P(x, D)u \| \le \text{const.} \, \| u \|_m \qquad u \in C^m(\overline{G}) \tag{9-8}$$

By means of this inequality we can define $P(x, D)u$ for $u \in H^m(G)$.

Our inequality is given by

Theorem 9-4 Under the hypotheses of Theorem 9-3 there is a constant C, such that

$$\|u\|_m \le C(\|P(x,D)u\| + \|u\|) \qquad u \in H \tag{9-9}$$

Notice the similarity between this inequality and inequalities (2-44) of Sec. 2-4, (6-131) of Sec. 6-7, and (8-48) of Sec. 8-4.

Our regularity theorem is given by

Theorem 9-5 Under the same hypotheses, if $f \in C^\infty(\overline{G})$, $u \in H$, and

$$(P(x,D)u, P(x,D)v) = (f, v) \qquad v \in \mathscr{D} \tag{9-10}$$

then $u \in \mathscr{D}$ and $P(x,D)u \in \mathscr{D}$. The same statement holds when $P(x,D)$ is replaced by $P'(x,D)$.

Theorem 9-4 will be proved in Sec. 9-5, and Theorem 9-5 in Sec. 9-3. Meanwhile, we show how they imply Theorem 9-3. First, we have

Corollary 9-6 Under the same hypotheses, there is a constant C, such that

$$\|u\|_m \le C(\|P'(x,D)u\| + \|u\|) \qquad u \in H \tag{9-11}$$

This follows from Theorem 9-4 and the observation that $P(x,D)$ is properly elliptic in $\overline{G}$, if and only if $P'(x,D)$ is. Thus, $P'(x,D)$ satisfies all of the hypotheses of Theorem 9-3 if the same is true of $P(x,D)$.

Corollary 9-7 Under the same hypotheses, if $u \in H$ and $P(x,D)u = 0$, then $u \in N$. Similarly, if $v \in H$ and $P'(x,D)v = 0$, then $v \in N'$.

PROOF If $u \in H$ and $P(x,D)u = 0$, then $(P(x,D)u, P(x,D)v) = (0, v)$ for all $v \in \mathscr{D}$. Thus, $u \in \mathscr{D}$ by Theorem 9-5. Consequently, $u \in N$. Similar reasoning gives the other statement.

In deriving further conclusions from Theorems 9-4 and 9-5 we shall make use of

Lemma 9-8 For any function v defined on G, set

$$\begin{aligned} \hat{v} &= v \text{ in } G \\ &= 0 \text{ outside } G \end{aligned}$$

If $v \in \mathscr{D}$, then $\hat{v} \in H^r$, and

$$|\hat{v}|_r = \text{const.} \|v\|_r^G \tag{9-12}$$

PROOF Set $v_\varepsilon = J_\varepsilon \hat{v}$. Then $v_\varepsilon \in S$ and $v_\varepsilon \to \hat{v}$ in L^2 as $\varepsilon \to 0$ (Theorem 2-3 of Sec. 2-2). Moreover, for $|\mu| \le r$, we have by integration by parts

$$D^\mu v_\varepsilon(x) = \int D^\mu j_\varepsilon(x-y)\hat{v}(y)\,dy = \int j_\varepsilon(x-y)D^\mu \hat{v}(y)\,dy$$

The boundary integrals, that would arise from the integration by parts, vanish because of the boundary conditions on v (see Eq. (1-27) of Sec. 1-3). Hence $D^\mu v_\varepsilon \to D^\mu \hat{v}$ in L^2. This shows that $\hat{v} \in H^r$ and that Eq. (9-12) holds (Lemma 8-28 of Sec. 8-8).

We can now prove conclusion 1 of Theorem 9-3. Consider N as a subspace of $L^2(G)$, and suppose $\{u_k\}$ is a sequence of functions in N, such that

$$\| u_k \| \le M$$

By Theorem 9-4

$$\| u_k \|_m \le CM$$

Consequently, by Lemma 9-8

$$| \hat{u}_k |_r \le C'$$

Let ζ be a function in C_0^∞, which equals one in G. By Theorem 8-18 of Sec. 8-6 $\{\zeta \hat{u}_k\} = \{\hat{u}_k\}$ has a subsequence converging in L^2. Thus, $\{u_k\}$ has a subsequence converging in $L^2(G)$. The result now follows from Theorem 8-17 of Sec. 8-5. Similar reasoning works for N'.

Next, let M be the set of all $u \in H$ such that $u \perp N$, and let M' be the set of all $v \in H$ such that $v \perp N'$ (in both cases the orthogonality is to be expressed by the $L^2(G)$ scalar product). We have

Corollary 9-9 There is a constant C, such that

$$\| u \|_m \le C \| P(x,D)u \| \qquad u \in M \tag{9-13}$$

$$\| v \|_m \le C \| P'(x,D)v \| \qquad v \in M' \tag{9-14}$$

PROOF Assume that (9-13) were false. Then there would be a sequence $\{u_k\}$ of functions in M, such that

$$\| u_k \|_m = 1 \qquad \| P(x,D)u_k \| \to 0 \qquad k \to \infty \tag{9-15}$$

By the reasoning given above, there is a subsequence (also denoted by $\{u_k\}$) which converges in $L^2(G)$. By inequality (9-9)

$$\| u_j - u_k \|_m \le C(\| P(x,D)(u_j - u_k) \| + \| u_j - u_k \|) \to 0 \quad \text{as} \quad j,k \to \infty$$

Thus, there is a $u \in H$ such that $u_k \to u$ in H. Clearly $u \in M$. Since $P(x,D)u_k \to P(x,D)u$, we have $P(x,D)u = 0$. Thus $u \in N$ (Corollary 9-7). This can happen only if $u = 0$. But $\| u \|_m = \lim \| u_k \|_m = 1$, by (9-15). This contradiction gives (9-13). A similar argument proves (9-14).

We are now in a position to prove conclusion 2 of Theorem 9-3. Suppose $u \in C^\infty(\overline{G})$ is a solution of Eqs. (9-3), (9-4). Then we have

$$(f, v) = (P(x, D)u, v) = (u, P'(x, D)v) \tag{9-16}$$

for all $v \in \mathscr{D}$. There are no boundary integrals because both u and v are in $\mathscr{D}$. In applying the integration by parts formula, each intermediate expression is of the form

$$\sum_{|\mu|+|\nu| \le m} (C_{\mu\nu} D^\nu u, D^\mu v)$$

Since either $|\mu| \le r$ or $|\nu| \le r$, no boundary integrals arise when integrating by parts (see Sec. 1-3). An inspection of Eq. (9-16) shows that $f \perp N'$ is a necessary condition for Eqs. (9-3), (9-4) to have a solution.

Next suppose $f \perp N'$. Note that there is a constant c, such that

$$c^{-1} \| v \|_m \le \| P'(x, D)v \| \le c \| v \|_m \qquad v \in M' \tag{9-17}$$

(Corollary 9-9 and inequality (9-8)). Since M' is a closed subspace of H, it itself is a Hilbert space with the norm of $H^m(G)$. We see from (9-17) that it remains a Hilbert space if we replace the scalar product of $H^m(G)$ with

$$(P'(x, D)u, P'(x, D)v) \tag{9-18}$$

Now (v, f) is a bounded linear functional on M'. For, by (9-17)

$$|(v, f)| \le \| v \| \ \| f \| \le c \| f \| \ \| P'(x, D)v \|$$

Thus, by the Fréchet–Riesz representation theorem (Theorem 1-5 of Sec. 1-5), there is a $g \in M'$, such that

$$(v, f) = (P'(x, D)v, P'(x, D)g) \qquad v \in M' \tag{9-19}$$

I claim that Eq. (9-19) holds not only for all $v \in M'$, but for all $v \in H$ as well. This is where the assumption $f \perp N'$ is used. Since N' is finite dimensional, it is a closed subspace of $L^2(G)$ (Lemma 7-12 of Sec. 7-4). By the projection theorem (Theorem 1-3 of Sec. 1-5), each $v \in H$ can be written in the form $v = v' + v''$, where $v' \in N'$ and $v'' \perp N'$. By definition $v'' \in M'$. Thus, if $f \perp N'$, we have by Eq. (9-19)

$$\begin{aligned}(f, v) = (f, v'') &= (P'(x, D)g, P'(x, D)v'') \\ &= (P'(x, D)g, P'(x, D)v)\end{aligned}$$

Thus (9-19) holds for all $v \in N$. Since we are assuming $f \in C^\infty(\overline{G})$, we can apply Theorem 9-5 to conclude that $u = P'(x, D)g$ is in $\mathscr{D}$. A simple argument now shows that u is a solution of Eqs. (9-3), (9-4). This completes the proof.

If we now interchange $P(x, D)$ and $P'(x, D)$, we get

Corollary 9-10 For $g \in C^\infty(\overline{G})$ there exists a solution $v \in \mathscr{D}$ of

$$P'(x, D)v = g \text{ in } G \tag{9-20}$$

if and only if $g \perp N$ (i.e., $g \in M$).

In proving the rest of Theorem 9-3 we make use of some further consequences of Theorem 9-4 and 9-5.

Lemma 9-11 If $u \in M$, and

$$(u, P'(x, D)v) = 0 \qquad v \in \mathscr{D} \tag{9-21}$$

then $u = 0$.

PROOF Let g be any function in $C^\infty(\overline{G})$. Then $g = g' + g''$, where $g' \in M$ and $g'' \in N$ (the projection theorem). Since $N \subset C^\infty(\overline{G})$, we have $g' \in C^\infty(\overline{G})$. By Corollary 9-10, there is a $v \in \mathscr{D}$ such that $P'(x, D)v = g'$. Since $u \perp N$, we have

$$(u, g) = (u, g') = (u, P'(x, D)v) = 0$$

by Eq. (9-21). Thus u is orthogonal to every $g \in C^\infty(\overline{G})$. Since $C^\infty(\overline{G})$ is dense in $L^2(G)$ (Lemma 4-17 of Sec. 4-6), we must have $u = 0$.

Lemma 9-12 If $u \in L^2(G)$ satisfies Eq. (9-21), then $u \in N$.

PROOF By the projection theorem we have $u = u' + u''$, where $u' \in M$ and $u'' \in N$. But

$$(u'', P'(x, D)v) = (P(x, D)u'', v) = 0$$

for $v \in \mathscr{D}$. Thus

$$(u', P'(x, D)v) = 0 \qquad v \in \mathscr{D}$$

By Lemma 9-11 this implies $u' = 0$. Hence $u = u'' \in N$, and the proof is complete.

Now we can give the proof of conclusion 3 of Theorem 9-3. Suppose u and f are as described there. Clearly $f \perp N'$. Thus, by conclusion 2, there is a function $u_0 \in \mathscr{D}$, such that $P(x, D)u_0 = f$. Integrating by parts, we have

$$(u_0, P'(x, D)v) = (f, v) \qquad v \in \mathscr{D}$$

Subtracting this from Eq. (9-5), we get

$$(u - u_0, P'(x, D)v) = 0 \qquad v \in \mathscr{D}$$

This implies that $u - u_0 \in N$ (Lemma 9-12). Since $N \subset \mathscr{D}$, we see that $u \in \mathscr{D}$ and it is a solution of Eq. (9-3). This completes the proof.

Similar reasoning gives conclusion 4. Thus, Theorem 9-3 is implied by Theorems 9-4 and 9-5.

9-3 LOCALIZATION

We now turn to the proof of Theorem 9-5. The idea of the proof is as follows. Differentiability is a local property. Thus, to show that $u \in C^\infty(\overline{G})$, it suffices to show that each point $x^0 \in \overline{G}$ is contained in a neighborhood $\mathcal{N}$, such that $u \in C^\infty(\mathcal{N} \cap \overline{G})$.

First, let us consider interior points. If u satisfies the hypotheses of Theorem 9-5, we have for $\varphi \in C_0^\infty(G)$

$$(u, P(x, D)P'(x, D)\varphi) = (P'(x, D)u, P'(x, D)\varphi) = (f, \varphi) \tag{9-22}$$

Clearly, $P(x, D)P'(x, D)$ is an elliptic operator of order $2m$ (check this!). Since elliptic operators are formally hypoelliptic (Lemma 3-2 of Sec. 3-1), every weak solution of Eq. (9-22) is in $C^\infty(G)$ (Theorem 3-1 of Sec. 3-1). Thus, every function satisfying the hypotheses of Theorem 9-5 is in $C^\infty(G)$. It remains, therefore, only to investigate boundary points.

Let x^0 be a point on ∂G. By the definition of smoothness given in Sec. 9-1, there are a neighborhood $\mathcal{N}$ of x^0 and a function Φ satisfying (a) to (d) given there. Now, Eq. (9-10) holds, a fortiori, for those $v \in \mathcal{D}$ which vanish outside $\mathcal{N}$. For $x \in \mathcal{N} \cap \overline{G}$ set $y = \Phi(x)$, $\tilde{f}(y) = Jf(x)$, $\tilde{u}(y) = u(x)$, where J is the Jacobian of the transformation. By hypothesis, J is in $C^\infty(\overline{\mathcal{N}})$ and is bounded away from zero.

Set
$$\tilde{P}(y, D_y) = J^{1/2}P(x, D_x) \tag{9-23}$$

where D_y means differentiation with respect to the coordinates y. It is easily checked that $\tilde{P}(y, D_y)$ is properly elliptic when $P(x, D)$ is. Thus, Eq. (9-10) implies

$$(\tilde{P}(y, D_y)\tilde{u}(y), \tilde{P}(y, D_y)w(y)) = (f, w) \qquad w \in \mathcal{D}_{1/2} \tag{9-24}$$

(see the definitions of $\mathcal{D}_R$ and H_R in Sec. 8-4). Let ζ be a function in C_0^∞, which equals one on $\sigma_{1/2}$ and vanishes outside $\sigma_{3/4}$. Then $\zeta\tilde{u}$ is in H_1. Moreover, Eq. (9-24) gives

$$(\tilde{P}(y, D_y)\zeta\tilde{u}, \tilde{P}(y, D_y)w) = (f, w) \qquad w \in \mathcal{D}_{1/2}$$

since functions in $\mathcal{D}_{1/2}$ vanish outside $\sigma_{1/2}$, and $\zeta u = u$ inside. We are now in a position to apply Theorem 8-24 of Sec. 8-7. Clearly $\tilde{f} \in C^\infty(\overline{\sigma}_1)$. Thus, it follows that there is an $R > 0$ such that $\tilde{u} \in C^\infty(\overline{\sigma}_R)$. Mapping back, we see that $u \in C^\infty(\mathcal{N}_1 \cap \overline{G})$, where $\mathcal{N}_1 = \Phi^{-1}(\sigma_R)$ in a neighborhood of x^0. Since x^0 was an arbitrary point of ∂G, it follows that $u \in C^\infty(\overline{G})$. Moreover, if we apply Corollary 8-27 of Sec. 8-7 to $\tilde{u}$, we see that all its derivatives of order less than r vanish on $\partial_0\sigma_R$. Mapping back, we see that all these derivatives of u vanish on $\mathcal{N}_1 \cap \partial G$. Thus, it follows that $u \in \mathcal{D}$.

Next, set $v = J^{1/2}P(x, D)u$. Then $v \in C^\infty(\overline{G})$, and Eq. (9-24) gives

$$(\tilde{v}, \tilde{P}(y, D_y)w(y)) = (f, w) \qquad w \in \mathcal{D}_{1/2}$$

It was shown in Sec. 8-2 that this implies that all derivatives of $\tilde{v}$ of order less than r vanish on $\partial_0\sigma_R$. Again, this implies that these derivatives of v vanish on $\partial G \cap \mathcal{N}_1$. Hence, $P(x, D)u$ is also in $\mathcal{D}$.

Since $P'(x, D)$ is properly elliptic in $\overline{G}$, and it has coefficients in $C^\infty(\overline{G})$, we see that the theorem holds with $P(x, D)$ replaced by $P'(x, D)$. This completes the proof.

9-4 SOME LEMMAS

Before proving Theorem 9-4, we give some technical lemmas which are useful in proving this type of inequality. As usual, we let Ω denote the half-space $t > 0$ in E^{n+1}.

Lemma 9-13 If $v \in H^{m,0}(\Omega)$ and vanishes outside σ_R, then there is a constant c, depending only on m, such that

$$c^{-1}|v|_{m,0} \leq \| v \|_m \leq c|v|_{m,0} \tag{9-25}$$

PROOF The inequalities follow immediately from inequalities (8-37) and (8-41) of Sec. 8-4.

Lemma 9-14 Let G and G_1 be bounded domains, and assume that there is a one-to-one mapping Φ of $\overline{G}$ on to $\overline{G}_1$, such that Φ and Φ^{-1} are infinitely differentiable. Then there is a constant c depending only on m, G, and Φ, such that

$$c^{-1}\| u \|_m^G \leq \| \tilde{u} \|_m^{G_1} \leq c\| u \|_m^G \qquad u \in H^m(G) \tag{9-26}$$

where $\tilde{u}$ is defined by $\tilde{u}(\Phi(x)) = u(x)$.

This lemma is a simple consequence of the definitions.

Lemma 9-15 Let G be a bounded smooth domain. Then, for each $m > 0$ and each $\varepsilon > 0$, there is a constant K, such that

$$\| u \|_{m-1} \leq \varepsilon \| u \|_m + K \| u \| \qquad u \in H^m(G) \tag{9-27}$$

In proving this lemma we shall make use of

Lemma 9-16 Let G be a bounded, smooth domain. Suppose $V_1, \ldots, V_q$ are open sets covering $\overline{G}$, i.e.,

$$\overline{G} \subset \bigcup_1^q V_k$$

Then there exists a C^∞ partition of unity, subordinate to this covering. This means that there are nonnegative functions $\zeta_1, \ldots, \zeta_q$ such that $\zeta_k \in C_0^\infty(V_k)$ and

$$\sum_{k=1}^q \zeta_k(x) = 1 \qquad x \in G$$

PROOF Since the V_k are open sets, we can find smaller open sets U_k, such that $\overline{U}_k \subset V_k$ and the U_k also cover $\overline{G}$ (check this). For each k, let φ_k be a nonnegative function in $C_0^\infty(V_k)$ which does not vanish in U_k (see Sec. 1-3).

We then set

$$\zeta_k = \frac{\varphi_k}{\sum_{j=1}^{q} \varphi_j} \qquad 1 \le k \le q \tag{9-28}$$

It is easily checked that the ζ_k have the required properties.

We can now give the

PROOF of Lemma 9-15 Since G is smooth, for each $x \in \partial G$ there are a neighborhood $\mathcal{N}$ and a mapping Φ satisfying (*a*) to (*d*) of Sec. 9-1. Since ∂G is bounded, it can be covered by a finite number of these neighborhoods, say $\mathcal{N}_1, \ldots, \mathcal{N}_q$. Together with a domain G_0, such that $\overline{G}_0 \subset G$, they form a finite open covering of $\overline{G}$ (prove this). By Lemma 9-16 there is a C^∞ partition of unity $\zeta_1, \ldots, \zeta_{q+1}$ subordinate to this covering (we take $\zeta_{q+1} \in C_0^\infty(G_0)$). Set $G_k = G \cap \mathcal{N}_k$, and let u be any function in $H^m(G)$. Set $u_k = \zeta_k u$. Then, we have

$$\begin{aligned} \| u \|_{m-1}^{G} &\le \sum \| u_k \|_{m-1}^{G_k} \le c \sum \| \tilde{u}_k \|_{m-1}^{\sigma_1} \\ &\le c_1 \sum \| \tilde{u}_k \|_{m-1}^{\sigma_1} \le c_2 \sum | u_k |_{m-1,0} \le \sum (\varepsilon | \tilde{u}_k |_{m,0} + K | u_k |_{0,0}) \\ &\le c_3 \sum (\varepsilon \| \tilde{u}_k \|_{m}^{\sigma_1} + K \| \tilde{u}_k \|^{\sigma_1}) \le c_4 \sum (\varepsilon \| u_k \|_{m}^{G_k} + K \| u_k \|^{G_k}) \\ &\le c_5 (\varepsilon \| u \|_{m}^{G} + K \| u \|^{G}) \end{aligned}$$

where we used Lemma 8-33 of Sec. 8-8 as well as Lemmas 9-13 and 9-14. We also made the harmless assumption that G is contained in Ω. The last inequality proves the lemma.

9-5 THE INEQUALITY

We can now give a proof of Theorem 9-4. Our first step is to show that it is really local in nature.

Lemma 9-17 A necessary and sufficient condition for inequality (9-9) to hold is that each point $x^0 \in \overline{G}$ has a neighborhood $\mathcal{N}$, such that

$$\| u \|_m \le C(\| P(x, D)u \| + \| u \|) \qquad u \in \mathcal{D} \cap C_0^\infty(\mathcal{N}) \tag{9-29}$$

PROOF It is clearly necessary. To show that it is sufficient, suppose each $x^0 \in \overline{G}$ has such a neighborhood. Since G is bounded, there are a finite number of these neighborhoods which cover $\overline{G}$, say $\mathcal{N}_1, \ldots, \mathcal{N}_q$. Let $\zeta_1, \ldots, \zeta_q$ be a C^∞ partition, subordinate to this covering (Lemma 9-16), and let u be any function in $\mathcal{D}$. Then $\zeta_k u \in \mathcal{D} \cap C_0^\infty(\mathcal{N}_k)$, and

$$\begin{aligned}\| u \|_m = \| \sum \zeta_k u \| &\le \sum \| \zeta_k u \| \\ &\le c_1 \sum (\| P(x, D)(\zeta_k u) \| + \| \zeta_k u \|) \\ &\le c_2 \sum \| \zeta_k P(x, D) u \| + K \| u \|_{m-1} \\ &\le \tfrac{1}{2} \| u \|_m + c_3(\| P(x, D)u \| + \| u \|)\end{aligned}$$

by (9-29) and Lemma 9-15. This proves (9-9) for $u \in \mathscr{D}$. By completion, it holds for all $u \in H$. This completes the proof.

It follows from Lemma 9-17 that, to prove Theorem 9-4, it suffices to show that each $x^0 \in \bar{G}$ has a neighborhood $\mathscr{N}$ such that (9-29) holds. First, assume that x^0 is an interior point. Then we can take $\mathscr{N}$ so small that it does not intersect the boundary. Then (9-29) follows from

Lemma 9-18 If $P(x, D)$ is elliptic at $x^0 \in G$ and has continuous coefficients, then there is a neighborhood $\mathscr{N}$, such that

$$\| u \|_m \le C(\| P(x, D)u \| + \| u \|) \qquad u \in C_0^\infty(\mathscr{N}) \tag{9-30}$$

PROOF By inequality (2-44) of Sec. 2-4, there is a constant c_4, such that

$$| v |_m \le c_4(| P(D)v |_0 + | v |_{m-1}) \qquad v \in S$$

where $P(D) = P(x^0, D)$. Since the norms $| \ |_m$ and $\| \ \|_m$ are equivalent on $C_0^\infty(G)$ (Lemma 9-8), this gives

$$\| u \|_m \le c_5(\| P(D)u \| + \| u \|_{m-1}) \qquad u \in C_0^\infty(G)$$

Now
$$P(x, D) = P(D) + \sum_{|\mu| \le m} b_\mu(x) D^\mu$$

where the $b_\mu(x)$ are continuous and vanish at x^0. Thus, there is a neighborhood $\mathscr{N}$, such that

$$\| P(x, D)u - P(D)u \| \le \tfrac{1}{2} c_5 \| u \|_m$$

Consequently $\quad \| u \|_m \le c_6(\| P(x, D)u \| + \| u \|_{m-1}) \qquad u \in C_0^\infty(\mathscr{N})$

Moreover, by Lemma 9-15 there is a constant K, such that

$$c_6 \| u \|_{m-1} \le \tfrac{1}{2} \| u \|_m + K \| u \|$$

Combining these inequalities, we get (9-30). This proves Lemma 9-18.

To complete the proof of Theorem 9-4, we need only consider boundary points. Let x^0 be a point on ∂G. By the definition of smoothness given in Sec. 9-1, there are a neighborhood $\mathscr{N}$ of x^0 and a function Φ satisfying (*a*) to (*d*) given there. Define $\tilde{P}(y, D_y)$ by Eq. (9-23). Then, by Theorem 8-6 of Sec. 8-4, there is a constant c_7, such that

$$| w |_{m,0} \le c_7(\| \tilde{P}(y, D_y)w \| + \| w \|) \qquad w \in \mathscr{D}_1$$

By Lemma 9-14, this implies

$$\| u \|_m \leq c_8(\| P(x, D)u \| + \| u \|) \qquad u \in \mathscr{D} \cap C_0^\infty(\mathscr{N})$$

which is the desired inequality. Thus, the proof of Theorem 9-4 is complete.

9-6 STRONGLY ELLIPTIC OPERATORS

An operator $P(x, D)$ is called *strongly elliptic* in a domain G if there is a smooth function $\gamma(x)$, such that

$$\operatorname{Re} \gamma(x) p(x, \xi) > 0 \tag{9-31}$$

for each $x \in \bar{G}$ and each real vector $\xi \neq 0$, where $p(x, \xi)$ is the principal part of $P(x, \xi)$. Clearly, every strongly elliptic operator is elliptic (see Sec. 3-1). What may not be so obvious is that

Lemma 9-19 Every strongly elliptic operator is properly elliptic.

PROOF Since every elliptic operator is properly elliptic when $n > 2$ (Lemma 9-1), we need only prove the lemma for $n = 2$. However, our proof will hold for arbitrary n. Fix $x \in \bar{G}$, and set

$$p(\xi) = \gamma(x) p(x, \xi)$$

We must show that m is even and that for each pair ξ, η of linearly independent real vectors, the polynomial in z

$$Q(z) = p(\xi + z\eta) \tag{9-32}$$

has exactly $m/2$ roots with positive imaginary parts. Put

$$p_1(\xi) = \operatorname{Re} p(\xi) \qquad p_2(\xi) = \operatorname{Im} p(\xi) \qquad \xi \in E^n$$

Then (9-31) says

$$p_1(\xi) > 0 \tag{9-33}$$

for each real vector $\xi \neq 0$. Since

$$p_1(-\xi) = (-1)^m p_1(\xi)$$

this shows that m is even. Moreover, (9-33) implies that the polynomial

$$Q_0(z) = p_1(\xi + z\eta)$$

has no real roots. Since its coefficients are real, its roots must come in conjugate pairs. Hence, it has exactly $m/2$ roots with positive imaginary parts. Next, consider the polynomials

$$Q_s(z) = p_1(\xi + z\eta) + isp_2(\xi + z\eta) \tag{9-34}$$

When s is real, $Q_s(z)$ has no real roots, by (9-33). The coefficients of these polynomials are continuous in s, and the coefficient of z^m is

$$p_1(\eta) + isp_2(\eta)$$

which is bounded away from zero for s real. Thus, the roots of $Q_s(z)$ converge to those of $Q_0(z)$ as $s \to 0$ (Lemma 4-14 of Sec. 4-4). Since $Q_s(z)$ can have no real roots, none of its roots can cross the real axis as $s \to 0$. Thus, the number of roots of $Q_s(z)$ which lie above the real axis must equal the number of roots of $Q_0(z)$ which lie above the real axis, for each real s. In particular, this is true for $s = 1$. Since $Q_0(z)$ has exactly $m/2$ roots above the real axis, the same must be true of $Q(z)$. This completes the proof.

We note that there are properly elliptic operators which are not strongly elliptic. A simple example is the operator corresponding to

$$P(x, \xi) = \xi_1^4 + \xi_2^4 - \xi_3^4 + i(\xi_1^2 + \xi_2^2)\xi_3^2 \tag{9-35}$$

which is properly elliptic in E^3, but not strongly elliptic.

The importance of strongly elliptic operators is shown by

Theorem 9-20 If $P(x, D)$ satisfies (9-31), there is a constant R, such that

$$[P(x, D) + \lambda\overline{\gamma(x)}]u = f$$

has a unique solution $u \in \mathscr{D}$, for each $f \in C^\infty(\bar{G})$ and each $\lambda > R$.

The proof of the theorem is based on the following lemma due to Gårding (1953).

Lemma 9-21 If $P(x, D)$ satisfies (9-31), then there are constants $c_0 > 0$ and K, such that

$$\text{Re}\,(\gamma(x)P(x, D)u, u) \geq c_0 \|u\|_r^2 - K\|u\|^2 \tag{9-37}$$

holds for all $u \in \mathscr{D}$.

We shall prove this inequality in Sec. 9-7. Now we show how it can be used to give the

PROOF of Theorem 9-20 Since $P(x, D)$ is properly elliptic, it suffices to show that $N = N' = \{0\}$ when $P(x, D)$ is replaced by $P(x, D) + \lambda\bar{\gamma}$ (Theorem 9-3). Since $\gamma(x)$ is a smooth function on $\bar{G}$ and it does not vanish there (otherwise (9-31) could not hold), there is a positive constant ρ, such that $|\gamma(x)| \geq \rho$ on G. Let λ be any number greater than $(K + 1)/\rho^2$, and set

$$P_1(x, D) = \gamma(x)P(x, D) + \lambda|\gamma(x)|^2$$

Then, by (9-37)

$$\text{Re}\,(P_1(x, D)u, u) \geq \|u\|_r^2 \qquad u \in \mathscr{D} \tag{9-38}$$

This shows that the only $u \in \mathscr{D}$ satisfying $P_1(x, D)u = 0$ is $u = 0$. Moreover, (9-38) gives

$$\operatorname{Re}(u, P_1'(x, D)u) \geq \|u\|_r^2 \qquad u \in \mathscr{D}$$

Thus, the only $u \in \mathscr{D}$ satisfying $P_1'(x, D)u = 0$ is $u = 0$. Let f be any function in $C^\infty(\overline{G})$. Then, by Theorem 9-3, there is a unique solution $u \in \mathscr{D}$ of

$$P_1(x, D)u = \gamma f$$

Dividing both sides by γ, we obtain Eq. (9-36). This completes the proof.

9-7 GÅRDING'S INEQUALITY

We now give a proof of Lemma 9-21. By integration by parts

$$(\gamma(x)P(x, D)u, v) = b(u, v) = \sum_{|\mu|,|\nu| \leq r} (b_{\mu\nu}(x)D^\nu u, D^\mu v) \tag{9-39}$$

for $u, v \in \mathscr{D}$. Note that no boundary terms arise, because at each step derivatives of order less than r are applied to v. Note also that

$$\operatorname{Re} \sum_{|\mu|=|\nu|=r} b_{\mu\nu}(x)\xi^{\mu+\nu} = \gamma(x)p(x, \xi) \tag{9-40}$$

Let x^0 be a point of $\overline{G}$, and set

$$b_0(u, v) = \sum_{|\mu|,|\nu| \leq r} (b_{\mu\nu}(x^0)D^\nu u, D^\mu v)$$

We have

Lemma 9-22 There are constants $c_0 > 0$ and K, such that

$$\operatorname{Re} b_0(u) \geq c_0 \|u\|_r^2 - K\|u\|^2 \qquad u \in \mathscr{D}$$

where $b_0(u) \equiv b_0(u, u)$.

Lemma 9-23 There is a neighborhood $\mathscr{N}$ of x^0, such that

$$|b(u) - b_0(u)| \leq \tfrac{1}{2}c_0 \|u\|_r^2 \qquad u \in C_0^\infty(\mathscr{N}) \tag{9-41}$$

where $b(u) = b(u, u)$.

Leaving the proofs of these lemmas for the moment, we show how they can be used to prove Lemma 9-21. By Lemma 9-23, each x^0 has a neighborhood $\mathscr{N}$ such that (9-41) holds. Since $\overline{G}$ is compact, it can be covered by a finite number of them, say $\mathscr{N}_1, \ldots, \mathscr{N}_q$ corresponding to the points $x^1, \ldots, x^q$. Let $\zeta_1, \ldots, \zeta_q$ be a C^∞ partition of unity, subordinate to this covering (Lemma 9-16). Set $u_k = \zeta_k u$. Since

$$\left(\sum_1^q \alpha_k\right)^2 \leq q \sum_1^q \alpha_k^2 \tag{9-42}$$

holds for real α_k, we have, by Lemmas 9-22 and 9-23

$$\operatorname{Re}\sum b(u_k) \geq \tfrac{1}{2}\sum c_k \| u_k \|_r^2 - \sum K_k \| u_k \|^2$$
$$\geq c_0 \| u \|_r^2 - K \| u \|^2 \tag{9-43}$$

where $c_0 = \min c_k/2q$ and K is sufficiently large. Next, note that

$$b(u) - \sum b(u_k) = \sum_{\substack{|\mu|,|\nu|\leq r \\ |\mu|+|\nu|<m}} (c_{\mu/\nu}(x)D^\nu u, D^\mu u) \tag{9-44}$$

This is another exercise in integration by parts. By Lemma 9-15, there is a constant C, such that

$$|b(u) - \sum b(u_k)| \leq \tfrac{1}{2}c_0 \| u \|_r^2 + C \| u \|^2$$

This, combined with (9-43), gives

$$\operatorname{Re} b(u) \geq \tfrac{1}{2}c_0 \| u \|_r^2 - (K + C) \| u \|^2$$

By Eq. (9-31), this is precisely what we want.

We now turn to the

PROOF of Lemma 9-22 It suffices to prove

$$\operatorname{Re} b_0(v) \geq c_0 |v|_r^2 - K \| v \|^2 \qquad v \in H^r \tag{9-45}$$

For, by Lemma 9-8, $\hat{v} \in H^r$ for each $v \in \mathscr{D}$ and Eq. (9-12) holds. Thus, (9-45) implies Lemma 9-22. Moreover, (9-45) is implied by

$$\operatorname{Re} b_0(v) \geq c_0 |v|_r^2 - K \| v \|^2 \qquad v \in S \tag{9-46}$$

since S is dense in H^r. It suffices, therefore, to prove (9-46). Now, (9-46) is equivalent to

$$c_0 \int (1 + |\xi|)^m |Fv|^2 \, d\xi \leq \int [\operatorname{Re}\sum b_{\mu\nu}(x^0)\xi^{\mu+\nu} + K] |Fv|^2 \, d\xi$$

By Eq. (9-40) and inequality (9-31) there is a constant $a > 0$, such that

$$\operatorname{Re} \sum_{|\mu|=|\nu|=r} b_{\mu\nu}(x^0)\xi^{\mu+\nu} \geq a|\xi|^m \tag{9-47}$$

On the other hand, there is a constant C, such that

$$\left| \sum_{|\mu|+|\nu|<m} b_{\mu\nu}(x^0)\xi^{\mu+\nu} \right| \leq C(1 + |\xi|)^{m-1}$$

Thus, everything reduces to showing that there are constants $c_0 > 0$ and K, such that

$$c_0(1 + |\xi|)^{m-1} \leq a|\xi|^m + K \qquad \xi \in E^n \tag{9-48}$$

This completes the proof.

PROOF of Lemma 9-23 We have

$$|b(u) - b(u_0)| \leq \max_{x \in \mathcal{N}} \sum |b_{\mu\nu}(x) - b_{\mu\nu}(x^0)| \, \|u\|_r^2$$

By the continuity of the coefficients, we can take $\mathcal{N}$ so small that this maximum is as small as desired.

9-8 STRONG AND WEAK SOLUTIONS

At the beginning of Chapter 5 we defined a strong solution of the Cauchy problem. Similarly, we define a strong solution of the Dirichlet problem. For $f \in L^2(G)$ we shall say that $u \in L^2(G)$ is a *strong solution* of Eqs. (9-3), (9-4), if there is a sequence $\{u_k\}$ of functions in $\mathcal{D}$, such that

$$u_k \to u \qquad P(x, D)u_k \to f \text{ in } L^2(G) \tag{9-49}$$

Unlike the case of the Cauchy problem, we have

Theorem 9-24 A function $u \in L^2(G)$ is a strong solution of Eqs. (9-3), (9-4) if and only if $u \in H$ and satisfies Eq. (9-3).

PROOF If $u \in H$, then there is a sequence of functions $\{u_k\}$ in $\mathcal{D}$ which converges to u in $H^m(G)$. By (9-8), $P(x, D)u_k \to P(x, D)u$ in $L^2(G)$. If u satisfies (9-3), we see that (9-49) holds. Conversely, suppose $\{u_k\} \subset \mathcal{D}$ satisfies (9-49). By (9-9)

$$\|u_j - u_k\|_m \leq C(\|P(x, D)(u_j - u_k)\| + \|u_j - u_k\|) \to 0$$

Thus, there is a $w \in H$, such that $u_k \to w$ in $H^m(G)$. In particular, u_k converges to w in $L^2(G)$. Thus, $u = w$ a.e. By the reasoning above, Eq. (9-3) holds.

Theorem 9-25 If $f \in L^2(G)$ and $f \perp N'$, then there is a $u \in M$ satisfying Eq. (9-3).

PROOF There is a sequence $\{f_k\}$ of functions in $C^\infty(\bar{G})$ which converges to f in $L^2(G)$. We write $f_k = f_k' + f_k''$, where $f_k' \perp N'$ and $f_k'' \in N'$. Since $N' \subset C^\infty(\bar{G})$, we have $f_k' \in C^\infty(\bar{G})$. Moreover

$$\|f_k' - f\|^2 + \|f_k''\|^2 = \|f_k - f\|^2 \to 0$$

Thus, $f_k' \to f$ in L^2. By Theorem 9-1, there is a function $u_k \in \mathcal{D}$, such that $P(x, D)u_k = f_k'$.

Again we write $u_k = u_k' + u_k''$, where $u_k' \in M$ and $u_k'' \in N$. Thus, we have $P(x, D)u_k' = f_k'$. By Corollary 9-9

$$\|u_j' - u_k'\|_m \leq C\|f_j' - f_k'\| \to 0$$

Since M is a closed subspace of $H^m(G)$, there is a $u \in M$, such that $u_k' \to u$ in $H^m(G)$. Clearly, u satisfies Eq. (9-3). This completes the proof.

As we did in the case of the Cauchy problem, we can define a weak solution for the Dirichlet problem. For $f \in L^2(G)$, we shall say that $u \in L^2(G)$ is a *weak solution* of Eqs. (9-3), (9-4), if

$$(u, P'(x, D)v) = (f, v) \qquad v \in \mathscr{D} \tag{9-50}$$

Since
$$(P(x, D)v, w) = (v, P'(x, D)w) \qquad v, w \in \mathscr{D} \tag{9-51}$$

by Eq. (9-16), it is easily seen that every strong solution is a weak solution. If $f \in C^\infty(\bar{G})$, Theorem 9-3 says that every weak solution is in $\mathscr{D}$; if f is only known to be in $L^2(G)$, we cannot say as much as this, but we have

Theorem 9-26 Every weak solution of Eqs. (9-3), (9-4) is a strong solution.

PROOF If u satisfies Eq. (9-50), we have $u = u' + u''$, where $u' \perp N'$, $u'' \in N$. Thus, u' is a weak solution as well. Now, Eq. (9-50) implies that $f \perp N'$. Thus, by Theorem 9-25, there is a $w \in M$, such that $P(x, D)w = f$. Thus, w is also a weak solution, and, consequently

$$(u' - w, P'(x, D)v) = 0 \qquad v \in \mathscr{D}$$

By Lemma 9-11, $u' = w$. Hence,

$$u = w + u'' \in H \quad \text{and} \quad P(x, D)u = P(x, D)(w + u'') = f$$

This completes the proof.

9-9 THE EXCEPTIONAL SET

Theorem 9-20 says that Eq. (9-36) has a unique solution $u \in \mathscr{D}$ for every $f \in C^\infty(\bar{G})$, provided $P(x, D)$ is strongly elliptic and λ is sufficiently large. If we apply Theorem 9-25, we have

Theorem 9-27 If $P(x, D)$ is strongly elliptic, then Eq. (9-36) has a unique solution $u \in H$ for each $f \in L^2(G)$, provided λ is sufficiently large.

We shall now show a bit more. We shall prove

Theorem 9-28 If $P(x, D)$ is strongly elliptic, then there exists at most a denumerable set Γ of complex numbers having no finite limit point, such that Eq. (9-36) has a unique solution $u \in H$ for each $f \in L^2(G)$, whenever λ is not in Γ.

In proving the theorem, we shall make use of the following two facts.

Lemma 9-29 Suppose S is a bounded linear operator from $L^2(G)$ into H. Then S is a compact operator from $L^2(G)$ into itself.

Lemma 9-30 If K is a compact linear operator from a Hilbert space to itself, then the set of points λ, for which

$$(I - \lambda K)u = 0 \tag{9-52}$$

has a nonzero solution, is denumerable and has no finite limit points.

Theorem 9-28 is a simple consequence of these lemmas. In fact, let R be the number given in Theorem 9-20. Then, the equation

$$[P(x, D) + 2R\overline{\gamma(x)}]u = f$$

has a unique solution $u \in H$ for each $f \in L^2(G)$. If we define u to be Sf, we see that S is a linear operator from $L^2(G)$ to H. It is bounded by Corollary 9-9. Thus, by Lemma 9-29, S is a compact operator on $L^2(G)$. Now, if

$$[P(x, D) + \lambda\bar{\gamma}(x)]u = 0$$

then

$$[I + (\lambda - 2R)S\bar{\gamma}(x)]u = 0$$

If $u \not\equiv 0$, then λ would have to belong to a denumerable set not having any finite limit points. The same reasoning shows that

$$[P'(x, D) + \bar{\lambda}\gamma(x)]v = 0$$

has a solution $v \not\equiv 0$, only for λ in such a set. We now apply Theorem 9-25 to reach the desired conclusion.

PROOF of Lemma 9-29 Let $\{u_k\}$ be a bounded sequence in $L^2(G)$. Then $\{Su_k\}$ is a bounded sequence in H. By Lemma 9-8, $\hat{v} \in H^r$ for each $v \in \mathscr{D}$ and Eq. (9-12) holds. Now, for each k, there is a $w_k \in \mathscr{D}$, such that

$$\| w_k - Su_k \|_m < \frac{1}{k}$$

Thus, $\{\hat{w}_k\}$ is a bounded sequence in H^r. As in the proof of Lemma 9-8, this implies that there is a subsequence $\{\hat{w}_{k_j}\}$ which converges in L^2. Clearly, this implies that $\{Su_{k_j}\}$ converges in $L^2(G)$.

PROOF of Lemma 9-30 Suppose the lemma were false. Then there would be a sequence $\{\lambda_k\}$ of distinct complex numbers, such that

$$|\lambda_k| \le C$$

and a sequence $\{u_k\}$ of nonzero elements, such that

$$(I - \lambda_k K)u_k = 0$$

First, I claim that the u_k are linearly independent (see Sec. 4-3). For otherwise there would be an integer j, such that

$$u_{j+1} = \alpha_1 u_1 + \cdots + \alpha_j u_j$$

while $u_1, \ldots, u_j$ are linearly independent. Set $\beta_k = 1/\lambda_k$ (note that none of the

λ_k can vanish). Then

$$(K - \beta_k)u_k = 0$$

Thus $$Ku_{j+1} = \alpha_1\beta_1 u_1 + \cdots + \alpha_j\beta_j u_j$$

while $$\beta_{j+1}u_{j+1} = \alpha_1\beta_{j+1}u_1 + \cdots + \alpha_j\beta_{j+1}u_j$$

Hence $$\alpha_1(\beta_{j+1} - \beta_1)u_1 + \cdots + \alpha_j(\beta_{j+1} - \beta_j)u_j = 0$$

Since $u_1, \ldots, u_j$ are linearly independent, all of the coefficients must vanish. Since the β_k are distinct, this means that all of the α_k must vanish. Thus $u_{j+1} = 0$, which is against our assumption. Thus, the u_k are linearly independent.

Let L_k be the subspace spanned by $u_1, \ldots, u_k$. Then for each k, L_k is a closed subspace (Lemma 7-12 of Sec. 7-4). Since the u_k are linearly independent, L_{k-1} is a proper subspace of L_k. Thus, there is an element $w_k \in L_k$, such that

$$\|w_k\| = 1 \qquad w_k \perp L_{k-1}$$

(Corollary 1-4 of Sec. 1-5). Now, K maps L_k into itself. For if

$$u = \alpha_1 u_1 + \cdots + \alpha_k u_k$$

then $$Ku = \alpha_1\beta_1 u_1 + \cdots + \alpha_k\beta_k u_k$$

Note also that

$$(K - \beta_k)u = \alpha_1\beta_1 u_1 + \cdots + \alpha_{k-1}\beta_{k-1}u_{k-1}$$

Thus, $K - \beta_k$ maps L_k into L_{k-1}. Now, if $j > k$, we have

$$\begin{aligned} K(w_j - w_k) &= (K - \beta_j)w_j - Kw_k + \beta_j w_j \\ &= \beta_j[w_j - \lambda_j(Kw_k - (K - \beta_j)w_j)] \end{aligned}$$

Now $Kw_k \in L_k \subset L_{j-1}$, while $(K - \beta_j)w_j$ is in L_{j-1}. Hence

$$\|K(w_j - w_k)\| \geq |\beta_j| \geq \frac{1}{C}$$

This shows that $\{Kw_j\}$ cannot have a convergent subsequence, contradicting the fact that K is assumed compact. Thus, no such sequence $\{\lambda_k\}$ can exist, and the proof is complete.

We shall discuss the exceptional set Γ in more detail in a later section.

PROBLEMS

9-1 If ∂G is smooth, show that $C^\infty(\bar{G})$ consists of functions which are the restrictions to G of functions in C_0^∞.

9-2 Prove Lemma 9-2.

9-3 Show that $P(x, D)$ is properly elliptic in $\overline{G}$, if and only if $P'(x, D)$ is.

9-4 Show that the product of elliptic operators is elliptic.

9-5 Prove Lemma 9-14.

9-6 Prove the first statement in the proof of Lemma 9-16.

9-7 Show that the functions given by Eq. (9-28) form a partition of unity, subordinate to the covering.

9-8 Show how to find the subdomain G_1 described in the proof of Lemma 9-15.

9-9 Show that the operator given by Eq. (9-23) is properly elliptic if and only if $P(x, D)$ is.

9-10 Show that the operator given by Eq. (9-35) is properly elliptic but not strongly elliptic.

9-11 Prove inequality (9-42).

9-12 Prove Eq. (9-44).

9-13 Prove inequality (9-47).

9-14 Prove inequality (9-48).

CHAPTER

TEN

GENERAL BOUNDARY VALUE PROBLEMS

10-1 FORMULATION OF THE PROBLEM

Since we have been so successful in solving the Dirichlet problem for a bounded domain, we wonder if it is possible to extend our results to more general boundary value problems. In this chapter we show that it is possible to extend some of the theorems, but in order to do so we must deal with several new questions.

The first question is how to express our boundary conditions. In Chapters 6 and 7 we used constant-coefficient partial differential operators to describe boundary conditions on a hyperplane. When we deal with curved surfaces, we must be careful. We shall assume that the boundary conditions are of the form

$$Q_j(x, D)u(x) = 0 \text{ on } \partial G \qquad 1 \le j \le r \tag{10-1}$$

where the Q_j are partial differential operators with coefficients defined in the whole of $\bar{G}$. If the coefficients are defined only in ∂G, this is equivalent to assuming that they can be continued smoothly into the interior. Of course, if the coefficients of the Q_j and ∂G are sufficiently smooth, this can always be done.

The next question is how to formulate the adjoint problem. For the Dirichlet problem this was simple; it consisted of the same boundary conditions for the formal adjoint $P'(x, D)$ of $P(x, D)$ (see Theorem 9.3 of Sec. 9-1). However, in the general case, it turns out that the adjoint problem need not have boundary conditions of the same form as Eq. (10-1), let alone the same operators. There are several ways of coping with this problem, and we have chosen the simplest. We find a subclass of boundary conditions of the form (10-1), having the property that the adjoint boundary conditions consist of boundary conditions of the same form, albeit usually involving different operators. We call this class *normal boundary conditions*. We have met a special case in Sec. 8-3. We discuss them more fully in Sec. 10-4. As in the case of the Dirichlet problem, the main theorem can be proved by means of an inequality and a regularity theorem. These are stated in Sec. 10-3, together with a theorem concerning adjoint boundary conditions. Their proofs are given in Secs. 10-4 to 10-6. In Secs. 10-2 to 10-6 everything is considered in the semisphere σ_R. The rest of the chapter then shows how this special case gives the complete solution.

We now state the main results. The hypotheses are as follows:

(*a*) G is a bounded domain with smooth boundary ∂G (see Sec. 9-1).

(*b*) $P(x, D)$ is a properly elliptic operator of order $m = 2r$ with coefficients in $C^\infty(\bar{G})$.

(*c*) $Q_1(x, D), \ldots, Q_r(x, D)$ are partial differential operators of orders $m_j < m$ with coefficients in $C^\infty(\bar{G})$, such that

(i) The m_j are distinct

(ii) If $q_j(x, D)$ is the principal part of $Q_j(x, D)$, then $q_j(\chi_0, \nu) \neq 0$ whenever $\chi_0 \in \partial G$ and $\nu \neq 0$ is orthogonal to ∂G at the point χ_0.

(*d*) If $\chi_0 \in \partial G$, $\xi \neq 0$ is parallel to ∂G at χ_0, and $\nu \neq 0$ is orthogonal to ∂G at χ_0, then the polynomials in z

$$q_1(\chi_0, \xi + z\nu), \ldots, q_r(\chi_0, \xi + z\nu)$$

are linearly independent modulo

$$p(z) = (z - \tau_1(\xi, \nu)) \cdots (z - \tau_r(\xi, \nu))$$

where the $\tau_k(\xi, \nu)$ are the roots of

$$p(\chi_0, \xi + z\nu) = 0 \tag{10-2}$$

having positive imaginary parts (see Sec. 6-7).

The boundary set $\{Q_j\}$ is called *normal* if the Q_j satisfy (*c*). We say that it *covers P* if it satisfies (*d*). We shall prove

Theorem 10-1 Under the above assumptions

1. There is a normal set $\{Q'_j\}$ of r boundary operators which covers the formal adjoint $P'(x, D)$ of $P(x, D)$, and such that $u \in C^\infty(\bar{G})$ satisfies Eq. (10-1) if and only if

$$(u, P'v) = (Pu, v) \tag{10-3}$$

holds for all v satisfying

$$Q'_j(x, D)v(x) = 0 \text{ on } \partial G \qquad 1 \leq j \leq r \tag{10-4}$$

Conversely, $v \in C^\infty(\bar{G})$ satisfies Eq. (10-4) if and only if Eq. (10-3) holds for all $u \in C^\infty(\bar{G})$ satisfying Eq. (10-1).

2. There is a constant C, such that

$$\| u \|_m \leq C(\| Pu \| + \| u \|) \tag{10-5}$$

holds for all $u \in C^\infty(\bar{G})$ satisfying Eq. (10-1).

3. Let N be the set of those $u \in C^\infty(\bar{G})$ satisfying Eq. (10-1) and $Pu = 0$, and N' the set of those $v \in C^\infty(\bar{G})$ satisfying Eq. (10-4) and $P'v = 0$. Then N and N' are finite dimensional.
4. For $f \in C^\infty(\bar{G})$ there is a $u \in C^\infty(\bar{G})$ satisfying Eq. (10-1) and $Pu = f$, if and only if $f \perp N'$. Similarly, for $g \in C^\infty(\bar{G})$ there is a $v \in C^\infty(\bar{G})$ satisfying Eq. (10-4) and $P'v = g$, if and only if $g \perp N$.
5. If $u \in L^2(G)$, $f \in C^\infty(\bar{G})$, and

$$(u, P'v) = (f, v) \tag{10-6}$$

holds for all $v \in C^\infty(\bar{G})$ satisfying Eq. (10-4), then $u \in C^\infty(\bar{G})$ (after correction in a set of measure 0) and satisfies Eq. (10-1) and $Pu = f$.

This chapter is devoted to proving Theorem 10-1. In Secs. 10-2 to 10-4 the problem is restricted to the semispheres σ_R. Then we show in Secs. 10-5 to 10-9 how to use the information obtained to deal with general domains. Many of the techniques were used in previous chapters, and there is not much new that is needed.

The results of this chapter are due to Schechter (1964 and 1959*b*).

10-2 THE PROBLEM IN σ_R

As we did in Chapter 8, we first study the problem in the region σ_R. Thus, we want to find a solution of

$$P(\chi, t, D)u(\chi, t) = f(\chi, t) \text{ in } \sigma_R \tag{10-7}$$

$$Q_j(\chi, 0, D)u(\chi, 0) = g_j(\chi) \qquad |\chi| < R \qquad 1 \leq j \leq r \tag{10-8}$$

As before, we assume that $P(\chi, t, D)$ is a properly elliptic operator of order $m = 2r$ (see Sec. 6-9). Again we are willing to take f and the g_j as infinitely differentiable functions. The best situation is when Eqs. (10-7), (10-8) have a solution for every choice of f and the g_j. Clearly, this cannot happen unless there exists a $u \in C^\infty(\bar{\sigma}_R)$ satisfying Eq. (10-8) for each choice of the g_j. Moreover, Lemma 8-4 of Sec. 8-3 gives a sufficient condition for this to happen, i.e., that the Q_j can be made part of a Dirichlet system. From the definition, we have

Lemma 10-2 The operators Q_j can be made part of a Dirichlet system if and only if their orders m_j are distinct, and the coefficient of $D_t^{m_j}$ in $Q_j(\chi, 0, D)$ does not vanish on $\partial_0 \sigma_R$.

A set of operators having this property is called *normal*. If we assume that the Q_j form a normal set, we can find a function u_0 satisfying Eq. (10-8) and subtract it. This will change the function f, but it will reduce the boundary conditions (10-8) to

$$Q_j(\chi, 0, D)u(\chi, 0) = 0 \qquad |\chi| < R \qquad 1 \le j \le r \tag{10-9}$$

If we can solve Eqs. (10-7), (10-9) with f replaced by $f - P(\chi, t, D)u_0$, we obtain a solution of Eqs. (10-7), (10-8) by adding u_0. Thus, it suffices to consider the problem (10-7), (10-9). We shall always assume that the orders m_j of the Q_j are less than m.

If we follow the methods of the preceding two chapters, we need to define a weak solution for the problem. To do this we need a counterpart of Eq. (8-13) of Sec. 8-2. The following theorem is used for this purpose. Let $\mathscr{C}_R$ denote the set of those functions in $C^\infty(\bar{\sigma}_R)$ which vanish near $\partial_1 \sigma_R$. We have

Theorem 10-3 Let $Q_1, \ldots, Q_r$ be a normal set of operators with orders less than m. Then there is a normal set $Q'_1, \ldots, Q'_r$, such that $v \in C^\infty(\bar{\sigma}_R)$ satisfies

$$(P(x, t, D)u, v) = (u, P'(x, t, D)v) \tag{10-10}$$

for all $u \in \mathscr{C}_R$ satisfying Eq. (10-9), if and only if

$$Q'_j(x, 0, D)v(x, 0) = 0 \qquad |x| < R \qquad 1 \le j \le r \tag{10-11}$$

This is an easy consequence of

Theorem 10-4 If $Q_1, \ldots, Q_m$ is a Dirichlet system or order m, then there is a Dirichlet system $Q'_1, \ldots, Q'_m$ of order m, such that

$$(P(x, t, D)u, v) - (u, P'(x, t, D)v) = \sum_{j=1}^{m} \int_{\partial_0 \sigma_R} Q_j(x, 0, D)u\, \overline{Q'_{m-j+1}(x, 0, D)v}\, dx \tag{10-12}$$

holds for $u \in \mathscr{C}_R$ and $v \in C^\infty(\bar{\sigma}_R)$.

Proof We may assume that Q_j is of the order $j - 1$. By Theorem 8-2 of Sec. 8-3

$$D_t^{j-1} = \sum_{k=1}^{j} \Lambda_{jk}(\chi, D_x) Q_k(\chi, 0, D) \qquad 1 \le j \le m \tag{10-13}$$

where the Λ_{jk} satisfy the conditions stipulated in Sec. 8-3. If we substitute this into Eq. (8-10) of Sec. 8-2, we find that the left-hand side of Eq. (10-12) equals

$$-i \sum_{j=1}^{m} \int_{\partial_0 \sigma_R} \sum_{k=1}^{j} \Lambda_{jk}(x, D_x) Q_k(x, 0, D)u\, \overline{N_j v}\, dx$$

$$= -i \sum_{j=1}^{m} \sum_{k=1}^{j} \int_{\partial_0 \sigma_R} Q_k(x, 0, D)u\, \overline{\Lambda'_{jk}(x, D_x) N_j v}\, dx$$

$$= \sum_{k=1}^{m} \int_{\partial_0 \sigma_R} Q_k(x, 0, D)u \, Q'_{m-k+1}(x, 0, D)v \, dx$$

where Λ'_{jk} is the formal adjoint of Λ_{jk}, and

$$Q'_{m-k+1}(\chi, t, D) = i \sum_{j=1}^{m} \Lambda'_{jk}(\chi, D_x) N_j \qquad 1 \leq k \leq m \tag{10-14}$$

Now, the N_j form a Dirichlet system. Moreover, the order of N_j is $m - j$. Thus, the order of Q'_j is $< j$. The coefficient of D_t^{j-1} in Q'_j is $\overline{\Lambda}_{m-j+1,m-j+1} a_{0,m}$, which is a nonvanishing function. Thus, the Q'_j form a Dirichlet system of order m, and the proof is complete.

We are now in a position to define a weak solution of problem (10-7), (10-9). Let V_R be the set of those $u \in \mathscr{C}_R$ which satisfy Eq. (10-9), and let V'_R be the set of those $v \in \mathscr{C}_R$ satisfying Eq. (10-11). Suppose $u, f \in C^\infty(\bar{\sigma}_k)$ are such that

$$(u, P'(\chi, t, D)v) = (f, v) \qquad v \in V'_R \tag{10-15}$$

Then I claim that u satisfies Eqs. (10-7) and (10-9). The first part is verified as in Sec. 8-2. To prove the second, let ζ be any function in $\mathscr{C}_R$. By Lemma 8-4 of Sec. 8-3 there is a $v \in \mathscr{C}_R$, such that

$$\begin{aligned} Q'_j(\chi, 0, D)v(\chi, 0) &= 0 & 1 \leq j \leq r \\ &= \bar{\zeta} Q_{m-j+1}(\chi, 0, D)u(\chi, 0) & r \leq j \leq m \end{aligned}$$

Substituting this into Eq. (10-12) and noting that $v \in V'_R$, we have

$$\sum_{j=1}^{r} \int_{\partial_0 \sigma_R} \zeta \, | Q_j(x, 0, D)u |^2 \, dx = 0$$

Since ζ was arbitrary, we see that u satisfies Eq. (10-9). Because of this we shall say that $u \in L^2(\sigma_R)$ is a weak solution of Eqs. (10-7), (10-9) if Eq. (10-15) holds. We have

Theorem 10-5 If a weak solution of Eqs. (10-7), (10-9) is in $C^\infty(\bar{\sigma}_R)$, then it is an actual solution.

Boundary operators satisfying the conclusions of Theorem 10-3 are called *adjoint to the Q_j relative to P*. We shall have more to say about them a bit later. For the moment, we note

Lemma 10-6 If $\{Q'_j\}$ is adjoint to $\{Q_j\}$ relative to P, then $\{Q_j\}$ is adjoint to $\{Q'_j\}$ relative to P'.

10-3 THE SOLUTION

The solutions in the last two chapters were each based upon an inequality and a regularity theorem. We can do the same here. First we state our hypotheses.

1. $P(0, 0, D)$ is properly elliptic (see Sec. 6-9) of order $m = 2r$.
2. $Q_1, \ldots, Q_r$ is a set of operators of orders $m_j < m$.
3. If $q_j(x, t, D)$ is the principal part of $Q_j(x, t, D)$, then for each $\xi \neq 0$ in E^n, the polynomials $q_j(0, 0, \xi, \tau)$ are linearly independent modulo
$$p_+(\xi, \tau) = (\tau - \tau_1(\xi)) \cdots (\tau - \tau_r(\xi)) \tag{10-16}$$
where the $\tau_k(\xi)$ are the roots of $p(0, 0, \xi, \tau)$ having positive imaginary parts.
4. The coefficients of P and the Q_j are in $C^\infty(\bar{\sigma}_R)$.
5. $P(x, t, D)$ is of order m in σ_R.

We shall say that the Q_j *cover* P if hypothesis 3 holds (see Theorem 6-15 of Sec. 6-7). We have

Theorem 10-7 If hypotheses 1 to 5 hold, then there are constants C and R, such that

$$|u|_{m,0} \leq C(\| P(x, t, D)u \| + \| u \| + \sum_{j=1}^{r} | Q_j(x, 0, D)u(x, 0)|_{m-m_j-1/2}) \quad u \in \mathscr{C}_R \tag{10-17}$$

Note that the $|\cdot|_{j,s}$ norm is taken over Ω while the $|\cdot|_s$ norm is taken over E^n, which is identified with $\partial\Omega$. Let $\hat{H}_R$ denote the completion of $\mathscr{C}_R$ with respect to the $|\cdot|_{m,0}$ norm. Our regularity theorem is

Theorem 10-8 If $f \in C^\infty(\bar{\sigma}_R)$, $u \in \hat{H}_R$, and

$$(P(x, t, D)u, P(x, t, D)v) + \sum_{j=1}^{r} (Q_j(x, 0, D)u(x, 0), Q_j(x, 0, D)v(x, 0))_{m-m_j-1/2}$$
$$= (f, v) \qquad v \in \mathscr{C}_R \tag{10-18}$$

then there is an $R' > 0$ such that $u \in C^\infty(\bar{\sigma}_{R'})$.

As usual, the conclusion of Theorem 10-8 holds after correction on a set of measure 0. We shall also need a theorem concerning adjoint boundary conditions.

Theorem 10-9 If the normal set $\{Q_j\}$ of boundary operators covers P, then every set $\{Q_j'\}$ adjoint to it, relative to P, covers its formal adjoint P'.

This theorem will be proved in Sec. 10-4; Theorem 10-8 will be proved in Sec. 10-5 and Theorem 10-7 in Sec. 10-6. Now let us show how they help solve our problem. As we did in Chapter 8, we can define $P(x, t, D)u$ for $u \in \hat{H}_R$ by means of inequality (8-31) given in Sec. 8-4. We can also define $Q_j(x, 0, D)u(x, 0)$ as an element of $H^{m-m_j-1/2}(E^n)$. This can be done by means of

Theorem 10-10 For $k \geq 1$

$$|v(x, 0)|^2_{s+k-1/2} \leq 2|v|^2_{k,s} \qquad v \in S(\Omega) \tag{10-19}$$

PROOF We use the same trick that was used in the proof of Lemma 6-17 of Sec. 6-7. By Lemma 6-14 of Sec. 6-6

$$|y(0)|^2 \leq 2 \sum_{j=0}^{1} \int_0^\infty |D_\tau^j y(\tau)|^2 \, d\tau \tag{10-20}$$

holds for $y \in H^1(0, \infty)$. Put $t = \tau/(1 + |\xi|)$ and $y(\tau) = Fv(\xi, t)$. Then $(1 + |\xi|)D_\tau y = D_t y$, and consequently

$$|Fv(\xi, 0)|^2 \leq 2 \sum_{j=0}^{1} (1 + |\xi|)^{1-2j} \int_0^\infty |D_t^j Fv(\xi, t)|^2 \, dt \tag{10-21}$$

If we multiply both sides by $(1 + |\xi|)^{2s+2k-1}$ and integrate with respect to ξ, we obtain the desired inequality.

Corollary 10-11 If $Q(x, t, D)$ is an operator of order $k < m$ with coefficients in $S(\Omega)$, then

$$|Q(x, 0, D)v(x, 0)|_{s+m-k-1/2} \leq C|v|_{m,s} \tag{10-22}$$

PROOF If $a(x) \in S(E^n)$ and $|\mu| \leq k$,

$$|a(x)D^\mu v(x, 0)|_{s+m-k-1/2} \leq C|D^\mu v|_{m-k,s} \leq C|v|_{m,s}$$

From this corollary we see that we can define $Q_j(x, 0, D)v(x, 0)$ as an element of $H^{m-m_j-1/2}(E^n)$ when v is in $H^{m,0}(\Omega)$. Let $\hat{N}_R$ be the set of those $u \in \hat{H}_R$ which satisfy Eq. (10-9) and

$$P(x, t, D)u = 0 \text{ in } \sigma_R \tag{10-23}$$

and let $\hat{N}'_R$ be the set of those $v \in \hat{H}_R$ which satisfy Eq. (10-11) and

$$P'(x, t, D)v = 0 \text{ in } \sigma_R \tag{10-24}$$

As in Chapter 8, we have

Corollary 10-12 For R sufficiently small, $\hat{N}_R$ and $\hat{N}'_R$ are finite dimensional.

The proof is the same as that of Theorem 8-5 of Sec. 8-4 (given at the end of Sec. 8-6).

Corollary 10-13 If $u \in \hat{N}_R$, then there is an $R' > 0$, such that $u \in C^\infty(\bar{\sigma}_{R'})$.

This is an immediate consequence of Theorem 10-8.

Corollary 10-14 If $v \in \hat{N}'_R$, then there is an $R' > 0$, such that $v \in C^\infty(\bar{\sigma}_{R'})$.

PROOF Since $P'(0, 0, D) = \overline{P}(0, 0, D)$, it is properly elliptic. The Q'_j cover it by Theorem 10-9. Thus, we may apply Corollary 10-13.

Theorem 10-15 If $f \in C^\infty(\bar{\sigma}_R)$ and $f \perp \hat{N}'_R$, then there is an $R' > 0$ and a function $u \in C^\infty(\bar{\sigma}_{R'})$, such that Eqs. (10-7), (10-9) hold in $\sigma_{R'}$.

PROOF We follow the proof of Theorem 8-26 of Sec. 8-7. Put

$$((u, v)) = (P'(x, t, D)u, P'(x, t, D)v) + \sum (Q'_j(x, 0, D)u(x, 0), Q'_j(x, 0, D)v(x, 0))_{m-m_j-1/2} \qquad (10\text{-}25)$$

Let M be the set of those $u \in \hat{H}_R$, such that $u \perp \hat{N}'_R$. By Theorem 10-7 and Corollary 10-12, we have

$$|u|^2_{m,0} \leq C((u, u)) \qquad u \in M \qquad (10\text{-}26)$$

(see the proof of Theorem 8-10 of Sec. 8-4). Thus, M is a Hilbert space with respect to the scalar product (10-25). Thus, there is a $w \in M$, such that

$$((w, v)) = (f, v) \qquad v \in M \qquad (10\text{-}27)$$

Since $f \perp \hat{N}'_R$, this holds for all $v \in \hat{H}_R$. Theorem 10-8 now tells us that $u \in C^\infty(\bar{\sigma}_{R'})$ for some $R' > 0$. Put $u = P'(x, t, D)w$. Then $u \in C^\infty(\bar{\sigma}_{R'})$, and u is a solution of Eq. (10-7) in $\sigma_{R'}$. Thus, Eq. (10-10) holds for all $v \in V'_R$. In view of Lemma 10-6, this implies that u satisfies Eq. (10-9).

10-4 THE ADJOINT SYSTEM

In this section we give the proof of Theorem 10-9. Two systems $Q_1, \ldots, Q_p$ and $\tilde{Q}_1, \ldots, \tilde{Q}_q$ will be called equivalent if $v \in C^\infty(\bar{\sigma}_R)$ satisfies

$$Q_j(x, 0, D)v(x, 0) = 0 \qquad 1 \leq j \leq p \qquad (10\text{-}28)$$

if and only if

$$\tilde{Q}_j(x, 0, D)v(x, 0) = 0 \qquad 1 \leq j \leq q \qquad (10\text{-}29)$$

We are going to show that equivalent normal sets contain the same number of operators (i.e., $p = q$), and that precisely the same orders are to be found among the operators of each set. In fact, we have

Lemma 10-16 Let m_j, $\tilde{m}_j$ denote the orders of the Q_j and $\tilde{Q}_j$, respectively. If the sets are normal and equivalent, then $p = q$ and the sets $\{m_j\}$ and $\{\tilde{m}_j\}$ are the same. Moreover, there is a set $\{\Lambda_{jk}(x, D_x)\}$ of operators such that

$$\tilde{Q}_j(x, 0, D) = \sum_{j=1}^{k} \Lambda_{jk}(x, D_x) Q_k(x, 0, D) \qquad (10\text{-}30)$$

and

(a) Λ_{jk} is a nonvanishing function for $\tilde{m}_j = m_k$.
(b) Order $\Lambda_{jk} \leq \tilde{m}_j - m_k$.

(As usual, (b) is to be interpreted to mean that $\Lambda_{jk} = 0$ if $\tilde{m}_j < m_k$.)

PROOF By Lemma 10-2, the Q_j are contained in a Dirichlet system $\{Q_j''\}$. By Corollary 8-3 of Sec. 8-3,

$$\tilde{Q}_j(x, 0, D) = \sum \Lambda_{jk}(x, D_x) Q_k''(x, 0, D)$$

where

(a') Λ_{jk} is a nonvanishing function for $\tilde{m}_j = m_k''$.
(b') Order $\Lambda_{jk} \leq \tilde{m}_j - m_k''$.

I claim that $\Lambda_{ik} = 0$ for each i, if Q_k'' is not one of the original Q_j. For if it did not vanish, there would be a $g \in C_0^\infty(\partial_0 \sigma_R)$, such that $\Lambda_{ik} g \neq 0$. By Lemma 8-4 of Sec. 8-3, there is a $v \in \mathscr{C}_R$, such that

$$\begin{aligned} Q_j''(x, 0, D) v(x, 0) &= 0 \qquad j \neq k \\ &= g \qquad j = k \end{aligned}$$

Since Q_k'' is not among the Q_j, v satisfies Eq. (10-28). But $\tilde{Q}_i(x, 0, D)v = \Lambda_{ik} g \neq 0$, contradicting the fact that the two sets are equivalent. We also see from this that, if Q_k'' is not one of the Q_j, then no $\tilde{Q}_i$ can have the same order as it. For then Λ_{ik} would be a nonvanishing function, contradicting what we have just shown. Thus, the $\tilde{m}_j$ are to be found among the m_k. Interchanging the two systems, we see that the m_j are only to be found among the $\tilde{m}_k$. This proves the lemma.

Next let $P(x, t, D)$ be a properly elliptic operator of order m. We have

Lemma 10-17 If a normal set $\{Q_j\}$ covers $P(x, t, D)$, then every equivalent normal set also covers it.

PROOF By definition, the polynomials $q_j(0, 0, \xi, \tau)$ are linearly independent modulo $p_+(\xi, \tau)$ (see Sec. 10-3). If $\{\tilde{Q}_j\}$ is any equivalent set, then Eq. (10-30) holds. Let $\lambda_{jk}(x, D_x)$ be the principal part of Λ_{jk}. Then Eq. (10-30) implies

$$\tilde{q}_j(0, 0, \xi, \tau) = \sum \lambda_{jk}(0, \xi) q_k(0, 0, \xi, \tau) \tag{10-31}$$

where

(a'') λ_{jk} is a nonzero constant for $\tilde{m}_j = m_k$.
(b'') Degree $\lambda_{jk}(0, \xi) \leq \tilde{m}_j - m_k$.

Suppose there are constants α_j, such that $\Sigma\, \alpha_j \tilde{q}_j(0, 0, \xi, \tau)$ is a multiple of $p_+(\xi, \tau)$. Then the same is true of

$$\sum_k \left(\sum_j \alpha_j \lambda_{jk}(0, \xi) \right) q_k(0, 0, \xi, \tau)$$

By hypothesis, this implies

$$\sum_j \alpha_j \lambda_{jk}(0, \xi) = 0 \text{ for each } k \tag{10-32}$$

By (a'') and (b'') this, in turn, implies that the α_j vanish. Hence, the $\tilde{Q}_j$ cover P as well.

The following is a trivial consequence of the definitions.

Lemma 10-18 If P is an operator of order $m = 2r$ and $Q_1, \ldots, Q_r$ is a normal set, then all sets $\{Q'_j\}$ which are adjoint to $\{Q_j\}$ relative to P are equivalent.

We can now give the

PROOF of Theorem 10-9 By Lemmas 10-17 and 10-18 it suffices to find one set adjoint to $\{Q_j\}$ relative to P which covers P'. By Lemma 10-2, the Q_j are contained in a Dirichlet system of order m. By Theorem 10-4, there is a Dirichlet system $\{Q'_j\}$ such that Eq. (10-12) holds. Replace u, v by $u(\lambda x, \lambda t)$, $v(\lambda x, \lambda t)$ and let $\lambda \to \infty$. This gives

$$(p(0,0,D)u, v) - (u, p'(0,0,D)v) = \sum_{j=1}^{m} \int_{\partial_0 \sigma_R} q_j(0,0,D)u\, \overline{q'_{m-j+1}(0,0,D)v}\, dx \tag{10-33}$$

Next, let $\xi \in E^n$, $\varphi \in C_0^\infty(E^n)$, and $g, h \in C_0^\infty(E^1)$. Put

$$u(x,t) = e^{i\xi x}\varphi(\lambda x) g(t) \qquad v(x,t) = e^{i\xi x}\varphi(\lambda x) h(t)$$

Substitute into Eq. (10-33) and let $\lambda \to 0$. We obtain

$$\int_0^\infty (p(0,0,\xi,D_t)g(t)\overline{h(t)} - g(t)\overline{p'(0,0,\xi,D_t)h(t)}\, dt$$
$$= \sum_{j=1}^{m} q_j(0,0,\xi,D_t)g(0)\overline{q'_{m-j+1}(0,0,\xi,D_t)h(0)} \tag{10-34}$$

Let $\xi \neq 0$ be fixed, and suppose that $h(t)$ is a solution of

$$p'(0,0,\xi,D_t)h(t) = 0 \qquad t > 0 \tag{10-35}$$

$$q'_j(0,0,\xi,D_t)h(0) = 0 \qquad 1 \le j \le r \tag{10-36}$$

By Theorem 6-8 of Sec. 6-4, there is a $g \in S(0, \infty)$, such that

$$p(0,0,\xi,D_t)g(t) = h(t) \qquad t > 0 \tag{10-37}$$

$$q_j(0,0,\xi,D_t)g(0) = 0 \qquad 1 \le j \le r \tag{10-38}$$

Substituting into Eq. (10-34) we get

$$\int_0^\infty |h(t)|^2\, dt = 0$$

which shows that $h(t) = 0$. Thus, every solution of Eqs. (10-35), (10-36) vanishes identically. By Sec. 6-4, the $q'_j(0, 0, \xi, \tau)$ are linearly independent modulo $p'_+(\xi, \tau)$. This shows that the Q'_j cover P'. Clearly, they are adjoint to $\{Q_j\}$ relative to P. Thus, the theorem is proved.

Corollary 10-19 A normal set covers P if and only if every normal set adjoint to it, relative to P, covers P'.

10-5 THE REGULARITY THEOREM

In this section we give the proof of Theorem 10-8. Fortunately, this was almost done in Sec. 8-9. In the present case we take $[u, v]$ as the left-hand side of Eq. (10-18), and we deal with $\mathscr{C}_R$ and $\hat{H}_R$ in place of $\mathscr{D}_R$ and H_R. The main step is to check that Lemma 8-34 of Sec. 8-9 holds under the new conditions, for, when we know that it does, the proof proceeds almost identically as it does there. In it, inequality (10-17) replaces Theorem 8-13 of Sec. 8-4.

In order to prove inequality (8-89) of Sec. 8-9, we add to the list of statements given there

5. For each real s

$$(\delta_i^h w(x, 0), v(x, 0))_s = (w(x, 0), \delta_i^{-h} v(x, 0))_s \tag{10-39}$$

6. For a smooth function $a(x)$ and real s

$$|(aw(x, 0), v(x, 0))_s - (w(x, 0), av(x, 0))_s| \le C|w|_{s-1}|v|_s \tag{10-40}$$

where the constant C depends only on a and s.

Before we prove 5 and 6, we show how they can be used to obtain the desired inequality. Note that it suffices to show (in the notation of Sec. 8-9) that

$$(aD^\rho \delta_i^h D_x^\mu \zeta u(x, 0), D^\sigma v(x, 0))_{k+1/2} \sim (aD^\rho u(x, 0), D^\sigma D_x^\mu \zeta \delta_i^{-h} v(x, 0))_{k+1/2} \tag{10-41}$$

for $|\rho|, |\sigma| < m - k$. By 5, 6 and Theorem 10-10 the left-hand side of (10-41) is

$$\begin{aligned}
&\sim (D^\rho \delta_i^h D_x^\mu \zeta u, \bar{a} D^\sigma v)_{k+1/2} = (D^\rho D_x^\mu \zeta u, \delta_i^{-h}(\bar{a} D^\sigma v))_{k+1/2} \\
&\qquad = (D^\rho D_x^\mu \zeta u, \bar{a}\, \delta_i^{-h} D^\sigma v + D^\sigma v(x_i^{-h})\, \delta_i^{-h} \bar{a})_{k+1/2} \\
&\sim (\delta_i^h a D^\rho D_x^\mu \zeta u, D^\sigma v)_{k+1/2} \\
&\sim (\delta_i^h a \zeta D^\rho D_x^\mu u, D^\sigma v)_{k+1/2} \\
&\sim (\delta_i^h \zeta D_x^\mu a D^\rho u, D^\sigma v)_{k+1/2} \\
&\sim (D_x^\mu a D^\rho u, \zeta\, \delta_i^{-h} D^\sigma v)_{k+1/2} \\
&\sim (D_x^\mu a D^\rho u, D^\sigma \zeta\, \delta_i^{-h} v)_{k+1/2} \\
&= (aD^\rho u, d_x^\rho D^\sigma \zeta\, \delta_i^{-h} v)_{k+1/2}
\end{aligned}$$

This proves that Lemma 8-34 holds under the new conditions. Now we can proceed as in the proof of Theorem 8-24 of Sec. 8-7.

It remains to prove 5 and 6. The former is simple since

$$F(\delta_j^h v) = \frac{1}{h}(e^{i\xi_j h} - 1)Fv \tag{10-42}$$

as is easily verified by the definition. To prove 6, put $\rho(\xi) = (1 + |\xi|^2)^{1/2}$, and, for s real, define the operator L^s by

$$FL^s u = \rho^s Fu$$

We have

Lemma 10-20 For each s, t, and $a \in S$

$$|L^s(av) - aL^s v|_t \le C_{s,t}|v|_{s+t-1} \int \rho(\xi)^{|s-1|+|t|+1}|Fa(\xi)|\,d\xi \tag{10-43}$$

PROOF By Theorem 2-10 of Sec. 2-5, we have

$$(2\pi)^n(L^s(av) - aL^s v) = F^{-1}\int [\rho(\xi)^s - \rho(\xi-\eta)^s]\,Fu(\xi-\eta)Fu(\eta)\,d\eta \tag{10-44}$$

Thus, the H^t norm of this is bounded by

$$\int |[\rho(\cdot)^s - \rho(\cdot - \eta)^s]\,Fu(\cdot - \eta)|_t|Fa(\eta)|\,d\eta$$

$$= \int |\rho(\cdot + \eta)^t[\rho(\cdot + \eta)^s - \rho(\cdot)^s Fu(\cdot)|_0|Fa(\eta)|\,d\eta \tag{10-45}$$

Now, by (3-48) and (3-49) of Sec. 3-4

$$|\rho(\xi+\eta)^s - \rho(\xi)^s| \le C\rho(\xi)^{s-1}\rho(\eta)^{|s-1|+1} \tag{10-46}$$

and

$$\rho(\xi+\eta)^t \le \rho(\xi)^t\rho(\eta)^{|t|} \tag{10-47}$$

Employing these estimates, we see that Eq. (10-45) is bounded by the right-hand side of (10-43). This completes the proof.

Corollary 10-21 For $a \in S$

$$|(au, v)_s - (u, \bar{a}v)_s| \le C|u|_{s-t}|v|_{s+t-1} \tag{10-48}$$

where the constant C depends only on s, t, and a.

PROOF The left-hand side of (10-48) is

$$|(L^{2s}(au) - aL^{2s}u|_{1-s-t}|v|_{s+t-1}$$

By Lemma 10-20, this is bounded by the right-hand side of (10-48).

To prove (10-40), we merely take $t = 0$ in (10-48).

10-6 THE INEQUALITY

Now we give a proof of Theorem 10-7. Because of the results already at our disposal, our job is relatively simple. First, we shall need

Lemma 10-22 For each integer $k \geq 1$, $s \geq 0$, and $\varepsilon > 0$, there is a constant K, such that

$$|u|_{k-1,s} \leq \varepsilon |u|_{k,s} + K|u|_{0,s} \qquad u \in H^{k,s} \tag{10-49}$$

PROOF This follows immediately from inequalities (6-129), (6-130) and Theorem 6-25 of Sec. 6-7.

We now give the

PROOF of Theorem 10-7 By Theorem 6-26 of Sec. 6-7, we have

$$|u|_{m,0} \leq C(\|P(0,0,D)u\| + \|u\| + \sum_{j=1}^{r} |Q_j(0,0,D)u(x,0)|_{m-m_j-1/2}) \tag{10-50}$$

Now, the coefficients of P and the Q_j are assumed smooth in σ_R, and in particular they can be made as close to their values at $(0,0)$ as desired, by taking R sufficiently small. Let $\varepsilon > 0$ be given. By Lemma 3-4 of Sec. 3-2 and inequality (10-49)

$$\|[P(x,t,D) - P(0,0,D)]u\| \leq \varepsilon |u|_{m,0} + K\|u\| \tag{10-51}$$

and $$\sum_{j=1}^{r} |[Q_j(x,0,D) - Q_j(0,0,D]u(x,0)|_{m-m_j-1/2}$$

$$\leq \varepsilon \sum_{j=0}^{m=1} |D_t^j u(x,0)|_{m-j-1/2} + K \sum_{j=0}^{m=1} |D_t^j u(x,0)|_{m-j-3/2} \tag{10-52}$$

Now, by Theorem 10-10

$$|D_t^j u(x,0)|_{m-j-1/2} \leq C|u|_{m,0}$$

and $$|D_t^j u(x,0)|_{m-j-3/2} \leq C|u|_{m-1,0}$$

Another application of Lemma 10-22 shows that the left-hand side of (10-52) is bounded by $\varepsilon|u|_{m,0} + K\|u\|$. If we now apply these inequalities to (10-50), we obtain inequality (10-17) with an additional term $C\varepsilon|u|_{m,0}$ added to the right-hand side. We take ε so small that $C\varepsilon < \frac{1}{2}$. This gives (10-17) for R sufficiently small.

10-7 THE GLOBAL ADJOINT OPERATORS

Now we turn to the problem in an arbitrary bounded domain G in E^n. We assume that the boundary ∂G of G is smooth (see Sec. 9-1). Let P be a properly elliptic operator in $\overline{G}$, and let $\{Q_j\}$ be a normal set of boundary operators which covers P

(see Sec. 10-1). As usual we assume that the coefficients of P and Q_j are in $C^\infty(\bar{G})$. As we did in Sec. 10-2, we shall call a set $\{Q'_j\}$ of boundary operators *adjoint to* $\{Q_j\}$ *relative to* P if $v \in C^\infty(\bar{G})$ satisfies

$$(Pu, v) = (u, P'v) \tag{10-53}$$

for all $u \in C^\infty(\bar{G})$ satisfying

$$Q_j u = 0 \text{ on } \partial G \text{ for all } j \tag{10-54}$$

if and only if

$$Q'_j v = 0 \text{ on } \partial G \text{ for all } j$$

Our first step is to prove

Theorem 10-23 If P and the Q_j satisfy the assumptions made above, then there exists a normal set of boundary operators adjoint to $\{Q_j\}$ relative to P. Moreover, every such set covers P'.

PROOF Actually, the main part of the proof of Theorem 10-23 has already been given. In fact, we can reduce it to Theorem 10-3 and Corollary 10-19. We can do this as follows. Let χ_0 be a fixed point on ∂G, and let D_χ be the row vector

$$D_\chi = -i\left[\frac{\partial}{\partial \chi_1}, \ldots, \frac{\partial}{\partial \chi_n}\right] \tag{10-55}$$

Let R be a rotation which takes the interior normal to ∂G at χ_0 into the positive χ_m-axis, and put $y = R(\chi - \chi_0)$. It can be checked easily that

$$D_\chi = D_y R \tag{10-56}$$

Put
$$\hat{P}(y, D_y) = P(\chi, D_\chi) = \sum a_\mu(R'y + \chi_0)(D_y R)^\mu \tag{10-57}$$

where R' is the inverse (adjoint) of R. Thus

$$\hat{p}(0, \xi) = \sum_{|\mu| = m} a_\mu(\chi_0)(\xi R)^\mu = p(\chi_0, \xi R) \tag{10-58}$$

Similarly, we put

$$\hat{Q}_j(y, D_y) = Q_j(\chi, D_\chi) \tag{10-59}$$

Thus
$$q_j(0, \xi) = q_j(\chi_0, \xi R) \tag{10-60}$$

Note that ξR is orthogonal to ∂G if and only if ξ is along the χ_m-axis. Next, let $z = \Phi(y)$ be the transformation given by

$$z_k = y_k \qquad 1 \le k \le n$$

$$z_n = y_n - \varphi(y_1, \ldots, y_{n-1})$$

where φ is the equation of the surface ∂G in a neighborhood of χ_0. Note that

$$\frac{\partial \Phi_j(0)}{\partial y_k} = \delta_{jk} \tag{10-61}$$

and

$$D_y = D_z \, d\Phi \tag{10-62}$$

where $d\Phi$ is the matrix with elements $\partial\Phi_j(y)/\partial y_k$. Put

$$\tilde{P}(z, D_z) = \hat{P}(y, D_y) \qquad \tilde{Q}_j(z, D_z) = \hat{Q}_j(y, D_y)$$

Thus

$$\tilde{P}(z, D_z) = \sum a_\mu(R'\Psi(z) + \chi_0)(D_z \, d\Phi \, R)^\mu$$

where $\Psi(z)$ is the inverse of $\Phi(y)$. By Eqs. (10-58), (10-60), and (10-61), we have

$$\tilde{p}(0, \xi) = p(\chi_0, \xi R) \qquad \tilde{q}_j(0, \xi) = q_j(\chi_0, \xi R) \tag{10-63}$$

This shows that $\tilde{P}$ and the $\tilde{Q}_j$ satisfy the hypotheses of Sec. 10-3. Thus we have shown that for each $\chi_0 \in \partial G$ there is a neighborhood η of it and $R > 0$, such that $\eta \cap \overline{G}$ can be mapped in a one-to-one C^∞ way on to $\bar{\sigma}_R$, so that the transformed operators satisfy hypotheses 1 to 5 of Sec. 10-3. Now, by Theorem 10-3 there is a boundary set $\{\tilde{Q}'_j\}$ in $\bar{\sigma}_R$ adjoint to $\{\tilde{Q}_j\}$ relative to $\tilde{P}$. Moreover, we can choose this set in a determined way. For instance, we can fill in the missing operators to make a Dirichlet system, by using only the operators D_t^k for the appropriate orders k (see Theorem 10-4). Since the set $\{Q_j\}$ is normal, the missing orders do not depend on χ_0. Thus, we can determine the $\tilde{Q}'_j$ uniquely in σ_R. Let Q'_j be the preimage of $\tilde{Q}'_j$ in $\eta \cap \overline{G}$. It is not difficult to show that the operator Q'_j, so defined in $\eta \cap \overline{G}$, coincides in any overlapping region with the corresponding operator defined in a neighborhood of another boundary point. Thus, Q'_j is well defined and infinitely differentiable in a neighborhood of ∂G. Multiplying it by a suitable function, which is one in a neighborhood of ∂G and vanishes in an interior region, will make its coefficients be in $C^\infty(\overline{G})$. By Corollary 10-19 the $\{\tilde{Q}'_j\}$ cover $\tilde{P}'$. The reasoning above then shows that the $\{Q'_j\}$ cover P'. By mapping neighborhoods of points of ∂G on to σ_R we see, by Corollary 10-19, that any normal boundary set covers P' as well. This completes the proof of Theorem 10-23.

10-8 THE BOUNDARY NORM

In order to prove a global counterpart of Theorem 10-7 we must introduce a boundary norm on ∂G. This can be done in several ways. We have chosen the one that is conceptually the easiest, but it lacks aesthetic appeal. If ∂G is smooth and bounded there are neighborhoods $\eta_1, \ldots, \eta_N$ such that $\overline{G} \cap \eta_k$ can be mapped on to $\bar{\sigma}_R$ in the way described in Sec. 10-7, and ∂G is contained in their union. By Lemma 9-16 of Sec. 9-4, there is a C^∞ partition of unity, subordinate to this covering. Let $\{\zeta_k\}$ be such a partition with $\zeta_k \in C^\infty(\overline{G} \cap \eta_k)$. For $u \in C^\infty(\overline{G})$, $\zeta_k u \in C^\infty(\overline{G} \cap \eta_k)$. Under the mapping it becomes a function $(\zeta_k u)^\sim$ in $\mathscr{C}_R$. For real s, we define

$$\langle u \rangle_s^2 = \sum_{k=1}^{N} |(\zeta_k u)^{\sim}(\chi, 0)|_s^2 \tag{10-64}$$

An important property of this norm is given by

Lemma 10-24 There is a constant C, such that

$$\langle u \rangle_{m-1/2} \leq C \| u \|_m \qquad u \in C^\infty(\bar{G}) \tag{10-65}$$

PROOF By Theorem 10-10

$$|(\zeta_k u)^{\sim}(\chi, 0)|_{m-1/2} \leq C_1 |(\zeta_k u)^{\sim}|_{m,0}$$

By Lemmas 9-13 and 9-14 of Sec. 9-4, this is bounded by

$$C_2 \| (\zeta_k u)^{\sim} \|_m \leq C_3 \| \zeta_k u \|_m \leq C_4 \| u \|_m$$

Summing over k we obtain inequality (10-65).

It is clear that Eq. (10-64) gives a norm. Lemma 10-24 allows us to define this norm for $s = m - \frac{1}{2}$, on functions which are in $H^m(G)$. We have

Theorem 10-25 Let P and the Q_j satisfy the hypotheses of Sec. 10-1 (the set $\{Q_j\}$ need not be normal). Then there is a constant C, such that

$$\| u \|_m \leq C\left(\| Pu \| + \sum_j \langle Q_j u \rangle_{m-m_j-1/2} + \| u \| \right) \qquad u \in H^m(G) \tag{10-66}$$

PROOF Let χ be a point of ∂G. It has a neighborhood M, such that $\bar{G} \cap M$ can be mapped into $\bar{\sigma}_R$ for some $R > 0$ and inequality (10-17) holds. By shrinking M if necessary, one can require that it be contained in one of the η_k used in defining the boundary norms. Since ∂G is bounded, it can be covered by a finite set $M_1, \ldots, M_k$ of such neighborhoods. Let $\omega_1, \ldots, \omega_k$ be a C^∞ partition of unity, subordinate to this covering (see Lemma 9-16 of Sec. 9-4). Under the mappings $\omega_k u, P$, and the Q_j are taken into $(\omega_k u)^{\sim}$, $\tilde{P}$, $\bar{Q}_j$ with $\tilde{P}$ and the $\tilde{Q}_j$ satisfying the hypotheses of Sec. 10-3 (see Sec. 10-7). Thus

$$\| (\omega_k u)^{\sim} \|_m \leq C\left(\| \tilde{P}(\omega_k u)^{\sim} \| + \| (\omega_k u)^{\sim} \| + \sum_j |\tilde{Q}_j(\omega_k u)^{\sim}(\chi, 0)|_{m-m_j-1/2} \right)$$

Mapping back to $M_k \cap \bar{G}$, we get

$$\| \omega_k u \| \leq C\left(\| P(\omega_k u) \| + \| \omega_k u \| + \sum_j \langle Q_j(\omega_k u) \rangle_{m-m_j-1/2} \right)$$

Applying Leibnitz's rule and Lemma 9-15 of Sec. 9-4, we have

$$\| u \|_m = \left\| \sum \omega_k u \right\|_m \leq \sum \| \omega_k u \|_m \leq C\left(\| Pu \| + \| u \| + \sum \langle Q_j u \rangle_{m-m_j-1/2} \right)$$

(See Lemma 3-4 of Sec. 3-2.) This completes the proof.

10-9 THE COMPACTNESS ARGUMENT

To show that N and N' are finite dimensional, we use a compactness argument similar to that of Sec. 9-2. We shall need

Lemma 10-26 If $m \geq 1$ and

$$\| u_k \|_m \leq C \qquad k = 1, 2, \ldots \tag{10-67}$$

then $\{u_k\}$ has a subsequence converging in $L^2(G)$.

PROOF There is an open covering $\eta_1, \ldots, \eta_N$ of ∂G, such that $\bar{G} \cap \eta_j$ can be mapped on to $\bar{\sigma}_R$ in the way described in Sec. 10-7. Together with an interior open set η_0, they cover $\bar{G}$. Let $\{\zeta_j\}$ be a partition of unity, subordinate to this covering (cf. Lemma 9-16 of Sec. 9-4). It suffices to show that, for each j, the sequence $\{\zeta_j u_k\}$ has a subsequence converging in $L^2(\bar{G})$. Since $\zeta_0 u_k$ has compact support, it follows, as in Sec. 9-2, that $\{\zeta_0 u_k\}$ has such a subsequence. For the other values of j, let $(\zeta_j u_k)^{\sim}$ be the image under the map on to σ_R. Then

$$\| (\zeta_j u_k)^{\sim} \|_m \leq C'$$

and, consequently

$$| (\zeta_j u_k)^{\sim} |_{m,0} \leq C''$$

in view of Lemma 9-13 of Sec. 9-4. Now, by Theorem 8-22 of Sec. 8-6, $(\zeta_j u_k)^{\sim}$ has a subsequence converging in $L^2(\sigma_R)$. Consequently $\{\zeta_j u_k\}$ has a subsequence converging in $L^2(G)$. This completes the proof.

Once we have Lemma 10-26 the finite dimensionality of N is simple to prove. In fact, let $\{u_k\}$ be a sequence of functions in N, such that $u_k \to u$ in $L^2(G)$. By Theorem 10-25, $\{u_k\}$ is a Cauchy sequence in $H^m(G)$. Thus $u \in H^m(G)$, $Pu = 0$, and u satisfies Eq. (10-1). Thus, the closure $\bar{N}$ of N in $L^2(G)$ consists of such functions (actually we shall show that $\bar{N} = N$, but we do not have to know this here). Let $\{u_k\}$ be a bounded sequence in $\bar{N}$. Then it satisfies inequality (10-67), by Theorem 10-25. In view of Lemma 10-26, it has a subsequence convergining in $L^2(G)$.

Since $\bar{N}$ is a Hilbert space, the desired result follows from Theorem 8-17 of Sec. 8-5. In the case of N', we use the fact that $\{Q'_j\}$ covers P' (Theorem 10-23), and proceed in the same way.

To complete the proof of Theorem 10-1 we follow the reasoning in Sec. 9-2. First, we note

Lemma 10-27 If $u \in H^m(G)$ satisfies Eq. (10-1) and $Pu = 0$, then $u \in N$.

PROOF We know by Theorem 3-1 of Sec. 3-1 that $u \in C^{\infty}(G)$. To show that it is in $C^{\infty}(\bar{G})$, it suffices to show that each $\chi^0 \in \partial G$ has a neighborhood η, such that $u \in C^{\infty}(\eta \cap \bar{G})$. Pick this neighborhood so small that $\eta \cap \bar{G}$ can be mapped into $\bar{\sigma}_R$ in the manner described in Sec. 10-7. Let $\tilde{u}$ denote the image of u under the map. Let $0 < R' < R$, and let $\zeta \in C^\infty(\bar{\sigma}_R)$ equal 1 in $\sigma_{R'}$ and vanish near $|\chi| = R$. If $v \in \mathscr{C}_{R'}$, we have

$$(P(\zeta\tilde{u}), Pv) + \sum \langle Q_j(\zeta\tilde{u}), Q_j v\rangle_{m-m_j-1/2} = 0$$

Thus, $\zeta\tilde{u}$ satisfies the hypotheses of Theorem 10-8. This shows that $\tilde{u} \in C^\infty(\bar{\sigma}_{R''})$ for some $R'' > 0$. Consequently, we have $u \in C^\infty(\bar{G} \cap \eta')$ for some neighborhood η'. This completes the proof.

Now we turn to the proof of conclusion 4 of Theorem 10-1. Let M denote the set of those $u \in H^m(G)$ satisfying Eq. (10-1) and $u \perp N$. Similarly, let M' denote the set of those $v \in H^m(G)$ satisfying Eq. (10-4) and $v \perp N'$. I claim

Lemma 10-28 There is a constant C, such that

$$\|u\|_m \leq C\|Pu\| \qquad u \in M \tag{10-68}$$

$$\|v\|_m \leq C\|P'v\| \qquad v \in M' \tag{10-69}$$

PROOF If (10-68) did not hold, there would be a sequence $\{u_k\} \subset M$, such that

$$\|u_k\|_m = 1 \qquad \|Pu_k\| \to 0 \tag{10-70}$$

By Lemma 10-26, there is a subsequence (also denoted by $\{u_k\}$) which converges in $L^2(G)$. By Theorem 10-25, this subsequence is a Cauchy sequence in $H^m(G)$. Since M is a closed subspace of $H^m(G)$, the limit u is in M. On the other hand $Pu = 0$, and consequently $u \in N$. Thus, we must have $u = 0$. But $\|u\|_m = 1$ by Eq. (10-70). This contradiction proves (10-68). Similar reasoning gives (10-69).

Lemma 10-29 If $u \in H^m(G)$, $f \in C^\infty(\bar{G})$, and

$$(Pu, Pv) + \sum \langle Q_j u, Q_j v\rangle_{m-m_j-1/2} = (f, v) \qquad v \in C^\infty(\bar{G}) \tag{10-71}$$

then $u \in C^\infty(\bar{G})$.

PROOF Since $P'P$ is elliptic and u is a weak solution of $P'Pu = f$, we see that $u \in C^\infty(G)$ (Theorem 3-1 of Sec. 3-1). To prove the theorem, it suffices to show that each $\chi^0 \in \partial G$ has a neighborhood η, such that $u \in C^\infty(\bar{G} \cap \eta)$. Following the proof of Lemma 10-27, we map $\bar{G} \cap \eta$ on to $\bar{\sigma}_R$ in the way described in Sec. 10-7. Then we see that the image $\tilde{u}$ of u satisfies the hypotheses of Theorem 10-8, from which we can draw the desired conclusion.

Now suppose $u \in C^\infty(\bar{G})$ satisfies (10-1) and $Pu = f$. Then, if $v \in N'$, we have

$$(f, v) = (Pu, v) = (u, P'v) = 0$$

by Eq. (10-3). Thus $f \perp N'$. Conversely, suppose $f \perp N'$. Then for $v \in M'$, (v, f) is a bounded linear functional (Lemma 10-28). By the Fréchet–Riesz representation theorem (Theorem 1-5 of Sec. 1-5), there is a $w \in M'$, such that

$$(v, f) = (P'v, P'w) \qquad v \in M' \tag{10-72}$$

It is also true for $v \in N'$. Since every function in $H^m(G)$ satisfying Eq. (10-4) is the sum of a function in M' and one in N', Eq. (10-72) holds for all such v. Applying Lemma 10-29 we conclude that $w \in C^\infty(\bar{G})$, and likewise for $u = P'w$. Applying 1 of Theorem 10-1, we see that u is a solution of Eq. (10-1) and $Pu = f$. A similar argument works for g.

In completing the proof of Theorem 10-1, we shall make use of

Theorem 10-30 If $f \in L^2(G)$ and $f \perp N'$, then there is a $u \in H^m(G)$ satisfying Eq. (10-1), such that $Pu = f$ and $u \perp N$.

PROOF There is a sequence $\{f_k\}$ of functions in $C^\infty(\bar{G})$ converging to f in $L^2(G)$ (Lemma 4-17 of Sec. 4-6). Since N is a closed subspace of $L^2(G)$, $f_k = f_k' + f_k''$, where $f_k'' \in N'$ and $f_k' \perp N'$. Now

$$\| f_k - f \|^2 = \| f_k' - f \|^2 + \| f_k'' \|^2$$

and consequently $f_k' \to f$ in $L^2(G)$. Since $N' \subset C^\infty(\bar{G})$, the same is true of the $\{f_k'\}$. By conclusion 4 of Theorem 10-1, there is a $u_k \in C^\infty(\bar{G})$ satisfying Eq. (10-1) and $Pu_k = f_k'$. We may take $u_k \in M$. By Lemma 10-28, we see that the $\{u_k\}$ form a Cauchy sequence in $H^m(G)$. Their limit u is clearly in M, and satisfies $Pu = f$. This completes the proof.

We now prove conclusion 5 of Theorem 10-1. Suppose $u \in L^2(G)$, $f \in C^\infty(\bar{G})$ and Eq. (10-6) holds for all $v \in C^\infty(\bar{G})$ satisfying Eq. (10-4). We put $u = u' + u''$, where $u'' \in N$ and $u' \perp N$. Clearly, u' satisfies Eq. (10-6) as well. Clearly $f \perp N'$. Thus, there is a $u_1 \in C^\infty(\bar{G}) \cap M$ satisfying $Pu_1 = f$. Hence, we have

$$(u' - u_1, P'v) = 0$$

for all $v \in C^\infty(\bar{G}) \cap M'$. There is a sequence $\{g_k\}$ of functions in $C^\infty(\bar{G})$ converging to $u' - u_1$ in $L^2(G)$. We may take $g_k \perp N$ (see the proof of Theorem 10-30). By conclusion 4 of Theorem 10-1, there is a $v_k \in C^\infty(\bar{G}) \cap M'$, such that $P'v_k = g_k$. Hence $(u' - u_1, g_k) = 0$ for each k. Taking the limit, we see that $u' = u_1 \in C^\infty(\bar{G})$. Since $u = u' + u''$ and $u'' \in N$, the result follows.

PROBLEMS

10-1 Prove Lemma 10-2.

10-2 Show that Theorem 10-4 implies Theorem 10-3.

10-3 Verify the last statement in the proof of Theorem 10-4.

10-4 Prove Lemma 10-6.

10-5 Prove Corollary 10-12.

10-6 Prove inequality (10-26).

10-7 Prove Eq. (10-31).

10-8 Show that Eq. (10-32) implies that the α_j vanish.

10-9 Derive Eqs. (10-33) and (10-34).

10-10 Fill in the details of the proof of Theorem 10-8 given in Sec. 10-5.

10-11 Prove Eq. (10-42).

10-12 Prove Eq. (10-56).

10-13 Prove Eqs. (10-61) and (10-62).

10-14 Show that $(\hat{P})' = (P')\hat{}$.

10-15 Show that $(\tilde{P})' - (P')\tilde{}$ is of lower order than P.

10-16 Show that Eq. (10-64) defines a norm.

10-17 Show that M and M' are closed subspaces of $H^m(G)$.

10-18 Prove inequality (10-69).

10-19 Why may we take $u_k \in M$ in the proof of Theorem 10-30?

BIBLIOGRAPHY

Agmon, S.: "Lectures on Elliptic Boundary Value Problems," van Nostrand, 1965.

Aronszajn, N.: On Coercive Integrodifferential Forms, *Kansas Report*, no. 14, pp. 94–106, 1954.

——— and A. N. Milgram: Differential Operators on Riemannian Manifolds, *Rend. Circ. Mat. Palermo*, vol. 2, pp. 1–61, 1953.

Berezanskii, Ju. M.: "Eigenspansions in Eigenfunctions of Selfadjoint Operators," American Mathematical Society, 1968.

Bers, L., John F. Fritz, and M. Schechter: "Partial Differential Equations," J. Wiley, 1964; American Mathemtical Society, 1974.

Browder, F. E.: On the Regularity Properties of Solutions of Elliptic Differential Equations, *Comm. Pure Appl. Math.*, vol. 9, pp. 351–361, 1956.

Carroll, R. W.: "Abstract Methods in Partial Differential Equations," Harper and Row, 1969.

Courant, R., and D. Hilbert: "Methods of Mathematical Physics II," Interscience, 1962.

Friedman, A.: "Generalized Functions and Partial Differential Equations," Prentice Hall, 1963.

———: "Partial Differential Equations," Holt, Reinhart, and Winston, 1969.

Friedrichs, K. O.: On the Differentiability of Solutions of Linear Elliptic Differential Equations, *Comm. Pure Appl. Math.*, vol. 6, pp. 299–326, 1953.

Gårding, L.: Dirichlet's Problem for Linear Partial Differential Equations, *Math. Scand.*, vol. 1, pp. 55–72, 1953.

———: Solution Directe du Problém du Cauchy pour les Equations Hyperboliques, *Colloques Internationaux du Centre Nat. de la Recherche Scient.*, vol. 71, pp. 71–90, 1956.

Gelfand, I. M., and G. E. Shilov: "Generalized Functions III," Academic Press, 1967.

Hellwig, G.: "Partial Differential Equations," Blaisdell, 1964.

Holmgren, E.: Über Systeme von linearen partiellen Differentialgleichungen, *Öfrersigt af Kongl. Vetenskaps-Akademien Förhandlinger*, vol. 58, pp. 91–103, 1901.

Hörmander, L.: On the Theory of General Partial Differential Operators, *Acta Math.*, vol. 94, pp. 161–248, 1955.

———: On the Interior Regularity of the Solutions of Partial Differential Equations, *Comm. Pure Appl. Math.*, vol. 11, pp. 197–218, 1958.

———: "Linear Partial Differential Operators," Springer, 1963.

Leray, J.: Hyperbolic Equations, Princeton Lecture Notes, 1954.

Lewy, H.: An Example of a Smooth Linear Partial Differential Equation without Solution, *Ann. Math.*, vol. 66, pp. 155–158, 1957.

Lions, J. L., and E. Magenes: "Problems aux Limites non Homogenes," I, II, III, Dunod, 1968.

Lopatinski, Y. B.: On a Method of Reducing Boundary Problems for a System of Differential Equations of Elliptic Type to Regular Integral Equations, (Russian) *Ukrain. Mat. Z.*, vol. 5, pp. 123–151, 1953.

Malgrange, B.: Sur une Classe di'operateurs Différentiels Hypoelliptiques, *Bull. Soc. Math. France*, vol. 85, pp. 283–306, 1957.

———: Operatori Differenziali Teorie delle Distribuzoni, *Centro Interzionale Matematico Estiro*, 2° Ciclo Soltino di Vallombrosa, 1–9, Settembre, 1961.

Miranda, C.: "Partial Differential Operators of Elliptic Type," Springer, 1970.

Mizohata, S.: "The Theory of Partial Differential Equations," Cambridge, 1973.

Necas, J.: "Les Methodes Directes en Theorie des Equations Elliptiques," Masson, Academia, 1967.

Nirenberg, L.: Remarks on Strongly Elliptic Partial Differential Equations, *Comm. Pure Appl. Math.*, vol. 8, pp. 649–675, 1955.

———: On Elliptic Partial Differential Equations, *Ann. Scuola Norm. Sup. Pisa*, vol. 13, pp. 116–162, 1959.

Peetre, J.: On Estimating the Solutions of Hypoelliptic Differential Equations near the Plane Boundary, *Math. Scand.*, vol. 9, pp. 337–351, 1961.

Peyser, G.: Energy Inequalities for Hyperbolic Equations in Several Variables with Multiple Characteristics and Constant Coefficients, *Trans. Amer. Math. Soc.*, vol. 180, pp. 478–490, 1963.

Schechter, M.: Solution of the Dirichlet Problem for Equations not necessarily Strongly Elliptic, *Bull. Amer. Math. Soc.*, vol. 64, pp. 371–372, 1958.

———: General Boundary Value Problems for Elliptic Partial Differential Equations, ibid., vol. 65, 70–72, 1959*a*.

———: Integral Inequalities for Partial Differential Operators and Functions Satisfying General Boundary Conditions, *Comm. Pure Appl. Math.*, vol. 12, pp. 37–66, 1959*b*.

———: Remarks on Elliptic Boundary Value Problems, ibid., vol. 12, pp. 561–578, 1959*c*.

———: On the Dominance of Partial Differential Operators, *Trans. Amer. Math. Soc.*, vol. 107, pp. 237, 1963; II, *Ann. Scand. Norm. Sup. Pisa*, vol. 18, pp. 255–282, 1964.

Schwartz, L.: Sir Alcuni Problemi della Teoria delle Equazioni Differenziali Lineari di Tipo Ellitico, *Rend. Sem. Mat. Fis. Milano*, vol. 27, pp. 3–41, 1958.

Spivak, M.: "Calculus on Manifolds," Benjamin, 1965.

Treves, F.: The Equation

$$\left[\frac{1}{4}\left(\frac{\partial^2}{\partial x^2} + \frac{\partial^2}{\partial y^2}\right) + (x^2 + y^2)\frac{\partial^2}{\partial t^2} + \left(x\frac{\partial}{\partial y} - y\frac{\partial}{\partial x}\right)\frac{\partial}{\partial t}\right]^2 u + \frac{\partial^2 u}{\partial t^2} = f$$

with real coefficients, is "without solutions," *Bull. Amer. Math. Soc.*, vol. 68, pp. 332, 1962.

———: "Basic Linear Partial Differential Equations," Academic Press, 1975.

TVSLB"O

INDEX